Advances in Astrobiology and Biogeophysics

More information about this series at http://www.springer.com/series/5118

Barbara Cavalazzi • Frances Westall
Editors

Biosignatures for Astrobiology

Springer

Editors
Barbara Cavalazzi
Dipartimento di Scienze Biologiche
Geologiche e Ambientali, Università di
Bologna
Bologna, Italy

Department of Geology
University of Johannesburg
Johannesburg, South Africa

Frances Westall
Centre de Biophysique Moléculaire
CNRS
Orléans, France

ISSN 1610-8957 ISSN 1613-1851 (electronic)
Advances in Astrobiology and Biogeophysics
ISBN 978-3-030-07147-9 ISBN 978-3-319-96175-0 (eBook)
https://doi.org/10.1007/978-3-319-96175-0

This Springer imprint is published by the registered company Springer Nature Switzerland AG
The registered company address is: Gewerbestrasse 11, 6330 Cham, Switzerland

Foreword

As a scientific discipline, astrobiology includes the study of life's origin, distribution, and fate in the Universe. By embodying the search for life, astrobiology includes research focused on the study of life's signatures within our Solar System and beyond. Recent discoveries suggest that the Universe is more amenable to life than previously recognized. Hence, the timeliness of *Biosignatures for Astrobiology*, a book edited by B. Cavalazzi and F. Westall, eloquently examines biosignatures in the context of the Earth and Solar System planetary bodies. Earth-like exoplanets are now being detected at a remarkable pace, which is likely to escalate as the Transiting Exoplanet Survey Satellite (TESS) comes online this year and the James Webb telescope to be launched next decade.

The number of interesting planetary targets for astrobiological exploration of extinct and extant life within our own Solar System has also increased as their geologically active nature has been revealed. While it has been known that Mars was potentially habitable early in its lifetime, with large oceans and lakes in abundance, the recent in situ discovery of complex organic molecules preserved in ancient sediments is consistent with the hypothesis that carbonaceous biosignatures could be preserved for long periods of time on the red planet. The recent finding that Mars releases methane seasonally is certainly indicative of the planet's dynamic nature on local scales, regardless of whether it was produced by microorganisms, which portends modern subsurface geochemical activity that could support life.

Evidence of the dynamic nature and habitability potential of other planetary moons in our Solar System has been captured in flyby mission images of geysers on the surfaces of Enceladus and Europa, young icy moons of Saturn and Jupiter, respectively. The incredible images captured by the Cassini–Huygens space probe of liquid methane lakes dotting the surface of Titan reminds us of how little we know about the inventory and distribution of organics in our Universe, a certainty substantiated by the incredible Hubble Space Telescope images of massive star-forming nurseries of the galaxy, regions of space replete with complex organic molecules.

Key to astrobiological exploration for extinct and extant life, whether beyond or within our Solar System, is the search for biosignatures. On Earth, astrobiologists conduct field and laboratory investigations, perform planetary simulations, and generate theoretical models in an attempt to understand how life originated and evolved on Earth. Given the uniqueness of Earth as the only known abode for life, along with the current trajectory of the future of our planet's climate, any improvements in understanding the early evolution of life and the potential for it to adapt to future terrestrial, low-Earth orbit, and eventually extraterrestrial environmental challenges may be essential for our survival as a species.

How life interacts with, and responds to, its environment to produce biosignatures that will be preserved on other worlds is an important thrust of astrobiology. Extremophiles are studied with evermore sophisticated technologies in microbially dominated habitats, in planetary conditions simulated in the laboratory, and in terrestrial samples of their fossilized habitats. These opportunities provide a baseline for testing hypotheses related to understanding what fraction of biosignatures become preserved in the geological record and how best to find, detect, and interpret them. Whether biosignatures of extinct or extant life can be distinguished from abiotic mimics is especially challenging, given the continuous rain of abiotically produced organic matter to planetary bodies throughout the Universe.

The search for biosignatures is compounded by the fact that they range in size from the atomic to planetary scales, and their age could virtually be any age during which life could have inhabited a planet and its biosignatures could have been preserved. On Earth, biosignatures of life were preserved in a variety of geological deposits throughout most of the planet's history. Hence, the amount of alteration that ancient biosignatures on Earth have received may not be directly comparable to that experienced by a similar type of biosignature preserved on another rocky planet like Mars, for example. The correct interpretation of any possible biosignature preserved in the geological record of a planet also requires an understanding of the processes that have altered it since its time of formation.

The book *Biosignatures for Astrobiology* brings together our current understanding of biosignatures with some of the most useful methodologies and technologies used in astrobiology search strategies in an easily readable and comprehensive manner. The book is suitable for the scientifically inclined layperson, the student of astrobiology, and the professional astrobiologist as well. All of the authors contributed background sections to help nascent readers of astrobiology literature grasp the main concepts. In its entirety, the book illustrates why the search for biosignatures beyond Earth is complicated, risky, and a compelling challenge for current and future generations of scientists worldwide. The book is laid out such that it lends itself to be read from cover to cover in sequence or one chapter at a time out of sequence. *Biosignatures for Astrobiology* will be useful for teaching an advanced course in the field or as a reference for practitioners.

The presence of life forms on Earth that have the capability to withstand the harshest of conditions, even those on nearby planets, underscores the hypothesis that life exists elsewhere in the Universe. As stated in the book's final chapter by Dunér, "So far, we have no conclusive evidence of the existence of extraterrestrial life. But could we ever be 100% sure that we are alone?" To be unequivocally sure that no life exists anywhere but on Earth is something we may never know. Yet we are confident that the innate curiosity of humans to know life's origins, and whether it will or does exist beyond Earth, is certain to drive astrobiological exploration far into the future so long as these questions remain unanswered. Even when answers are forthcoming, the field of astrobiology will endure. Surely, if life exists in at least one other place in the Universe, the possibility that it occurs in a third locality is almost a certainty.

Environmental Molecular Sciences Laboratory, Pacific Northwest National Laboratory, Richland, WA, USA — Sherry L. Cady
Bay Area Environmental Research Institute, Moffett Field, CA, USA — Rocco L. Mancinelli
August 27, 2018

Preface

This book on *Biosignatures for Astrobiology* has had a long germination. It started with an article that we wrote on *Biosignatures in Rocks* for the *Encyclopedia of Geobiology* published by Springer in 2009, and early suggestions from Ramon Khanna at the 13th meeting of European Astrobiology Network Association (EANA) in 2013 hosted in Edinburgh.

With the ongoing Mars Science Laboratory Mission to Gale Crater and the ExoMars Trace Gas Orbiter (2016) and lander/rover (2020), as well as the future JUICE and Europa Clipper Mission to Jupiter and its satellites and the various telescopes (HUBBLE, Spitzer, Kepler, TESS, JWST, WFIRST) searching for exoplanets, the time is appropriate to review biosignatures of relevance for astrobiology, addressing all aspects of the discipline. Our book thus aims at capitalising on the latest advances to provide overviews in all the domains of astrobiology.

Since the paradigm-changing discovery of the first exoplanet orbiting a main-sequence star and the finding of highly controversial biosignatures in the Martian meteorite ALH84001 by David McKay and his colleagues in 1996, biosignature research across the board in astrobiology has made enormous advances. Whether or not ALH84001 or other Martian meteorites harbour traces of life, the very idea that microbial fossils could be preserved in a meteorite from Mars sparked a huge interest in biosignature research covering the minimum size of microorganisms, extremophiles, through to the fossilised traces of life, and astrobiology missions to Mars.

We address the signatures of life with respect to life on Earth as well as life elsewhere in the Solar System (i.e. Mars) and exoplanets. For our purposes, we are assuming that the extraterrestrial life forms, hopefully encountered in the future (even if in fossil form), are based on carbon molecules and water as a solvent. Therefore, in order to place the topic of the biosignatures in context, we start with a couple of chapters that "set the scene". André Brack starts Chap. 1 with an overview of the chemical signatures at the origin of life in which he underlines the basic precepts of carbon and life and the carbonaceous signatures of life, namely an overrepresentation of organics and long strands of homochiral sequences. Since it is widely believed that the majority of the carbon required for the emergence of life on Earth came from extraterrestrial sources, and that the flux of this kind of carbon continues to this day, albeit at a much lower rate than on the early Earth or the early planets and satellites, André Brack's chapter is followed by a contribution from Eric Quirico and Lydie Bonal who, in Chap. 2, review the present state of knowledge on the composition, structure, and formation and evolution of the exogenous organics accreted by the Earth on their original asteroidal or cometary parent bodies. The importance of this knowledge is put into perspective when one considers the fact that, despite the harsh radiation and oxidising environment reigning at the surface of Mars for more than 3 billion years that effectively destroys the more volatile fraction of any organic matter, abiotic (meteoritic) or potentially biogenic, recent, hard-won results from the SAM instrument on the Curiosity rover on Mars do, indeed, show that organic molecules are present in the Martian surface.

Both the MSL and the ExoMars missions hope to find traces of life on Mars, more likely fossil life than extant life. The molecular compounds detected in Gale Crater are important in their own right since it is to be expected that, at a minimum, extraterrestrial carbon should be present at the surface. The continued hope is to find signatures of life. David J. Des Marais and Linda L. Jahnke address biosignatures of cellular components and metabolic activity in Chap. 3. They review life's basic capabilities of energy harvesting, metabolism, and self-replication, which can create objects, substances, and patterns—biosignatures—that indicate their biological origins. They conclude that the simultaneous presence of multiple biosignature objects, substances, and patterns in a demonstrably habitable earlier environment constitutes the most compelling evidence of past life.

Although it is widely believed that the likelihood of extant life forms at the surface of Mars is very low, the idea that life could subsist in the Martian subsurface is gaining credence. In this perspective, Frédéric Gaboyer, Gaëtan Burgaud, and Virginia Edgcomb in Chap. 4 describe very slow living extremophiles found up to several kilometres deep in subsea sediments and emphasise their relevance for the search of biosignatures in the Martian subsurface. The search for life in situ on another planet requires an approach that incorporates systematic preparation in terms of ground- and space-based studies before a mission. Jean-Pierre de Vera and colleagues from the BIOMEX and BIOSIGN experiments provide an overview of the necessary steps in order to search for life in situ on another planet or moon in Chap. 5 and show results obtained from research performed in the field, in the lab, and in space to help enhance knowledge of the traces and signatures of life, and how to recognise life itself. The different kinds of mineralogical traces that can be produced by microbial life forms are described in Chap. 6 by Karim Benzerara, Sylvain Bernard, and Jennyfer Miot. They review the manner in which many organisms impact mineral nucleation and growth, thus producing biominerals with specific chemical, structural, and textural properties that can provide clues to their biogenicity. Taking the concept of mineralisation and the preservation of microorganisms further, Fances Westall, Keyron Hickman-Lewis, and Barbara Cavalazzi make an overview of biosignatures in deep time, concentrating specifically on the oldest preserved biosignatures in well-preserved, although moderately metamorphosed, rocks up to 3.5 billion years old from the greenstone belts of Barberton in South Africa and the Pilbara in Australia. They show that the earliest preserved traces of life record an already thriving and, for an anaerobic world, evolved microbial ecosystem that included anoxygenic photosynthesis.

Indeed, carbonates as biominerals were described by David McKay and co-authors (1996) in the ALH84001 meteorite from Mars as one of the criteria in their interpretation of fossil life in the meteorite. Fractures within the meteorite contain rosette-shaped, aqueously deposited carbonate containing small ovoid to filamentous-shaped objects that the McKay team believed to be microbial nanofossils. Associated with the carbonates are minute magnetite crystals that were interpreted as biominerals produced by magnetotactic bacteria. Harry Y. McSween in Chap. 8 makes a critical appraisal of the original claims and the mountain of experimental data that ensued the 1996 publication, concluding that the evidence for biogenicity is weak.

Continuing on the theme of minerals, John Robert Brucato and Teresa Fornaro look at the role of mineral surfaces in prebiotic processes and life detection investigations focusing mainly on Mars exploration in Chap. 9. They show how molecule–mineral interactions provide important support for space missions aimed at searching for past or present signs of life in the form of molecular biomarkers within rocks.

One way of studying organic molecules as prebiotic or biosignatures in rocks is to look at the effects of photochemistry in space on astrobiologically relevant substrates. These molecules formed part of the organic inventory that was used for the prebiotic processes leading to the emergence of life. In Chap. 10, Avinash Dass, Hervé Cottin, and André Brack review the history of experiments to expose organic molecules to space radiation, observing the kinds of changes that occur in them.

The search for extraterrestrial life concerns not only our nearest planetary neighbour, Mars, but also rocky, Earth-like (or other habitable but not so Earth-like) planets in general. John Lee Grenfell in Chap. 11 provides a brief overview of potential biosignatures of relevance to remote observation and reviews knowledge of the main processes which influence biosignatures in an exoplanetary context, looking specifically at atmospheric model studies for Earth-like planets which predict climate, photochemistry, and potential spectral signals of biosignature species.

Just such a potential atmospheric signature for life has been found on Mars over the last couple of decades. Methane has been measured in the Martian atmosphere from ground-based telescopes, from Martian orbiters, and now in situ by the Curiosity rover. Franck Lefèvre reviews the history of the observations and evaluates their implications for the origin of methane as a possible biogas in Chap. 12.

One of the instruments in the Pasteur payload of the ExoMars 2020 rover is a Raman spectrometer, and another Raman spectrometer will fly on the Mars 2020 caching mission. Frédéric Foucher in Chap. 13 presents an overview of the different types of biosignatures that can be detected and/or characterised using Raman spectroscopy, including organic molecules, microfossils, biominerals, or even living cells.

The search for life on Mars is not new. The two 1976 Viking landers on Mars were the first dedicated astrobiology missions to the red planet. Jorge L. Vago,

Frances Westall, and Barbara Cavalazzi in Chap. 14 review the ambiguous results from this mission and introduce the objectives of the ExoMars 2020 rover and its approach to the search for past or present life on the planet.

The final part consists of a philosophical appraisal of the notion of biosignatures by David Dunér in Chap. 15 examines the human search, understanding, interpretation of biosignature natures, the concepts of conceptualisation, analogy, perception, and the semiotics of biosignatures.

Bologna, Italy Barbara Cavalazzi
Orléans, France Frances Westall

Acknowledgements

The idea for a book on biosignatures was born during the 15th meeting of the European Astrobiology Network Association (EANA) in 2015 hosted at the ESTEC-European Space Agency (ESA).

This editorial project was supported by the INACMa project under Grant Agreement FP7-PEOPLE-2013-CIG no. 618657 (to BC) and the MASE project supported by European Community's Seventh Framework Programme (FP7/2007-2013) under Grant Agreement no. 607297.

We are very thankful to all the authors who have done the most important work and contributed in preparing this volume of *Biosignatures for Astrobiology* and enabled to produce a very good scientific and updated review on the subject.

We would also like to thank all the declared and anonymous reviewers who dedicated their time and expertise to criticise, evaluate and improve all the chapters:

Conel M. O'D. Alexander, André Brack, Sherry L. Cady, Andrew D. Czaja, Mohit Melwani Deswani, Charlene Estrada, Frédéric Gaboyer, Axel Kleidon, Petr Vítek, R. Todd Clancy, Francois Raulin and Andreas Teske.

A sincere thank you to Keyron Hickman-Lewis for his assistance in English language editing. We are grateful to Ramon Khanna for his encouragement and support of our contribution to Springer series. We also thank the Editorial Office for their kindness and immense patience with the edition of the book.

BC would like to offer extra special thanks to Prof. Roberto Barbieri. Without his insight this book may not have been possible. His mentorship, support and encouragement throughout my research and career have been and still are a great privilege.

During a long career studying biosignatures and astrobiology, FW has benefitted from interactions with many colleagues in many disciplines. In particular, she acknowledges the contributions of the first two presidents of EANA, André Brack and Gerda Horneck, to the field.

We kindly thank our families and friends in accommodating long writing and editing sessions in off-working hours.

We covered many of the fundamentals in our *Biosignatures for Astrobiology*; however, we would like to emphasise that if we are far away from being complete—always there are questions left without response—we would like to stimulate further discussions and open new perspectives in the research for life.

Bologna, Italy
Orléans, France
May 2018

Barbara Cavalazzi
Frances Westall

Contents

Part I Biosignatures on Earth

1 Chemical Biosignatures at the Origins 3
André Brack
1.1 Introduction 3
1.2 Chemical Prerequisites 4
1.2.1 Liquid Water 4
1.2.2 Organic Molecules 5
1.3 Chemical Biosignatures 11
1.3.1 Over Representation of Organics 12
1.3.2 One-Handedness 13
1.4 Conclusions 17
References 17

2 Organic Matter in Interplanetary Dusts and Meteorites 23
Eric Quirico and Lydie Bonal
2.1 Introduction 23
2.2 Cosmomaterials Inherited by Telluric Planets 24
2.2.1 Past and Present Flux of Extraterrestrial Matter 24
2.2.2 Cosmomaterials in Earth Collections 26
2.2.3 Post-Accretion Processes on the Parent Body 27
2.3 Diversity and Complexity of Organics in Meteorites 28
2.3.1 Soluble Organic Matter 28
2.3.2 Insoluble Organic Matter 31
2.4 Organic Matter in Stratospheric IDPs and AMMs 35
2.5 Origin and Formation of Organics in Chondrites and Dust 37
2.6 Dust and Meteorites at the Surface of Mars 39
2.6.1 Flux of Exogenous Matter on Mars 39
2.6.2 Exogenous Organics on Mars: Detection and Preservation 40
References 43

3 Biosignatures of Cellular Components and Metabolic Activity 51
David J. Des Marais and Linda L. Jahnke
3.1 Introduction 51
3.1.1 Biosignatures 51
3.1.2 Concepts of Life 52
3.1.3 Attributes of Life 54
3.2 Organic Molecular Biosignatures 55
3.2.1 Membrane Polar Lipids 56
3.2.2 Cyclic Triterpenoids 58
3.2.3 Modern Microbial Mat Ecosystems as Analogs for Life on the Early Earth 59
3.2.4 Linking Geological and Biological Records 62
3.3 Stable Isotope Abundance Patterns as Biosignatures 65
3.3.1 Stable Isotope Basics 65
3.3.2 Carbon Compounds and Microbial Metabolism 65
3.3.3 Reduced Carbon and Carbonates in Sedimentary Rocks 71
3.3.4 Isotopic Patterns Arising from Biological Redox Reactions 76
3.4 Final Comments 78
References 78

4 The Deep Subseafloor and Biosignatures 87
Frédéric Gaboyer, Gaëtan Burgaud, and Virginia Edgcomb
4.1 The Deep Biosphere: An Unseen World of Contrasting Habitats 88
4.1.1 Definition 88
4.1.2 Environmental Conditions of Subsurface Environments 89
4.2 Tools to Detect Subsurface Biosignatures 92
4.2.1 Contamination Issues 92
4.2.2 Sensitive Analytical Methods 92
4.3 Microbiology of Subsurface Sediments Using the Canterbury Basin (CB) as a Case Study 94
4.3.1 Drilling Expeditions and Environmental Context 94
4.3.2 Cell Abundance 94
4.3.3 Evolution of Diversity with Depth 95
4.4 Activity *Versus* Viability 98
4.4.1 The Canterbury Basin Subseafloor Case Study 99
4.5 Possible Metabolisms 100
4.5.1 The Canterbury Basin Subseafloor Case Study 101
4.6 Physiological Potential 102
4.6.1 The Canterbury Basin Subseafloor Case Study 104
References 106

5 A Systematic Way to Life Detection: Combining Field, Lab and Space Research in Low Earth Orbit 111
Jean-Pierre de Vera and The Life Detection Group of BIOMEX/ BIOSIGN
5.1 Introduction 111
5.2 Promising Locations in the Solar System for Life Detection: Mars, Europa and Enceladus 112
5.2.1 The Search for Habitable Worlds 113
5.2.2 Promising Potentially Habitable Targets of Relevance to Life Detection Missions 113
5.3 Finding Stable and Space-Resistant Biomolecules as Potential Biosignatures 114
5.3.1 The Content of a Life Detection Database 114
5.3.2 Low Earth Orbit Missions Supporting Exploration Missions to Search for Life 115
5.4 Relevance to Biosignature, Bio-Trace and Bio-Fingerprint Research 115
5.5 Experimental Approach 116
5.6 Analysis Strategy and Biosignature Database Concept 117
References 120

6 Mineralogical Identification of Traces of Life 123
Karim Benzerara, Sylvain Bernard, and Jennyfer Miot
6.1 Introduction 123
6.2 How Does Life Influence the Formation of Minerals? 125
6.2.1 Impact of Life on Solution Saturation 125
6.2.2 Impact of Organic Molecules on Crystal Nucleation and Growth 128
6.2.3 Extracellular Versus Intracellular Biomineralization 129
6.3 Examples of Biominerals with Particular Physico-Chemical and Structural Properties 130
6.3.1 Genetic Control Over Biomineral Properties: Example of Magnetite Formed by Magnetotactic Bacteria (MTB) 130
6.3.2 Properties of Extracellular and Cell Wall-Associated Iron Biominerals 131
6.4 The Loss of the Biological Information Contained by Biominerals Upon Ageing 132
6.4.1 Structural Alteration of Biominerals in Natural Settings 133
6.4.2 (Geo)chemical Alteration of Biominerals in Natural Settings 133
6.4.3 Constraining Biomineral Ageing Processes in the Laboratory 134

6.5 The Thin Frontier Between Biominerals and Abiotic Minerals . . . 135
6.5.1 Biogenic and Abiotic Framboidal Pyrite 136
6.5.2 Twisted Stalks 137
6.6 Conclusions 138
References 139

7 Biosignatures in Deep Time 145
Frances Westall, Keyron Hickman-Lewis, and Barbara Cavalazzi
7.1 Introduction 146
7.2 Ancient Biosignatures from the Early-Mid Archaean Eon 147
7.2.1 Graphitic Biosignatures from Highly Metamorphosed Sediments from the 3.95 Uivak-Iquluk Gneiss, Labrador and Greenland 147
7.2.2 Biosignatures from the 3.5–3.0 Ga Pilbara Greenstone Belt 150
7.2.3 Biosignatures of the 3.5–3.2 Ga Barberton Greenstone Belt 155
7.3 Significance of Early Earth Microorganisms for the Search for Extraterrestrial Life 158
7.4 Conclusions 160
References 161

Part II Biosignatures in Space

8 The Search for Biosignatures in Martian Meteorite Allan Hills 84001 167
Harry Y. McSween Jr.
8.1 Introduction 168
8.2 Petrologic Context for Claims of Biologic Activity 169
8.3 Biogenic Carbonates? 170
8.4 Biogenic Magnetite and Sulphide? 172
8.5 Biological Organic Matter? 174
8.6 Microfossils and Related Biogenic Structures? 176
8.7 Summary and Lessons Learned 178
References 179

9 Role of Mineral Surfaces in Prebiotic Processes and Space-Like Conditions 183
John Robert Brucato and Teresa Fornaro
9.1 Introduction 183
9.2 Role of Minerals in the Prebiotic Context 187
9.2.1 Adsorption Processes 187
9.2.2 Catalysis 191

9.3 Role of Minerals in the Context of Space Missions 192
9.3.1 Mission Instruments to Detect Biomarkers in Space 192
9.3.2 Future Exploration of Mars 195
9.3.3 Laboratory Studies to Support Life Detection Investigations 197
9.4 Conclusions 199
References 200

10 Photochemistry and Photoreactions of Organic Molecules in Space 205
Avinash Vicholous Dass, Hervé Cottin, and André Brack
10.1 Why Are We Interested in Radiation Chemistry for Biosignatures? 206
10.2 Modes of Production and Delivery of Organic Material 206
10.2.1 Endogenous Delivery 206
10.2.2 Exogenous Delivery 207
10.3 Radiation Studies on Earth *versus* Space 208
10.4 Space Experiments Relevant to Exposure of Organics 209
10.5 Conclusions and Perspectives 219
References 220

11 Exoplanetary Biosignatures for Astrobiology 223
John Lee Grenfell
11.1 Introduction 223
11.2 Challenges of Defining Life and Its Needs 225
11.3 Preparing a Planetary Target List for Biosignature Assessment 225
11.3.1 Steps to Forming an Earth-Like Planet 225
11.3.2 Habitability and the Habitable Zone (HZ) 226
11.3.3 HZ Around M-Dwarf Stars 227
11.3.4 Exoplanet Missions Providing Planetary Targets 227
11.4 Remote Biosignature Techniques 228
11.5 Spectroscopic Exoplanetary Biosignatures 228
11.5.1 Reflectivity Increase with Wavelength 228
11.5.2 Isotopic Ratios 229
11.5.3 Gas-Phase Species as Biosignatures 230
11.5.4 Atmospheric Redox Disequilibrium as a Biosignature 239
11.5.5 Additional Atmospheric Biosignatures 240
11.6 Earth-Like Planets Without Life ('Dead Earths') 240
11.7 Discussion and Conclusions 241
References 243

Part III Biosignatures, Instruments and Missions

12 The Enigma of Methane on Mars . . . 253
Franck Lefèvre
12.1 Observations . . . 253
12.2 Methane Production . . . 257
12.3 Methane Loss . . . 258
12.4 Open Questions and Future Measurements . . . 263
References . . . 264

13 Detection of Biosignatures Using Raman Spectroscopy . . . 267
Frédéric Foucher
13.1 Introduction . . . 267
13.2 Raman Effect and Instrumentation . . . 268
13.3 Biosignatures Raman Detection . . . 272
13.3.1 Organic Molecules . . . 272
13.3.2 Pigments . . . 274
13.3.3 Carbonaceous Matter and Microfossils . . . 274
13.3.4 Minerals and Microfossils . . . 277
13.3.5 Living Organisms . . . 278
13.4 Summary . . . 279
References . . . 280

14 Searching for Signs of Life on Other Planets: Mars a Case Study . . . 283
Jorge L. Vago, Frances Westall, Barbara Cavalazzi, and The ExoMars Science Working Team
14.1 Introduction . . . 284
14.2 Geological Context and Biosignatures . . . 287
14.2.1 Morphological Biosignatures . . . 287
14.2.2 Chemical Biosignatures . . . 291
14.3 Conclusion . . . 295
References . . . 296

Part IV Biosignatures in a Philosophical Perspective

15 The History and Philosophy of Biosignatures . . . 303
David Dunér
15.1 Introduction . . . 304
15.2 The Concept of Life . . . 305
15.2.1 The Definition of Definition . . . 307
15.2.2 The Categorization of Our Categories . . . 308
15.2.3 Empirical Ignorance . . . 309
15.3 The Analogy of the Earth-Twins . . . 310
15.3.1 The Mountains of the Moon . . . 312
15.4 Epistemic Perception . . . 315
15.4.1 The Atmosphere of Venus . . . 316
15.4.2 The Canals of Mars . . . 319

15.5 Semiotics of Biosignatures 321
15.5.1 Icons of Life 324
15.5.2 Indices of Life 326
15.5.3 Symbols of Life 329
15.6 Conclusion 331
References 332

Index 339

Contributors

Karim Benzerara IMPMC, CNRS UMR 7590, Sorbonne Universités, MNHN, UPMC, IRD UMR 206, Paris, France

Sylvain Bernard IMPMC, CNRS UMR 7590, Sorbonne Universités, MNHN, UPMC, IRD UMR 206, Paris, France

Lydie Bonal Université Grenoble Alpes, Grenoble, France

Institut de Planétologie et d'Astrophysique de Grenoble, Grenoble, France

André Brack Centre de Biophysique Moléculaire, CNRS, Orléans, France

John Robert Brucato INAF – Astrophysical Observatory of Arcetri, Florence, Italy

Gaëtan Burgaud Université de Brest, Laboratoire Universitaire de Biodiversité et Ecologie Microbienne (EA3882), ESIAB, Plouzané, France

Barbara Cavalazzi Dipartimento di Scienze Biologiche, Geologiche e Ambientali, Università di Bologna, Bologna, Italy

Department of Geology, University of Johannesburg, Johannesburg, South Africa

Hervé Cottin Laboratoire Interuniversitaire des Systèmes Atmosphériques (LISA), UMR CNRS 7583, Université Paris Est Créteil et Université Paris Diderot, Institut Pierre Simon Laplace, Paris, France

Avinash Vicholous Dass Centre de Biophysique Moléculaire, CNRS, Orléans, France

Jean-Pierre de Vera German Aerospace Center (DLR), Institute of Planetary Research, Management and Infrastructure, Astrobiological Laboratories, Berlin, Germany

David J. Des Marais NASA-Ames Research Center, Mountain View, CA, USA

David Dunér History of Science and Ideas, Lund University, Lund, Sweden

Cognitive Semiotics, Lund University, Lund, Sweden

Virginia Edgcomb Woods Hole Oceanographic Institution, Woods Hole, MA, USA

Teresa Fornaro Carnegie Science, Geophysical Laboratory, Washington, DC, USA

Frédéric Foucher Centre de Biophysique Moléculaire-CBM, CNRS, Orléans, France

Frédéric Gaboyer Centre de Biophysique Moléculaire, CNRS, Orléans, France

John Lee Grenfell Department of Extrasolar Planets and Atmospheres (EPA), Institute for Planetary Research (PF), German Aerospace Centre (DLR), Berlin, Germany

Keyron Hickman-Lewis Centre de Biophysique Moléculaire, CNRS, Orléans, France

Università di Bologna, Bologna, Italy

Linda L. Jahnke NASA-Ames Research Center, Mountain View, CA, USA

Franck Lefèvre LATMOS, CNRS/Sorbonne Université, Paris, France

Harry Y. McSween University of Tennessee, Knoxville, TN, USA

Jennyfer Miot IMPMC, CNRS UMR 7590, Sorbonne Universités, MNHN, UPMC, IRD UMR 206, Paris, France

Eric Quirico Université Grenoble Alpes, Grenoble, France

Institut de Planétologie et d'Astrophysique de Grenoble, Grenoble, France

Jorge L. Vago European Space Agency, ESTEC, Noordwijk, The Netherlands

Frances Westall Centre de Biophysique Moléculaire, CNRS, Orléans, France

Part I
Biosignatures on Earth

Chapter 1
Chemical Biosignatures at the Origins

André Brack

Abstract Chemists searching for chemical biosignatures begin to define the chemical prerequisites for the emergence of life, a process based on organized molecules capable of self-reproduction and also with the capability of evolution. It is generally accepted that these prerequisites are liquid water and organic molecules, i.e. molecules that contained carbon and hydrogen atoms associated with atoms of oxygen, nitrogen and sulphur. This is not just an anthropocentric point of view, since water and carbon chemistry have very specific peculiarities. Two different kinds of chemical biosignatures are considered: an overrepresentation of organics and a long strand of homochiral sequences.

1.1 Introduction

From a chemical point of view, it is difficult to define life (Luisi 1998). Perhaps the most general working definition is that adopted in October 1992 by the NASA Exobiology Programme: "Life is a self-sustained chemical system capable of undergoing Darwinian evolution" (Joyce 1995). The concept of evolution implies that the chemical system transfers its information fairly faithfully but, in so doing, makes a few random errors. These may potentially lead to a higher complexity or efficiency, and possibly to a better adaptation to changes under existing environmental constraints. Life is expected to appear as an open chemical system capable of self-reproduction, i.e. making more of itself by itself, and to be capable of evolving, and can thus be defined as a sort of chemical automaton (Brack and Troublé 2010).

Chemists tackle three different aspects of the origins of life: the origin of terrestrial life, the possibility of an alien life on Earth, and the possible emergence of an extraterrestrial form of life. For each of these three fundamental questions, it is necessary to define the biological prerequisites before searching for any chemical biosignatures.

A. Brack (✉)
Centre de Biophysique Moléculaire, CNRS, Orléans, France
e-mail: brack@cnrs-orleans.fr

B. Cavalazzi, F. Westall (eds.), *Biosignatures for Astrobiology*, Advances in Astrobiology and Biogeophysics, https://doi.org/10.1007/978-3-319-96175-0_1

1.2 Chemical Prerequisites

It is generally believed that the requisite for the an emergence of life is the simultaneous presence of liquid water and organic molecules, i.e. molecules that contained carbon and hydrogen atoms associated with oxygen, nitrogen and sulphur, as in the case of present life. This is not just an anthropocentric guideline, since water and carbon chemistry have a number of peculiarities when compared, for example, to silicon-based biochemistry in non-aqueous solvents (Bains 2004).

1.2.1 Liquid Water

As parts of an open system, the constituents of a living system must be able to diffuse at a reasonable rate. A solid-state life is generally discarded, the constituents being unable to migrate or to be easily exchanged. A gaseous phase would allow fast diffusion of the parts but the limited inventory of stable volatile organic molecules would constitute a severe restriction. A liquid phase offers the best environment for the diffusion and the exchange of dissolved organic molecules. Besides liquid water, other solvents can be considered such as liquid ammonia, hydrogen sulphide, and sulphur oxide, together with hydrocarbons, organic acids and/or alcohols. Compared to any of these possible solvents, liquid water exhibits many promising specificities. Liquid water is a fleeting substance that can persist only above 0 °C and under a pressure higher than 6 mbars. The freezing point of water can be depressed by adding salts (brines). For instance, the 5.5% by weight salinity of the Dead Sea depresses the freezing point of seawater by about 3 °C. Large freezing point depressions are observed for 15% LiCl (23.4 °C) and for 22% NaCl (19.2 °C). Monovalent and divalent salts are essential for terrestrial life because they are required as co-catalysts in many enzymatic activities. Usually, the tolerated salt concentrations are quite low (<0.5%) because high salt concentrations disturb the networks of ionic interactions that shape biopolymers and hold them together. However, both eukaryotic and prokaryotic salt-loving microorganisms—known as extreme halophiles—tolerate a wide range of salt concentrations (1–20%) and some prokaryotes have managed to thrive in hypersaline biotopes (such as sabkhas, salt-lakes) containing up to 25–30% sodium chloride.

Water is a good solvent thanks to its hydrogen bonds. According to its molecular weight, water should be a gas under standard terrestrial conditions by comparison with CO_2, SO_2 or H_2S. Its liquid state is due to its ability to form hydrogen bonds. This is not restricted to water molecules since alcohols exhibit a similar behaviour, however, the polymeric network of water molecules *via* H-bonds is so tight that the boiling point of water is raised from 40 °C, a temperature inferred from the boiling point of the smallest alcohols, to 100 °C. Biopolymers, such as nucleic acids, proteins and membranes, contain C_xH_yO, N, S-groups and C_xH_y-groups (hydrocarbon groups). Groups like C_xH_yO, N, S, especially those bearing ionisable groups such as –COOH or $–NH_2$, form hydrogen bonds with water molecules and therefore display an affinity for water. They are soluble in water and hydrophilic. The large dipole moment of water (1.85 debye) favours the dissociation of the ionisable groups while the high dielectric constant ($\varepsilon = 80$) prevents recombination of the ions, the attraction forces for ion re-association being proportional to $1/\varepsilon$. This is also true for metallic ions, which are associated with the biopolymers. C_xH_y-groups cannot form hydrogen bonds with water molecules and thus water molecules tend to escape. They are insoluble in water and hydrophobic. These two groups co-exist in biopolymers and this co-existence drives the conformation (geometry) of the biopolymers in water, i.e. into forms such as helices, β-sheets, micelles, vesicles or liposomes. Water participates in the production of clays, which probably played an important role in the emergence of life. It stabilises the biopolymer conformation by hydrophobic clustering and is also a good heatsink.

Liquid water was almost permanently present at the surface of the Earth thanks to both the size of the planet and its distance to the Sun (Pinti 2005). If the planet were happened to be much smaller, like Mercury or the Moon, it would not have been able to retain any atmosphere and, therefore, no ocean of liquid water. If the planet were too close to the star, the mean temperature would have risen due to starlight intensity. Any seawater present would evaporate delivering large amounts of water vapour to the atmosphere thus contributing to the greenhouse effect. Such a positive feedback loop could lead to a runaway greenhouse: all of the surface water would be transferred to the upper atmosphere where photo-dissociation by ultraviolet light would break the molecules into hydrogen, which escapes into space, and oxygen, which would be recombined into the crust. The Earth hosted permanent liquid water thanks to its constant greenhouse atmosphere, however, water risked provoking its own disappearance. The atmospheric greenhouse gas CO_2 normally dissolves in the oceans and is eventually trapped as insoluble carbonates through rock weathering. This negative feedback is expected to lower the surface pressure and temperature to an extent that water would be largely frozen. On Earth, active plate tectonics and volcanism recycled the carbon dioxide by breaking down subducted carbonates.

1.2.2 Organic Molecules

Life is autocatalytic in essence and must be able to evolve. To evolve, i.e. improving its efficiency of self-reproduction and increasing its diversity, the molecules bearing hereditary memory must reach a certain level of complexity. This can be best achieved with a scaffolding of polyvalent atoms. In chemists' hands, carbon chemistry is very productive in this respect. Another clue in favour of carbon is provided by radio astronomers: about 110 carbon-containing molecules, up to $HC_{10}CN$, have been identified in the interstellar medium, whereas while only 11 silicon-based molecules, up to SiH_4, have been detected (Wikipedia).

Charles Darwin was the first to envision an organic approach to the origin of life. In February 1871, he wrote in a private letter to Joseph Hooker: "If (and oh, what a big if) we could conceive in some warm little pond, with all sorts of ammonia and phosphoric salts, light, heat, electricity, etc., present that a protein compound was chemically formed, ready to undergo still more complex changes, at the present day such matter would be instantly devoured or adsorbed, which would not have been the case before living creatures were formed".

1.2.2.1 Production of Organics in the Atmosphere

The simplest sources of carbon susceptible to building up the prebiotic organic molecules are gaseous, i.e. carbon dioxide (CO_2) and monoxide (CO) for the oxidised forms and methane (CH_4) for the reduced forms. Oparin (1924) suggested that the reduced small organic molecules needed for primitive life were formed in a primitive atmosphere dominated by methane. His idea was tested in the laboratory by Miller who exposed a mixture of methane, ammonia, hydrogen, and water to spark and silent electric discharge (Miller 1953). In this initial experiment, Miller obtained three amino acids (glycine, alanine and β-alanine) *via* the intermediary formation of hydrogen cyanide and aldehydes. More generally, simple gaseous molecules, like CH_4, H_2, NH_3, and H_2O, require a supply of energy (UV, heat, electric discharges, cosmic rays, shock waves) to react with each other. They generate compounds like formaldehyde and hydrogen cyanide, which store chemical energy in their double and triple chemical bonds, respectively. Chang (1993) reviewed the possible sources of atmospheric synthesis including electric effects, solar UV and impact shocks.

Miller's laboratory synthesis of amino acids occurs efficiently when a reducing gas mixture containing significant amounts of hydrogen is used. However, the true composition of the primitive Earth's atmosphere is not known. The dominant view is that the primitive atmosphere consisted mainly of CO_2, N_2, and H_2O, along with small amounts of CO and H_2 (Kasting and Brown 1998; Catling and Kasting 2007). Only small yields of amino acids are formed in such a mixture (Schlesinger and Miller 1983; Miller 1998). More recent studies show that the low yields previously reported appear to be the outcome of the oxidation of organic compounds during hydrolytic workup by nitrite and nitrate produced in the reactions. The yield of amino acids is greatly increased when oxidation inhibitors, such as ferrous iron, are added prior to hydrolysis, suggesting that endogenous synthesis from neutral atmospheres may be more important than previously thought (Cleaves et al. 2008). Additionally, twenty-two amino acids and five amines were obtained when re-analysing archived Miller's archived samples obtained by lightning applied to volcanic gases. The volcanic apparatus experiment suggests that, even if the overall atmosphere was not reducing, localized prebiotic synthesis could have occurred in volcanic plumes (Johnson et al. 2008). Stanley Miller also sparked a gaseous mixture of CH_4, NH_3, and H_2O, while intermittently adding the plausible prebiotic condensing reagent cyanamide. For unknown reasons, an analysis of the samples was not reported. After his death, the archived samples were analysed for amino acids, dipeptides, and diketopiperazines by liquid chromatography, ion mobility spectrometry, and mass spectrometry. A dozen amino acids, ten glycine-containing dipeptides, and three glycine-containing diketopiperazines were detected. Miller's experiment was repeated and aqueous heating experiments indicate that Strecker synthesis intermediates play a key role in facilitating polymerization (Parker et al. 2014).

The escape of hydrogen from the early Earth's atmosphere has recently been re-evaluated (Tian et al. 2005). It likely occurred at rates two orders of magnitude more slowly than previously thought. The balance between slow hydrogen escape and volcanic outgassing could have maintained a hydrogen mixing ratio of more than 30%, thus producing more amino acids than previously thought.

Intense bombardment probably caused some chemical reprocessing of the Earth's primitive atmosphere by impact shock chemistry (Brack 2009). An indication of the number and timing of the impacts onto the early Earth can be obtained by comparison with the crater record of the Moon, which records impacts from the earliest history of the Solar System (Ryder 2003). Because of the larger size of the Earth and its greater gravitational pull, about 20 times as many impacts would have occurred on the early Earth as on the Moon. Computer modelling of the resulting impact shock chemistry shows that the nature of the atmosphere strongly influences the shock products (Fegley et al. 1986). A neutral CO_2-rich atmosphere produces CO, O_2, H_2 and NO, whereas a reducing CO-rich atmosphere yields primarily CO_2, H_2, CH_4, HCN, NH_3, and H_2CO. The last three compounds are particularly interesting for prebiotic chemistry since they can lead to amino acids *via* Strecker synthesis. However, a CO-rich primitive atmosphere probably has no counterpart in prebiotic reality. In laboratory experiments, a gas mixture of methane, ammonia and water subjected to shock heating followed by rapid thermal quenching yielded the amino acids glycine, alanine, valine and leucine (Bar-Nun et al. 1970). Here again, the gas mixture used does not represent a realistic primitive atmosphere, which was dominated by CO_2. Laboratory simulations of shocks were also run with a high-energy laser. CH_4-containing mixtures generated hydrogen cyanide and acetylene but no organics could be obtained with CO_2-rich mixtures (McKay and Borucki 1997).

Hydrogen cyanide was produced in the laboratory by the impact of a polycarbonate projectile and graphite through N_2-rich atmosphere. A significant fraction (>0.1 mol%) of the vaporized carbon was converted into HCN and cyanide condensates, even when the ambient gas contains as much as a few hundred mbar of CO_2 (Kurosawa et al. 2013).

1.2.2.2 Submarine Hydrothermal Systems

The reducing conditions in hydrothermal systems may have been an important source of biomolecules on the primitive Earth (Baross and Hoffman 1985; Holm 1992; Holm and Andersson 1998, 2005). The reducing environment results from the flow of substances dissolved in seawater through inorganic compounds present in very hot crustal material that reduces compounds in seawater. These reduced compounds flow out of the hydrothermal system and the resulting inorganic sulphides formed precipitate when they mix with the cold (4 °C) ocean water. For example, hydrocarbons containing 16–29 carbon atoms have been detected in the Rainbow ultramafic hydrothermal system, Mid-Atlantic Ridge (Holm and Charlou 2001). Hydrothermal vents are often disqualified as efficient reactors for the synthesis of bioorganic molecules due to their high temperature. Experiments exploring the potential for amino acid synthesis at high temperature from synthetic seawater solutions of varying composition have been conducted (Aubrey et al. 2009). The synthesis of amino acids was examined as a function of temperature, heating time, starting material composition and concentration. Using very favourable reactant conditions (high concentrations of reactive, reduced species), small amounts of a limited set of amino acids can be generated at moderate temperature conditions (~125–175 °C) over short heating times of only a few days, but even these products are significantly decomposed after exposure times of approximately one week. Therefore, although amino acids can be generated from simple, likely environmentally available precursors under submarine hydrothermal system conditions, their equilibrium at high temperatures favours net amino acid degradation rather than synthesis, and that synthesis at lower temperatures may be more favourable. However, the products that are synthesized in hot vents are rapidly quenched in the surrounding cold water thanks to the good heat conductivity of water and may therefore be preserved (Ogata et al. 2000).

1.2.2.3 Delivery of Extraterrestrial Organic Matter

The Earth has experienced a large range of impactors ranging from the huge Mars-sized impactor that created the Moon to cosmic dust less than 1 μm in size. A great number of organic molecules, including amino acids, have been found in carbonaceous chondrites. Micrometeorite collection and analysis from the Greenland and Antarctic ice sheets suggests that the Earth accreted large amounts of complex organic molecules of extraterrestrial origin. Intense bombardment probably also caused some chemical reprocessing of the Earth's primitive atmosphere.

Comets—Comets are, as is known thus far, the planetary objects richest in organic compounds. Ground-based observations have detected hydrogen cyanide and formaldehyde in the coma of comets. In 1986, on-board analyses performed by the two Russian missions Vega 1 and 2, as well as observations obtained by the European mission Giotto and the two Japanese missions Suisei and Sakigake, demonstrated that Comet Halley shows substantial amounts of organic material. On average, dust particles ejected from the nucleus of Comet Halley contain 14% of organic carbon by mass. About 30% of cometary grains are dominated by the light elements C, H, O, and N, and 35% are close in composition to the carbon-rich meteorites. Many chemical species of interest for astrobiology were detected in Comet Hyakutake in 1996, including ammonia, methane, acetylene, acetonitrile, and hydrogen isocyanide. In addition, the study of Comet Hale-Bopp in 1997 led to the detection of methane, acetylene, formic acid, acetonitrile, hydrogen isocyanide, isocyanic acid, cyanoacetylene, formamide and thioformaldehyde.

The Stardust mission collected samples of Comet Wild 2 and returned them to Earth in January 2006 for laboratory analysis. Unexpectedly, most of the comet's rocky matter formed inside the Solar System at extremely high temperature. The grains contain organic functions (alcohol, ketone, aldehyde carboxylic acids, amides, nitrile). The protein-building amino acid glycine has also been discovered. Cometary grains may therefore appear as an important source of organic molecules delivered to the primitive Earth (Ehrenfreund and Charnley 2000; Despois and Cottin 2005). However, it is unlikely that whole comets could have safely delivered organics to the Earth. They would either have exploded while crossing the atmosphere or when impacting the Earth's surface.

The ESA Rosetta robotic spacecraft performed the most detailed study of a comet ever attempted (Glassmeir et al. 2007). Launched in March 2004, the spacecraft reached Comet 67P/Churyumov–Gerasimenko in August 2014. It consisted of two main elements: the Rosetta space probe orbiter and the Philae robotic lander. The orbiter featured eleven instruments. Among them, ROSINA (Rosetta Orbiter Spectrometer for Ion and Neutral Analysis) measured the deuterium-to-hydrogen ratio of water vapour emanating from the comet and found it to be more than three times greater than for Earth's oceans (Altwegg et al. 2015). The discovery fuels the debate on the origin of Earth's oceans. The instrument also made the first measurement of molecular nitrogen for a comet, providing clues about the temperature of the environment in which the comet formed (Rubin et al. 2015). Moreover, ROSINA detected volatile glycine, accompanied by methylamine and ethylamine, in the coma, confirming the results of the Stardust mission. Together with the detection of phosphorus and a multitude of organic molecules, this result demonstrates that comets could have played a crucial role in the emergence of life on Earth (Altwegg et al. 2016).

The Philae robotic probe landed on Comet 67P/Churyumov–Gerasimenko on 12 November 2014, achieving the first-ever soft landing on a comet nucleus. It hosted nine instruments. Among them, COSAC (Cometary Sampling and Composition experiment) was designed to detect and identify complex organic molecules from their elemental and molecular composition. SD2 (Sample and Distribution Device) was designed to drill more than 20 cm into the surface, collect samples and deliver them to different ovens or for microscope inspection. When the Philae lander successfully touched down on the comet, it unfortunately bounced twice to finally settling in a location and layout preventing the solar panels from charging the batteries. After a year of silence from the craft, the European Space Agency began to power down the systems.

Philae nonetheless provided interesting results for astrobiology. Just after the first comet touchdown, COSAC mass spectrometer took a spectrum in sniffing mode of the ejected material. The spectrum displayed a suite of 16 organic compounds, including methyl isocyanate, acetone, propionaldehyde, and acetamid (Goesmann et al. 2015).

Meteorites—Carbonaceous chondrites also delivered organic materials to the early Earth. They contain between 1.5% and 4% carbon, for the most part as organic materials. One hundred kilograms of the Murchison meteorite, a CM2 type carbonaceous chondrite that fell in Australia in 1969, have been extensively analysed

(Pizzarello 2007; Pizzarello and Shock 2010). Murchison organic materials are generally classified according to their solubility in water and organic solvents. Insoluble and soluble components represent respectively 70% and 30% of the total carbon components. The insoluble organic material is a poorly identified macromolecular material of complex composition with average elemental abundances of $C_{100}H_{46}N_{10}O_{15}S_{4.5}$. NMR, IR and pyrolysis analyses suggest the presence of aromatic ring clusters bridged by aliphatic chains, with peripheral branching and functional groups. The insoluble organic material releases a variety of aromatic and heteroatomic hydrocarbons as well as a suite of alkyl dicarboxylic acids up to C18 chain length under conditions similar to those of hydrothermal vents (Yabuta et al. 2007).

The soluble organic compounds of the Murchison meteorite represent a diverse and abundant group of organics that vary from small water-soluble compounds such as amino acids and polyols up to 30 carbon-long hydrocarbons. This diversity has been analysed in detail for the amino acids. The total number of meteoritic amino acids is about one hundred. All the possible α-amino alkylamino acids up to seven-carbon have been identified as well as large abundances of N-substituted, cyclic, β-, γ-, δ-, and ε-amino acids. Eight protein-building amino acids (glycine, alanine, proline, leucine, isoleucine, valine, aspartic acid and glutamic acid) have been found. Nucleic acid bases, purines and pyrimidines have also been noted in the Murchison meteorite (Stoks and Schwartz 1982; Callahan et al. 2011). No ribose, the sugar moiety linking together nucleic acid building blocks, has been detected in meteorites. Vesicle-forming fatty acids have been extracted from different carbonaceous meteorites (Deamer 1985, 1998).

A combination of high-resolution analytical methods, comprising organic structural spectroscopy FTICR/MS, UPLC-QTOF-MS and NMR, applied to the organic fraction of Murchison extracted under mild conditions allowed chemists to extend its indigenous chemical diversity to tens of thousands of different molecular compositions, and likely millions of diverse structures (Schmitt-Kopplin et al. 2010).

Most of the amino acids detected in the carbonaceous chondrites are chiral but present as racemates, i.e. L- and D-enantiomers are present in equal proportions. However, enantiomeric excesses have been detected, as described in Sect. 1.3.2.1. The discovery of a large number of meteorites since 1969 has provided new opportunities to search for organic compounds in CM type carbonaceous chondrites (Pizzarello et al. 2001; Glavin et al. 2006; Pizzarello and Shock 2010).

Micrometeorite. Micrometeorite collections in the Greenland and Antarctica ice sheet (Fig. 1.1) show that the Earth captures interplanetary dust as micrometeorites at a rate of about 20,000 tonnes per year. About 99% of this mass is carried by micrometeorites in the 50–500 μm size ranges. This value is about 2000 times higher than the most reliable estimate of the meteorite flux (about 10 tonnes per year). The amazing dominance of micrometeorites within impactors already supports their possible role in delivering complex organics to the early Earth between 4.1 to 3.9 Ga ago, when the flux of impacting objects was probably enhanced by several orders of magnitude. Antarctic micrometeorite flux measurements suggest that a huge mass ($\sim 5 \times 10^{24}$ g) of micrometeorites was accreted by the Earth during ~200 Ma of the late heavy bombardment (Maurette 1998, 2006).

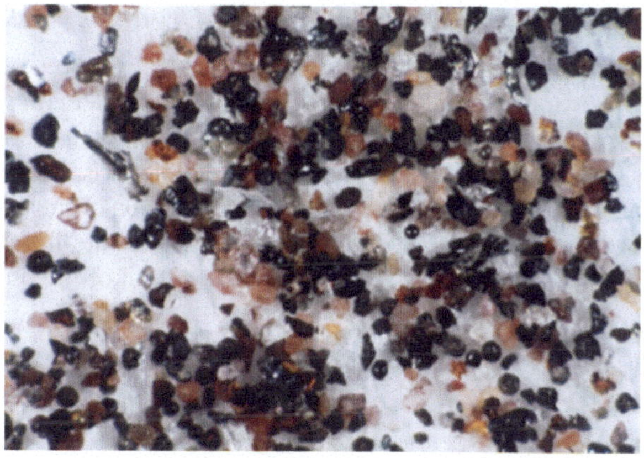

Fig. 1.1 Micrometeorites, 50–100 μm in size, collected in Antarctica ice. Source: Image courtesy of M Maurette

At least about 20 wt.% of the micrometeorites survives unmelted upon atmospheric entry. As their insoluble organic fraction represents about 2.5 wt.% of carbon, this amounts to a total mass of carbon of ~2.5×10^{22} g on the early Earth surface. This is equivalent to a ~30 m thick global layer deposited during ~200 Ma (Maurette and Brack 2006). This delivery represents more carbon than that present in the biomass of the present day Earth (10^{18} g). One amino acid, α-amino isobutyric acid, has been identified in Antarctic micrometeorites (Brinton et al. 1998; Matrajt et al. 2004). These grains also contain a high proportion of metallic sulphides, oxides and clay minerals, a rich variety of inorganic catalysts which could have promoted reactions of the carbonaceous material leading to the origin of life. Analysis of the dust grains collected by the Cosmic Dust mission supports a cometary origin for the micrometeorites collected in Antarctica.

A collection of CONCORDIA Antarctic micrometeorites recovered from ultraclean snow close to Dome C provided the most unbiased collection of large cosmic dust available. Many similarities can be found between Antarctic micrometeorites and Wild 2 samples, in terms of chemical, mineralogical, and isotopic compositions, and in the structure and composition of their carbonaceous matter (Dobrica et al. 2013). The cometary origin has been confirmed by a zodiacal cloud model based on the orbital properties and lifetimes of comets and asteroids, and constrained by Infrared Astronomical Satellite observations of thermal emission, but is also qualitatively consistent with meteor observations, with spacecraft impact experiments, and with the properties of recovered micrometeorites (Nesvorny et al. 2010).

1.3 Chemical Biosignatures

Two different kinds of chemical biosignatures can be considered: an overrepresentation of organics and long strands of homochiral sequences, such as homochiral poly amino acids.

1.3.1 *Over Representation of Organics*

Any extraterrestrial explorer searching for life on Earth would probably be struck by the overabundance of organics by comparison to the composition of the mantle. Carbon, which constitutes 17.9% of the biomass, accounts for only 0.094% of the mantle. However, the presence of active carbon chemistry alone—or even associated with liquid water—does not necessarily generate life as illustrated by the case of Titan and also by the search for an alien form of life on Earth.

1.3.1.1 Titan

The Voyager 1 mission in 1980 first revealed the composition of Titan's atmosphere: 90% molecular nitrogen and about 1–8% methane. Further, a great number of trace constituents were observed in the form of hydrocarbons, nitriles, and oxygen-containing compounds, mostly CO and CO_2. Titan is the only other object in our Solar System to bear any resemblance to our own planet in terms of atmospheric pressure (1.5 bar) and carbon/nitrogen chemistry. It represents, therefore, a natural laboratory to study the formation of complex organic molecules on a planetary scale and over geological timescales.

The ISO satellite has detected tiny amounts of water vapour in the higher atmosphere, but Titan's surface temperature (94 K) is much too low to allow the presence of liquid water. Although liquid water is totally absent, the satellite provides a unique milieu to study, in situ, the products of the fundamental physical and chemical interactions driving planetary organic chemistry.

In 2004, the NASA/ESA Cassini-Huygens spacecraft launched in October 1997 arrived in the vicinity of Saturn and performed several flybys of Titan, taking spectroscopic, imaging, radar, and other measurements. On January 14, the European instrumented descent probe penetrated the atmosphere and systematically studied the organic chemistry in Titan's geofluid. For 150 min, in situ measurements provided analyses of the organics present in the air, in the aerosols, and at the surface. The GC-MS of the Huygens probe measured the chemical composition and the isotopic abundances from an altitude of 140 km down to the surface (Niemann et al. 2005; Israël et al. 2005). The main findings were:

1. nitrogen and methane are the main constituents of the atmosphere;
2. the isotopic ratio $^{12}C/^{13}C$ suggests a permanent supply of methane in the atmosphere;
3. the surface is "wetted" by liquid methane and rich in organics (cyanogen, ethane);
4. the presence of ^{40}Ar suggests the existence of internal geological activity.

Unfortunately, the Huygens probe had no specific instruments to detect any potential autocatalytic systems leading to non-stochastic chemicals such as homochiral molecules (Brack and Spach 1987).

1.3.1.2 Alien Life on Earth

Life, being basically an autocatalytic chemical system, should be a reproducible process, if the starting conditions are identical. Therefore, the emergence of one or more types of alien terrestrial life, including a mirror-image life, coexisting with known life, should be conceivable (Davies and Lineweaver 2005). So far, no fossils of such a terrestrial alien life have been found. As for the emergence of a modern alien life, the insatiable appetite of bacteria, which have squatted in all habitable terrestrial sites, would probably prevent any de novo origin of life.

1.3.2 One-Handedness

The carbon atom occupies the centre of a tetrahedron and when the four substituents at the four summits are different, the carbon atom becomes one-handed and shows two mirror images, a left-handed form and a right-handed form.

Present terrestrial life uses proteins, which catalyse biochemical reactions, nucleic acids, which carry genetic information, and phospholipids, which form the semi-permeable membranes around cells. Most of the constituents, amino acids, sugars and lipids, contain at least one asymmetric carbon atom. Life is one-handed so only one enantiomer of each chiral biomolecule is present in living systems. For example, each amino acid, with the exception of glycine, exists in two enantiomeric forms, L and D, but proteins use only L forms (Fig. 1.2).

Proteins adopt asymmetrical rigid geometries, α-helices and β-sheets, which play a key role in the catalytic activity. Homochirality is now believed to be not just a consequence of life, but also a prerequisite for life, because stereoregular structures such as protein β-sheets, for example, do not form with mixtures of monomers of both handednesses (Brack and Spach 1981). The use of one-handed biomonomers also sharpens the sequence information of the biopolymers. For a polymer made of n units, the number of sequence combinations will be divided by 2^n when the system uses only homochiral monomers. Taking into account the fact that enzyme chains are generally made of hundreds of monomers, the tremendous gain in simplicity offered by the use of monomers restricted to.

Any chemical reaction performed in a symmetrical environment that forms chiral molecules yields a racemic mixture i.e. a mixture of equal quantities of right- and left-handed enantiomers. Theoretical models show that autocatalytic systems fed with both left- and right-handed molecules must become one-handed in order to survive. The problem of the generation of homochirality thus has two parts, the origin of a prevalent enantiomer and its further amplification until the appearance of life. The amplification of enantiomeric excesses of a few percent has been well documented in the laboratory using crystals, crystallization processes or biopolymers. Shibata et al. (1998) have clearly demonstrated that an excess of one enantiomer can be induced by the presence of few per cent of a chiral initiator and this

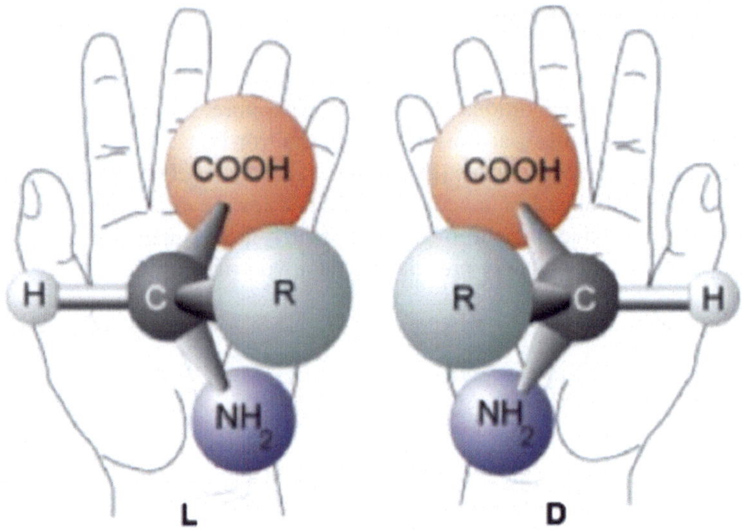

Fig. 1.2 L-form and D-form, mirror image enantiomers of a generic chiral amino acid. Source: credit ESA

excess can be dramatically amplified by asymmetry autocatalysis. The source of the chiral initiator could be extraterrestrial, as already mentioned.

Theoretical models for the existence of the slightly prevalent enantiomer excess on Earth can be divided into two classes, those which call for a chance mechanism and those which call for a determinate mechanism resulting from an asymmetrical environment.

1.3.2.1 Enantiomeric Excess Via a Chance Mechanism

The proponents of the chance mechanism hypothesis argue that the notion of equimolarity of a racemic mixture is relative, and random fluctuations may favour one enantiomer over the other. For instance, for a population of ten million molecules, which is about the amount of chiral constituents of the smallest living cell, the probability to find an excess of 0.02% or more in one enantiomer, is about 50% (Spach and Brack 1988). In a rather simple kinetic model proposed by Frank (1953), an open flow reactor, running in far from equilibrium conditions, is fed by prochiral compounds and forms two enantiomers reversibly and auto catalytically. If the two enantiomers can react to form an irreversible combination flowing out the reactor (by precipitation, for instance), and if certain conditions of fluxes and concentrations are reached, the racemic production may become metastable and the system switches permanently toward the production of either one or the other enantiomer, depending on a small excess in one enantiomer. The Frank model has inspired a further model described as a theoretical framework, based on the stereo-selective reactivity of pre-existing chiral monomeric building blocks (polymerization, epimerization, and depolymerisation) maintained out of equilibrium by a continuous energy income, *via* an activation reaction. It permits the self-conversion of all monomeric subunits into a single chiral configuration (Plasson et al. 2004). A model featuring enantiomer cross-inhibition and chiral bias has been used to study the diffusion equations controlling the spatiotemporal development of left- and right-handed domains in the context of autocatalytic polymerization reaction networks (Gleiser 2007). A fully self-contained model of homochirality has been proposed that contains the effects of both polymerization and dissociation (Brandenburg et al. 2005).

Spontaneous resolution on crystallization represents the most effective means of chiral symmetry breaking by chance mechanism. Kondepudi et al. (1990) demonstrated a total spontaneous resolution by crystallization. Sodium chlorate crystals are optically active although the molecules of the compound are not chiral. When crystallized from an aqueous solution while the solution is not stirred, statistically equal numbers of L and D crystals were found. When the solution was stirred, almost all the $NaClO_4$ crystals (99.7%) in a given sample had the same chirality, either L or D. Quartz, known as an asymmetric adsorbent and a catalyst, may have undergone such a spontaneous resolution during crystallization when it cooled down on the primitive Earth. Close examination of over 2700 natural quartz crystals gave 49.83% L and 50.17% D, i.e. an almost even distribution. Beautiful enantio-separations have been obtained by the Lahav and colleagues with crystals, crystallites and crystalline self-assemblies at the water surface (Weissbuch et al. 1988, 1997; Zepik et al. 2002; Rubinstein et al. 2005).

Enantio-enrichment of a variety of amino acids has also been obtained by sublimation of near-racemic samples (Fletcher et al. 2007; Perry et al. 2007; Tarasevych et al. 2013, 2015).

1.3.2.2 Determinate Mechanisms

In terms of determinate mechanisms, parity non-conservation has raised many hopes and caused many disappointments. There is an extremely small parity-violating energy difference in favour of L-amino acids in water and in favour of D-sugars (Nordén et al. 1985). The energy difference is about 3×10^{-19} eV corresponding to one part in 10^{-17} for the excess of L-molecules in a racemic mixture at thermodynamic equilibrium at ambient temperature. This fundamental asymmetry of matter has been examined from various aspects such as circularly polarized photons emitted by the slowing down of longitudinally polarized electrons (Bremsstrahlung), inducing degradation reactions or stereo selective crystallization of racemic mixtures. No experiment has convincingly supported these theoretical considerations for the origin of a dominant enantiomer on Earth. Either the results were shown to be artefacts or so weak that they are doubtful (Mac Dermott 1995).

Other chiral force fields that could have been acting on the Earth surface have been considered. Asymmetric synthesis and degradation have been achieved using circularly polarized light. On Earth, the classical electromagnetic interactions such as circularly polarized light or other fields would probably never result in a very high yield of optically pure compounds. They would also probably cancel on a time and space average.

The possibility to induce a chiral effect by submitting a suitable chemical reaction to a magnetic field first attempted by Pasteur has received experimental support in Rikken and Raupach (2000). The authors used a chiral chromium complex, which is unstable in solution and spontaneously dissociates and re-associates. The dissociation is accelerated by the absorption of light. In the presence of an unpolarised laser beam travelling parallel to a static magnetic field, a small excess of one enantiomer is produced and maintained. On reversing the magnetic field direction, the mirror-image enantiomer is obtained. Magnetochiral photochemistry therefore appears to be a possible source for biological homochirality. This is especially pertinent to hypotheses suggesting that complex organic molecules could evolve in the ice mantles of dust grains in interstellar space, because magnetic fields and unpolarised light are more common in the cosmos than circularly polarized light.

1.3.2.3 Extraterrestrial Homochirality

Most of the amino acids detected in carbonaceous chondrites are asymmetric, with L- and D-enantiomers present in equal proportions. However, Cronin and Pizzarello (Cronin and Pizzarello 1997; Pizzarello and Cronin 2000) found small L-enantiomer excesses in six α-methyl-α-amino alkanoic acids from the Murchison (2.8–9.2%) and Murray (1.0–6.0%) carbonaceous chondrites. Enantiomeric excesses up to 18% have been measured for isovaline, 2-methyl-2-aminobutyric acid. These amino acids are either unknown or rare in the terrestrial biosphere and cannot therefore be attributed to terrestrial contamination (Pizzarello 2007). In addition, the indigeneity of D- and L-isovaline enantiomers is supported by carbon and hydrogen isotopic data (Pizzarello et al. 2003; Pizzarello and Huang 2005). The Renazzo-type (CR) chondrites found in Antarctica revealed natal enantiomeric excesses of up to 60% (Pizzarello et al. 2012).

The excess of one-handed amino acids found in the Murchison meteorite may result from the processing of the organic mantles of interstellar grains by circularly polarized synchrotron radiation from the neutron star remnant of a supernova (Bonner 1991). Strong infrared circular polarization, resulting from dust scattering in reflection nebulae in the Orion OMC-1 star-formation region, has been observed (Bailey et al. 1998) and circular polarization at shorter wavelengths might have been important in inducing chiral asymmetry in interstellar organic molecules that could have been subsequently delivered to the early Earth (Bailey 2001).

1.4 Conclusions

On Earth, life probably appeared about 4 billion years ago, when some assemblages of organic molecules in a liquid water medium began to transfer their chemical information and to evolve by making a few accidental transfer errors. The number of molecules required for those first assemblages is still unknown. The problem is that, on Earth, those molecules have been erased. If life started on Earth with the self-organisation of a relatively small number of molecules, its emergence must have been quick; therefore the chances for the appearance of life on any appropriate celestial bodies are real. On the contrary, if the process required thousands of different molecules, the event risks being unique and restricted to the Earth. The discovery of a second independent genesis of life on a body presenting environmental conditions similar to those which prevailed on the primitive Earth, such as Mars (Westall et al. 2015), would strongly support the idea of a rather simple genesis of terrestrial life. More than just a societal wish, the discovery of a second genesis of life is a scientific need for the study of the origin of life. It will demonstrate that life is not a magic one-shot process but rather a common phenomenon. Many scientists are convinced that microbial life is not restricted to the Earth but such conviction now needs to be supported by facts. So far, chemists have been able to propose requirements for the origin of terrestrial, alien terrestrial and extraterrestrial life and put forward a certain number of chemical biosignatures. However, only joint efforts between astronomers, planetologists, geologists and geochemists will be needed to legitimize those prerequisites and evaluate chemical biosignatures.

Acknowledgements The occasion is given here to thank all the exo/astrobiologists I have met for the quality of their company while enjoying good science, good food, and good wine. They are too numerous to be quoted individually, but they know who they are, for sure. I thank also CNES and ESA for their constant support.

References

Altwegg K, Balsiger H, Bar-Nun A et al (2015) 67P/Churyumov-Gerasimenko, a Jupiter family comet with a high D/H ratio. Science 347:1261952

Altwegg K, Balsiger H, Bar-Nun A et al (2016) Prebiotic chemicals-amino acid and phosphorus-in the coma of comet 67P/Churyumov-Gerasimenko. Sci Adv 27:e1600285

Aubrey AD, Cleaves HJ, Bada JL (2009) The role of submarine hydrothermal systems in the synthesis of amino acids. Orig Life Evol Biosph 39:91–108

Bailey J (2001) Astronomical sources of circularly polarized light and the origin of homochirality. Orig Life Evol Biosph 31:167–183

Bailey J, Chrysostomou A, Hough JH et al (1998) Circular polarization in star formation regions: implications for biomolecular homochirality. Science 281:672–674

Bains W (2004) Many chemistries could be used to build living systems. Astrobiology 4:137–167

Bar-Nun A, Bar-Nun N, Bauer SH et al (1970) Shock synthesis of amino acids in simulated primitive environments. Science 168:470–473
Baross JA, Hoffman SE (1985) Submarine hydrothermal vents and associated gradient environment as sites for the origin and evolution of life. Orig Life Evol Biosph 15:327–345
Bonner WA (1991) The origin and amplification of biomolecular chirality. Orig Life Evol Biosph 21:59–111
Brack A (2009) Impacts and origins of life. Nat Geosci 2:8–9
Brack A, Spach G (1981) Enantiomer enrichment in early peptides. Orig Life 11:135–142
Brack A, Spach G (1987) Search for chiral molecules and optical activity in extraterrestrial systems. Example of Titan. Biosystems 20:95–98
Brack A, Troublé M (2010) Defining life: connecting robotics and chemistry. Orig Life Evol Biosph 40:31–136
Brandenburg A, Andersen AC, Nilsson M (2005) Dissociation in a polymerization model of homochirality. Orig Life Evol Biosph 35:507–521
Brinton KLF, Engrand C, Glavin DP et al (1998) A search for extraterrestrial amino acids in carbonaceous Antarctic micrometeorites. Orig Life Evol Biosph 28:413–424
Callahan MP, Smith KE, Cleaves JC II et al (2011) Carbonaceous meteorites contain a wide range of extraterrestrial nucleobases. Proc Natl Acad Sci U S A 108:13995–13998
Catling D, Kasting JF (2007) Planetary atmospheres and life. In: Sullivan WT III, Baross JA (eds) Planets and life. Cambridge University Press, Cambridge, pp 91–116
Chang S (1993) Prebiotic synthesis in planetary environments. In: Greenberg JM, Mendoza-Gomez CX, Pirronello V (eds) The chemistry of life's origin. Kluwer Academic, Dordrecht, pp 259–300
Cleaves HJ, Chalmers JH, Lazcano A et al (2008) A reassessment of prebiotic organic synthesis in neutral planetary atmospheres. Orig Life Evol Biosph 38:105–115
Cronin JR, Pizzarello S (1997) Enantiomeric excesses in meteoritic amino acids. Science 275:951–955
Davies PCW, Lineweaver CH (2005) Finding a second sample of life on earth. Astrobiology 5:154–163
Deamer DW (1985) Boundary structures are formed by organic components of the Murchison carbonaceous chondrite. Nature 317:792–794
Deamer DW (1998) Membrane compartments in prebiotic evolution. In: Brack A (ed) The molecular origins of life: assembling pieces of the puzzle. Cambridge University Press, Cambridge, pp 189–205
Despois D, Cottin H (2005) Comets: potential sources of prebiotic molecules. In: Gargaud M, Barbier B, Martin H, Reisse J (eds) Lectures in astrobiology. Springer, Berlin, pp 289–352
Dobrica E, Engrand C, Duprat J, Gounelle M, Leroux H, Quirico E, Rouzaud J-N (2013) Connection between micrometeorites and Wild 2 particles: from Antarctic snow to cometary ices. Meteorit Planet Sci 44:1643–1661
Ehrenfreund P, Charnley SB (2000) Organic molecules in the interstellar medium, comets and meteorites. Annu Rev Astron Astrophys 38:427–483
Fegley B Jr, Prinn RG, Hartman H et al (1986) Chemical effects of large impacts on the earth's primitive atmosphere. Nature 319:305–308
Fletcher SP, Jagt RBC, Feringa BL (2007) An astrophysically-relevant mechanism for amino acid enantiomer enrichment. Chem Commun 25:2578–2580
Frank FC (1953) On spontaneous asymmetric synthesis. Biochim Biophys Acta 11:459–463
Glassmeir K-H, Boehnhardt H, Koschny D et al (2007) The Rosetta mission: flying towards the origin of the solar system. Space Sci Rev 128:1–21
Glavin DP, Dworkin JP, Aubrey A et al (2006) Amino acid analyses of Antarctic CM2 meteorites using liquid chromatography-time of flight-mass spectrometry. Meteorit Planet Sci 41:889–902
Gleiser M (2007) Asymmetric spatiotemporal evolution of prebiotic homochirality. Orig Life Evol Biosph 37:235–251
Goesmann F, Rosenbauer H, Bredehöft JH et al (2015) Organic compounds on comet 67P/Churyumov-Gerasimenko revealed by COSAC mass spectrometry. Science 349:aab0689

Holm NG (1992) Marine hydrothermal systems and the origins of life. Orig Life Evol Biosph 22:1–191
Holm NG, Andersson EM (1998) Organic molecules on the early earth: hydrothermal systems. In: Brack A (ed) The molecular origins of life: assembling pieces of the puzzle. Cambridge University Press, Cambridge, pp 86–99
Holm NG, Andersson EM (2005) Hydrothermal simulation experiments as a tool for studies of the origin of life on Earth and other terrestrial planets: a review. Astrobiology 5:444–460
Holm NG, Charlou J-L (2001) Initial indications of abiotic formation of hydrocarbons in the Rainbow ultramafic hydrothermal system, Mid-Atlantic ridge. Earth Planet Sci Lett 191:1–8
Israël G, Szopa C, Raulin F et al (2005) Evidence for the presence of complex organic matter in Titan's aerosols by in situ analysis. Nature 438:796–799
Johnson AP, Cleaves HJ, Dworkin JP et al (2008) The Miller volcanic spark discharge experiment. Science 322:404
Joyce GF (1995) The RNA world: life before DNA and protein. Cambridge University Press, Cambridge, pp 139–151
Kasting JF, Brown LL (1998) The early atmosphere as a source of biogenic compounds. In: Brack A (ed) The molecular origins of life: assembling pieces of the puzzle. Cambridge University Press, Cambridge, pp 35–56
Kondepudi DK, Kaufman RJ, Singh N (1990) Chiral symmetry breaking in sodium chlorate crystallization. Science 250:975–976
Kurosawa K, Sugita S, Ishibashi K et al (2013) Hydrogen cyanide production due to mid-size impacts in a redox-neutral N_2-rich atmosphere. Orig Life Evol Biosph 43:221–245
Luisi PL (1998) About various definitions of life. Orig Life Evol Biosph 28:613–622
Mac Dermott A (1995) Electroweak enantioselection and the origin of life. Orig Life Evol Biosph 25:191–199
Matrajt G, Pizzarello S, Taylor S et al (2004) Concentration and variability of the AIB amino acid in polar micrometeorites: implications for the exogenous delivery of amino acids to the primitive Earth. Meteorit Planet Sci 39:1849–1858
Maurette M (1998) Carbonaceous micrometeorites and the origin of life. Orig Life Evol Biosph 28:385–412
Maurette M (2006) Micrometeorites and the mysteries of our origins. Springer, Berlin
Maurette M, Brack A (2006) Cometary petroleum in Hadean time? Meteorit Planet Sci 41:5247
McKay CP, Borucki WJ (1997) Organic synthesis in experimental impact shocks. Science 276:390–392
Miller SL (1953) The production of amino acids under possible primitive Earth conditions. Science 117:528–529
Miller SL (1998) The endogenous synthesis of organic compounds. In: Brack A (ed) The molecular origins of life: assembling pieces of the puzzle. Cambridge University Press, Cambridge, pp 59–85
Nesvorny D, Jenniskens P, Levison HF et al (2010) Cometary origin of the zodiacal cloud and carbonaceous micrometeorites. Implications for hot debris disks. Astrophys J 713:816–836
Niemann HB, Atreya SK, Bauer SJ et al (2005) The abundances of constituents of Titans' atmosphere from the GCMS instrument on the Huygens probe. Nature 438:779–784
Nordén B, Liljenzin J-O, Tokay RK (1985) Stereoselective decarboxylation of amino acids in the solid state, with special reference to chiral discrimination in prebiotic evolution. J Mol Evol 21:364–370
Ogata Y, Imai E-I, Honda H et al (2000) Hydrothermal circulation of seawater through hot vents and contribution of interface chemistry to prebiotic synthesis. Orig Life Evol Biosph 30:527–537
Oparin AI (1924) Proikhozndenie Zhizni Izd. Moskowski Rabochi
Parker ET, Zhou M, Burton AS et al (2014) A plausible simultaneous synthesis of amino acids and simple peptides on the primordial Earth. Angew Chem Int Ed 53:8132–8136

Perry RH, Wu C, Nefliu M et al (2007) Serine sublimes with spontaneous chiral amplification. Chem Commun 10:1071–1073

Pinti DL (2005) The origin and evolution of the oceans. In: Gargaud M, Barbier B, Martin H, Reisse J (eds) Lectures in astrobiology. Springer, Heidelberg, pp 83–112

Pizzarello S (2007) The chemistry that preceded life's origin: a study guide from meteorites. Chem Biodivers 4:680–693

Pizzarello S, Cronin JR (2000) Non-racemic amino acids in the Murray and Murchison meteorites. Geochim Cosmochim Acta 64:329–338

Pizzarello S, Huang Y (2005) The deuterium enrichment of individual amino acids in carbonaceous meteorites: a case for the presolar distribution of biomolecules precursors. Geochim Cosmochim Acta 69:599–605

Pizzarello S, Shock E (2010) The organic composition of carbonaceous meteorites: the evolutionary story ahead of biochemistry. Cold Spring Harb Perspect Biol 2:a002105

Pizzarello S, Huang Y, Becker L et al (2001) The organic content of the Tagish Lake meteorite. Science 293:2236–2239

Pizzarello S, Zolensky M, Turk KA (2003) Non-racemic isovaline in the Murchison meteorite: chiral distribution and mineral association. Geochim Cosmochim Acta 67:1589–1595

Pizzarello S, Schrader DL, Monroe AA et al (2012) Large enantiomeric excesses in primitive meteorites and the diverse effects of water in cosmochemical evolution. Proc Natl Acad Sci USA 109:11949–11954

Plasson R, Bersini H, Commeyras A (2004) Recycling frank: spontaneous emergence of homochirality in noncatalytic systems. Proc Natl Acad Sci USA 101:16733–16738

Rikken GLJA, Raupach E (2000) Enantioselective magnetochiral photochemistry. Nature 405:932–935

Rubin M, Altwegg K, Balsiger H et al (2015) Molecular nitrogen in comet 67P/Churyumov-Gerasimenko indicates a low formation temperature. Science 348:232–235

Rubinstein I, Kjaer K, Weissbuch I et al (2005) Homochiral oligopeptides generated via an asymmetric induction in racemic 2D crystallites at the air–water interface; the system ethyl/thio-ethyl esters of long-chain amphiphilic α-amino acids. Chem Commun 43:5432–5434

Ryder G (2003) Bombardment of the Hadean Earth: wholesome or deleterious? Astrobiology 3:3–6

Schlesinger G, Miller SL (1983) Prebiotic syntheses in atmospheres containing CH_4, CO, and CO_2. I. Amino acids. J Mol Evol 19:376–382

Schmitt-Kopplin P, Gabelica Z, Gougeon RD et al (2010) High molecular diversity of extraterrestrial organic matter in Murchison meteorite revealed 40 years after its fall. Proc Natl Acad Sci USA 107:2763–2768

Shibata T, Yamamoto J, Matsumoto N et al (1998) Amplification of a slight enantiomeric imbalance in molecules based on asymmetry autocatalysis. J Am Chem Soc 120:12157–12158

Spach G, Brack A (1988) Chemical production of optically pure systems. In: Marx G (ed) Bioastronomy. The next steps. Kluwer Academic, Dordrecht, pp 223–231

Stoks PG, Schwartz AW (1982) Basic nitrogen-heterocyclic compounds in the Murchison meteorite. Geochim Cosmochim Acta 46:309–315

Tarasevych AV, Sorochinsky AE, Kukhar VP et al (2013) Partial sublimation of enantioenriched amino acids at low temperature. Is it coming from the formation of a euatmotic composition1 of the gaseous phase? J Org Chem 78:10530–10533

Tarasevych AV, Sorochinsky AE, Kukhar VP et al (2015) High temperature sublimation of a-amino acids: a realistic prebiotic process leading to large enantiomeric excess. Chem Commun 51:7054–7057

Tian F, Toon OB, Pavlov AA et al (2005) A hydrogen-rich early atmosphere. Science 308:1014–1017

Weissbuch I, Addadi L, Leiserowitz L et al (1988) Total asymmetric transformations at interfaces with centrosymmetric crystals: role of hydrophobic and kinetic effects in the crystallization of the system glycine/α-amino acids. J Am Chem Soc 110:561–567

Weissbuch I, Berfeld M, Bouwman W, Kjaer K, Als-Nielsen J, Lahav M, Leiserowitz L (1997) Separation of enantiomers and racemate formation in two-dimensional crystals at the water surface from racemic-amino acid amphiphiles: design and structure. J Am Chem Soc 119:933–942

Westall F, Foucher F, Bost N et al (2015) Biosignatures on Mars: what, where, and how? Implications for the search for Martian Life. Astrobiology 15:998–1029

Wikipedia: http://en.wikipedia.org/wiki/List_of_molecules_in_interstellar_space

Yabuta H, William LB, Cody GD et al (2007) The insoluble carbonaceous material of CM chondrites: a possible source of discrete compounds under hydrothermal conditions. Meteorit Planet Sci 42:37–48

Zepik H, Shavit E, Tang M et al (2002) Chiral amplification of oligopeptides in two-dimensional crystalline self-assemblies on water. Science 295:1266–1269

Chapter 2
Organic Matter in Interplanetary Dusts and Meteorites

Eric Quirico and Lydie Bonal

Abstract Asteroids and comets have continuously delivered organics to Earth and telluric planets since their formation ~4.55 Ga ago. Characterizing these organics and investigating their origin constitutes a major goal of astrobiology and planetary sciences. This chapter reviews past and current knowledge on the nature of the exogenous organics accreted by the Earth, their composition and structure and different issues regarding their origin and subsequent evolution through the effects of secondary processes in/on their original asteroidal or cometary parent bodies. Tertiary processes such as weathering and, in the case of dust, the heating and oxidation effects during atmospheric entry, are also discussed. The last section focuses on the nature and preservation of organics at the surface of Mars, in the context of the ExoMars and Mars2020 missions.

2.1 Introduction

Asteroids and comets have continuously delivered organics to Earth and telluric planets since their formation ~4.55 Ga ago. Characterizing these organics and investigating their origin constitutes a major goal of astrobiology and planetary sciences. First and foremost, extraterrestrial organics could have been a source of prebiotic molecules from which life arose on Earth. In any case, they were a significant source of carbon to primitive telluric planets. However, they also testify to a complex abiotic chemistry and shed light on the astrophysical locations and conditions that led to the emergence of organic complexification. In this respect, they provide insightful evidence that helps investigate new paradigms in planetary habitability.

E. Quirico (✉) · L. Bonal
Université Grenoble Alpes, Grenoble, France

Institut de Planétologie et d'Astrophysique de Grenoble, Grenoble, France
e-mail: eric.quirico@univ-grenoble-alpes.fr; lydie.bonal@univ-grenoble-alpes.fr

B. Cavalazzi, F. Westall (eds.), *Biosignatures for Astrobiology*, Advances in Astrobiology and Biogeophysics, https://doi.org/10.1007/978-3-319-96175-0_2

The inventory of the past and present delivery of extraterrestrial organics is a major field of investigation and relies on the determination of (1) the flux of cosmomaterials over time, including size and mass distributions; (2) their composition; (3) the effect of atmospheric entry; and (4) alteration processes at the surface of the planet (Earth or Mars). The objective of this chapter is to review past and current knowledge on these different issues, with emphasis on the organic inventory of the main cosmomaterials available on Earth and the detection and characterization of exogenous organic matter at the surface of Mars. The following paragraph concerns the cosmomaterials available in laboratories, such as meteorites collected on the ground, and micrometeorites and stratospheric Interplanetary Dust Particles (IDPs) from the Earth's atmosphere. We then review the organic content in primitive meteorites and focussing on recent analytical developments and characterization, subsequently addressing organic matter in stratospheric IDPs and Antarctic micrometeorites (AMMs) that represent the dominant source of extraterrestrial organics on Earth and Mars. Finally, we discuss the origin of organic matter in chondrites and dust, and address the issue of the preservation and detection of allochthonous extraterrestrial organic matter at the surface of Mars, in the context of the Mars 2020 and ExoMars space missions. Given the broad scope of the present chapter, several points could not be exhaustively addressed. We refer the reader to excellent reviews by Sephton (2002) (overview of organics in chondrites), Burton et al. (2012) (amino acids and nucleobases in chondrites) and Alexander et al. (2017) (insoluble organic matter in chondrites).

2.2 Cosmomaterials Inherited by Telluric Planets

2.2.1 Past and Present Flux of Extraterrestrial Matter

Love and Brownlee (1993) estimated the pre-atmospheric dust flux (mass range 10^{-9} to 10^{-4} g) based on direct measurements of hypervelocity impact craters on the Long Duration Exposure Facility satellite. The meteoroid mass distribution peaks near 1.5×10^{-5} g—or roughly 200 μm in diameter—and the small particle mass accretion rate is 40 $\pm$ 20 ktons per year (Love and Brownlee 1993). Estimates of post-atmospheric flux are, however, slightly higher, but may not be significantly different, with values of 78 $\pm$ 30 ktons per year (Gabrielli et al. 2004) and of 64 $\pm$ 20 ktons per year (Lanci and Kent 2006). Most meteoritic material reaching the Earth system is in the form of meteoritic smoke made of nanometer-sized particles produced through recondensation of meteor ablated products (Hunten et al. 1980; Love and Brownlee 1991). Lanci and Kent (2006) used magnetic measurements of Greenland ices to estimate meteoritic smoke concentration and deduced an accretion rate assuming a given snow deposition rate. The estimates of Gabrielli et al. (2004) were based on measured concentrations of platinum and iridium Greenland ices during the Holocene epoch. Meteoritic smoke represents the main input of extraterrestrial matter to Earth but the comparison to pre-atmospheric flux shows that only a small fraction of extraterrestrial materials (and of the organics potentially within) survives entry to the Earth's atmosphere.

Micrometeorites represent the major fraction of recoverable solid mass accreted to the Earth, with values of 2.7 ± 1.4 ktons per year estimated by Taylor et al. (1998) based on counting statistics of the South Pole water well (SPWW) collection and assuming that a similar fraction of unmelted micrometeorites is present in the SPWW and in Antarctic blue ice. The present flux of meteorites was determined by Halliday et al. (1984, 1989) to be 83 meteorites of mass equal or greater than 1 g per 10^6 km^2 per year based on direct observations of fireballs by the Canadian camera network between 1974 and 1985 and on some assumptions made to estimate the proportion of surviving material. Bland et al. (1996) calculated between 36 and 116 meteorites >10 g per 10^6 km^2 and per year based on counting statistics of meteorites on the ground of hot deserts and on an estimate of their preservation over time. This represents a total flux between 2900 and 7300 kg per year in the 10 g to 1 kg size range over the last 50,000 years.

Now let us look at whether the composition and intensity of the extraterrestrial flux to the Earth has changed over time. Access to several meteorite collection sites with slow weathering rates and the possibility of dating the terrestrial residence times of meteorites provides a time window of up to several millions years (e.g., terrestrial ages of up to two million years for meteorites in Antarctica; Jull 2006). A review by Zolensky et al. (2006a) concluded that there has been a significant variation of the flux of extraterrestrial material over time, with the present day flux being lower than the flux over the past few millions years by a factor of up to two. A higher past flux of extraterrestrial material to Earth has also been suggested from the study of fossil meteorites, which are traceable back to 470 Ma (Schmitz et al. 2001). Ordinary chondrites are the dominant form in both the modern and past flux of meteorites falling to Earth. Nevertheless, Zolensky et al. (2006a) also pointed out a variation in the types of meteorites, reaffirmed recently by Gattacceca et al. (2011), who focused on the San Juan meteorite field of the Atacama Desert (Chile). With terrestrial ages up to more than 40 ky included in that collection, they showed an overabundance of H chondrites and a shortage of LL chondrites in comparison with the population of modern falls. This suggests that there are short-term variations in the composition of meteorite flux, in agreement with observations made based on the Antarctic collection (e.g., Harvey and Cassidy 1989).

Collections of IDPs contain only contemporary material, but some chemical tracers of oceanic sediments (e.g., ^3He; Farley 1995) make it possible to determine the flux of dust back through time. This is also possible from the study of micro-xenoliths (i.e., inclusions having a different origin from the host meteorites) that can be considered as fossil micrometeorites (Gounelle et al. 2003; Briani et al. 2012). Gounelle et al. (2003) noted that the ancient micrometeorite flux is dominated by C2-like matter, comparable to the present micrometeorite flux. Further investigations would be required to characterize a potential compositional variation.

The prerequisite conditions to find meteorites on planetary surfaces are (1) the presence of a sufficiently dense atmosphere to decelerate the projectiles to a low impact speed to prevent total vaporization and melting of material and (2) slow weathering to allow accumulation over time. These conditions exist in at least certain locations on Earth and Mars.

2.2.2 Cosmomaterials in Earth Collections

Cosmomaterials available for laboratory studies include extraterrestrial rocks and interplanetary dust particles collected on the ground as micrometeorites (e.g., in the blue ices and snowfields of Antarctica) and in the atmosphere of Earth as stratospheric IDPs. They also comprise materials collected in situ on the Moon, in the coma of comet Wild 2 (the Stardust mission) and on the asteroid Itokawa (the Hayabusa-1 mission). Meteorites are rocks derived from asteroids, although a cometary origin has been proposed for several of them (Gounelle et al. 2006). They are named *falls* when their fall is certified by an eyewitness and the material is recovered shortly afterwards. Should this not be the case, they are considered as *finds*. Although fresh falls are the preferred objects, as they largely escape terrestrial organic contamination and oxidation, they are rare. Indeed, collections mostly contain finds collected in Antarctica and to a lesser extent in hot deserts. Meteorites (in particular metal phases and organics) are better preserved in Antarctica than in hot deserts (Ash and Pillinger 1995; Alexander et al. 2007).

Meteorites are subdivided into two main groups: originating from undifferentiated (chondrite) or differentiated (achondrite, stony-irons and irons) parent bodies. The second group may contain some carbonaceous material but mostly in inorganic form (e.g., graphite, diamonds). Thus, hereafter we focus on chondrites. Chondrites are classified into three categories, ordinary, carbonaceous and enstatite meteorites, according to their bulk composition, mineralogy and petrology (Krot et al. 2014). Among the meteorites collected in Antarctica, 87% are classified as ordinary, 5% as carbonaceous and 1% as enstatite chondrites, the rest comprising not chondritic objects (Grady 2000). This distribution does not reflect the relative abundances of their presumed parent bodies: S-type (for OCs), C-type (for CM carbonaceous chondrites) and possibly D- and P-type asteroids (likely unrepresented by chondrites, but possibly by micrometeorites) (DeMeo and Carry 2013). This suggests that the dynamics of delivery to Earth-crossing orbits might favour OCs over other types, however carbonaceous chondrite material is also expected to be more friable than OC material and could therefore be more represented in micrometeorite collections.

Stratospheric IDPs and micrometeorites are micrometer-sized particles (2–60 μm for IDPs, Rietmeijer 1998; 30–1000 μm for Antarctic micrometeorites (AMMs), Genge et al. 2008), collected in the high stratosphere and in Antarctic and Greenland ice and snow and oceanic sediments, respectively. IDPs are classified according to their bulk composition (chondritic *versus* non-chondritic, carbon abundance), their morphology (compact or porous) and the dominant silicate, such as olivine and pyroxenes (anhydrous IDPs) or phyllosilicates (hydrated IDPs). Compact porous

CP-IDPs generally have an anhydrous mineralogy and are believed to have a cometary origin (Rietmeijer 1998; Bradley 2014). In contrast, hydrated compact IDPs resemble CI, CM and CR chondrites and may be related to asteroids. However, the source of a large fraction of these particles is still controversial because of the anhydrous mineralogy of some asteroids, the probable existence of a comet-asteroid continuum, the hotly discussed origin of zodiacal cloud dusts, and the effects of atmospheric heating that remain only partly understood (e.g., Dermott et al. 2002; Genge et al. 2008; Nesvorný et al. 2010; Vernazza et al. 2012).

Antarctic micrometeorites are classified into four groups: fine-grained, crystalline, scoriaceous and cosmic spherules (Engrand and Maurette 1998; Genge et al. 2008). This classification mostly reflects the degree of heating during atmospheric entry. The respective abundances of the four groups in the CONCORDIA collection (Duprat et al. 2007) are 28%, 8%, 35% and 29%, respectively. The majority of the fine-grained class is comprised of 20% of fine-grained compact types and 8% of fine-grained fluffy types, among which 4% are the rare Ultra Carbonaceous AMMs (UCAMMs). UCAMMs share similarities with anhydrous stratospheric IDPs and cometary grains (Dobrica et al. 2009).

IDPs and AMMs have spiralling orbits due to Poynting-Robertson drag, limiting gravitational bias. They more likely sample friable and fragile objects that spread into micrometer-sized dust. Fine-grained and scoriaceous AMMs are carbon-rich and are related to carbonaceous chondrites, though they are not strictly similar to CI, CM and CR matrices. A fraction of cosmic spherules may be related to ordinary chondrites (e.g., Suavet et al. 2011; Rudraswamy et al. 2016).

2.2.3 *Post-Accretion Processes on the Parent Body*

Post-accretion—or secondary—processes refer to physical and chemical modifications occurring during and/or after parent body accretion. This includes geological processes such as aqueous alteration (AA), thermal metamorphism (TM), hydrothermalism, shock metamorphism, and space weathering of the surface and subsurface generated by micrometeorite impacts and solar wind irradiation (Huss et al. 2006). All of these processes have potential effects on the formation or evolution of organic matter.

The extent of thermal metamorphism and aqueous alteration is evaluated by the so-called petrologic type of a meteorite (see review by Huss et al. 2006). A number from 3 to 6 reflects increasing thermal metamorphism and from 3 to 1 an increasing degree of aqueous alteration. In practice, TM and AA are not independent and several chondrites show evidence of the combined effects of both these processes (e.g., the type 3 COs and oxidized CVs). Some C1 and C2 chondrites have experienced short periods of thermal metamorphism, possibly generated by impacts or solar heating (Tonui et al. 2014).

2.3 Diversity and Complexity of Organics in Meteorites

The search for organics started soon after meteorites were recognized as extraterrestrial rocks (Berthelot 1868). However, this field of research remained uncertain until the 1970s, when a new era was opened by the discovery of extraterrestrial amino acids and other compounds in the Murchison fresh fall (a CM2 chondrite, Kvenvolden et al. 1970). Earlier studies were limited by analytical capacities and by the difficulty to discriminate the contribution of terrestrial contamination.

Organic matter in meteorites is generally described as *soluble* and *insoluble* fractions. Soluble organic matter (SOM) comprises molecules that are extracted using common solvents (water, methanol, dichloromethane, toluene, etc). Insoluble organic matter (IOM) is the organic residue recovered after SOM extraction and acid (HF, HCl or CsF) digestion of most minerals. IOM is a polyaromatic organic solid that resembles, in many respects, type III terrestrial kerogens. The respective abundance of SOM *versus* IOM is unclear and depends upon the chemical class and secondary processes. Older publications (e.g., Hayes 1967) reported a value of 30 wt.%, but they mention neither the experimental protocols nor the chondrites that were used to derive this value.

2.3.1 Soluble Organic Matter

The Murchison CM2 chondrite has been extensively investigated, due to both its high available mass (100 kg) and the fact that it has undergone very minor terrestrial contamination. It is certainly the most well known chondrite in terms of soluble organics and, as such, is often used as a proxy for comparison with other chondrites. The fraction of molecules with C1–C10 carbon atoms can be classified into 25 groups (Fig. 2.1). Species with C, H and O atoms are the dominant fraction (72 wt.%), followed by C, H, O, N compounds (20 wt.%), C, H, O, S compounds (7 wt.%), and others (1 wt.%). Carboxylic acids are the main compounds, accounting for 43% of the whole SOM content. Gaseous species, amino acids, sugar-related compounds, urea, dicarboximides and sulfonic acids comprise 38%, while other compounds account for 19%.

Amino acids are of primary interest in astrobiology and have been extensively studied by GC-MS and HPLC combined with fluorescence detection (FD) and mass spectrometry (MS), as well as with dedicated setups for analysing chirality and stable isotopic composition (see Burton et al. 2012 for a thorough review). So far, 80 amino acids have been detected in Murchison, with the following characteristics: (1) carbon numbers ranging between 2 and 9; (2) abundance decreasing with increasing number of carbon atoms; (3) all isomers present for C3–C5 acids; (4) mono-amino acids in the major form with few di-amino acids and N, N acids;

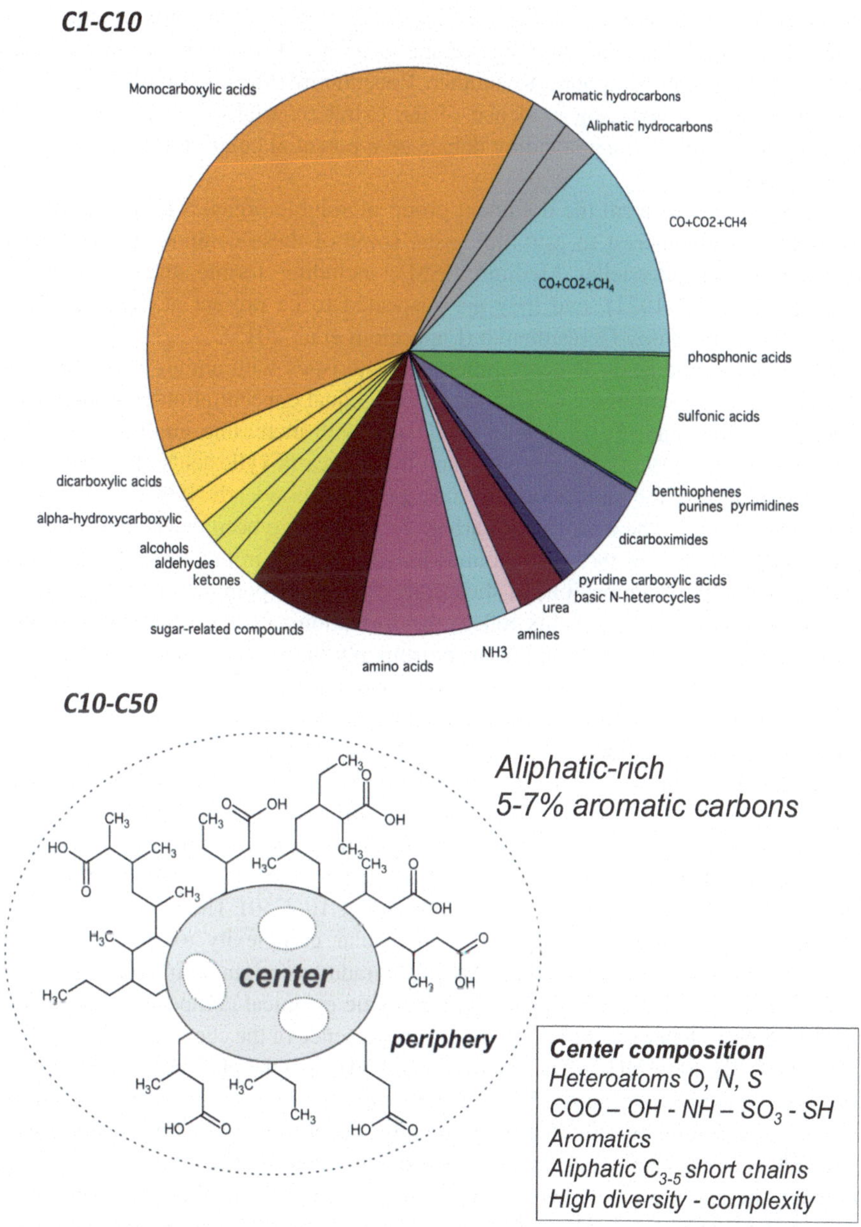

Fig. 2.1 Top: Abundances of molecular groups (C1–C10) present in SOM extracts of the Murchison CM2 chondrite (plotted with data compiled in Sephton et al. 2002). Bottom: SOM model based on NMR and FT-ICR analysis of methanolic extracts. Source: Adapted from Hertkorn et al. (2015) and Schmitt-Koplin et al. (2010)

(5) racemic composition with some exceptions for most of the amino acids; and (6) D- and ^{15}N-enrichments showing average solar $^{13}C/^{12}C$ ratios but with significant variations from one compound to another. Racemic mixtures and the D- and ^{15}N-enrichments are convincing evidence of the extraterrestrial origin of these compounds and ended the long-standing debate on a potential contribution of terrestrial contamination.

Carboxylic acids form the dominant group of soluble organics (Sephton 2002). Apart from their interest to astrobiologists, some of these compounds have been detected in the Interstellar Medium (ISM), including formic and acetic acids (Zuckerman et al. 1971), and they are suspected to be present at the surface of comet 67P/Churyumov-Gerasimenko (Capaccioni et al. 2015).

Monocarboxylic acids share common characteristics with amino acids, such as (1) all isomers present in the C3–C5 range, with equal concentration of straight and branched chains, and (2) decreasing abundance with increasing number of carbon atoms (Yuen et al. 1984). Dicarboxylic and hydroxycarboxylic acids have also been detected, related to the most abundant corresponding monocarboxylic acids (Peltzer and Bada 1978). In the bulk rocks, carboxylic acids have been detected as carboxylates and a fraction of these compounds may be present in this form. Carbon and hydrogen isotopic compositions of short carboxylic acids point to an extraterrestrial origin, and high D-enrichments reflect the formation of either precursors or the whole molecule under cold conditions, presumably in the ISM. Other molecules of interest to astrobiology are aldehydes, sugar precursors, alcohols and nucleobases. The reader is referred to Sephton (2002) and Burton et al. (2012) for thorough reviews of these compounds.

Schmitt-Kopplin et al. (2010) reported the first high-resolution FT-ICR spectroscopy of SOM extracted from Murchison with different polar and apolar solvents. The thorough analysis of spectra collected from methanolic extracts with negative and positive electron spray ionization (ESI) led to the identification of 10,299 compositions over the mass range 150–1000 m/z (C10–C50). The main conclusions of this study are the following: (1) a molecular complexity exceeding that of terrestrial organic matter, including that of degradative origin as in soils and sediments; (2) a wide range of aliphatic and aromatic chemical composition including contributions of heteroatoms (O, N, S); (3) the presence of the chemical groups CH_2, COO, OH, SH, SO_3, NH and very weak N and NH_2; (4) the presence of nitrogen as amide or heterocyclic groups; (5) all isomers of each composition; and (6) a global abiotic sulfurization process that occurred subsequent to the formation of the CHO and CHON compounds, presumably in the parent body, by fluid circulation during aqueous alteration. The whole elemental composition of the methanolic extract was estimated to be $C_{100}H_{155}O_{20}N_3S_3$ and the total number of soluble molecules in Murchison is estimated to be several million. This is strikingly larger than the ~500 species detected in the C1–C50 range and is evidence of the amazing complexity of SOM.

1D and 2D H and ^{13}C NMR, combined with FT-ICR, provided further insights into the chemical structure of Murchison SOM (Hertkorn et al. 2015). This study proposed a model structure consisting of a molecular center with high chemical diversity and complexity onto which chains formed by aliphatic and carboxylic groups are branched. This peripheral network is readily evidenced in NMR spectra thanks to its simple structure.

In contrast, the molecular center displays high molecular diversity and complexity, combining aromatic species and the heteroatoms O, N and S through a complex linkage that includes C3–C5 aliphatic groups with a statistical branching ratio. Overall, Murchison SOM is a highly aliphatic material with only 5–7% of aromatic species, unlike IOM which is highly aromatic (>60% of aromatic carbons). Note that the polar fraction analysed by ESI FT-ICR contains a large aromatic abundance (~27%) with respect to the entirety of SOM (Schmitt-Kopplin et al. 2010). A long-standing issue has been the origin and formation processes of SOM. The isotopic composition, chirality and high number of isomers for each elemental composition point to a complex chemistry that does not select specific chemical routes. This chemistry presents similarities with ISM chemistry, in particular at low temperatures in order to account for D-enrichments. The detection of amino acids and/or their isomers in comets Wild 2 and 67P/Churyumov-Gerasimenko (Elsila et al. 2009; Altwegg et al. 2016) thus support a pre-accretion origin. Experimental simulations, such as synthesis in cold plasma or radiolysis of ices, show that amino acids can indeed be formed in the ISM or under protosolar disk conditions (Elsila et al. 2007; Horst et al. 2012). The contribution of aqueous chemistry in the parent body has been proposed to account for the synthesis of amino acids through Strecker synthesis (see review by Burton et al. 2012). The location and physical conditions of this chemistry, either in the ISM or in the protosolar disk, is not presently clear.

2.3.2 *Insoluble Organic Matter*

The first analysis of IOM (often referred to as 'organic polymers' at the time) reported a highly aromatic and condensed material, containing carbonyl, aliphatic and carboxylic chemical groups (Bandurski and Nagy 1976 and references therein). This study also noted similarities with coals, kerogens, humic acid, lignine and laboratory polymers formed by Fischer-Tropsch reactions.

In the following years, several degradative techniques of analysis were successfully applied: gradual, stepwise and flash pyrolysis, hydrous pyrolysis, and combustion and oxidation (Hayatsu et al. 1977, 1980; Hayatsu and Anders 1981; Komiya and Shimoyama 1996; Alexander et al. 1998; Sephton et al. 2000; 2004; Sephton and Gilmour 2001; Wang et al. 2005; Remusat et al. 2005a,b). These experiments revealed highly substituted small-sized polyaromatic units (1–4 rings), an aliphatic linkage composed of short chains, bridged with oxygenated species, such as ester and ether groups, confirming the presence of carboxyl and ketone groups. A ^{13}C NMR study by Cronin et al. (1987) pointed to the presence of aromatic/olefinic and aliphatic carbons in IOMs extracted from Orgueil, Murchison and Allende. Subsequent CP-MAS ^{13}C NMR analyses by Gardinier et al. (2000) revealed and determined the abundance of several chemical groups such as CH_3 branched on aliphatics and aromatics, CH_2, aliphatic carbons bonded to heteroatoms, protonated and non-protonated aromatic carbons, carboxyls, and carbonyls. The aromatic abundance was found to range between 61–67% for Murchison and 69–78% for Orgueil. The single pulse (SP) and CP-MAS measurements of Cody et al. (2002) confirmed the aromatic content of Murchison IOM, the presence of oxygenated species and the highly branched aliphatic linkage. They also ruled out the presence of large polyaromatic units. Infrared spectroscopy pointed to the presence of aromatic carbons, carbonyl, hydroxyl and aliphatic groups and determined the CH_2/CH_3 ratio (Hayatsu et al. 1977; Gardinier et al. 2000; Kebukawa et al. 2011; Orthous-Daunay et al. 2013). Insights into the speciation of minor elements were provided by ^{15}N NMR and UV resonant Raman spectroscopy, which identified pyrrholes, indole and carbazole groups in the CI1 Orgueil and cyanide in the CI1 Alais meteorites, respectively (Remusat et al. 2005b; Dobrica et al. 2009). A broad range of sulfur speciation was determined by XANES spectroscopy (Remusat et al. 2005b; Orthous-Daunay et al. 2010).

Raman micro-spectroscopy is vibrational spectroscopy operating at the micrometric scale. As a technique that is relatively easy to implement and essentially non-destructive, it has been used in several studies and can be combined with Secondary Ion Mass Spectroscopy (SIMS) and FTIR measurements. A typical Raman spectrum of IOM consists of two peaks at ~1600 and 1350 cm^{-1}, termed the G- and D-bands, respectively. These features provide information on the structure of the carbonaceous material in terms of the degree of disorder and are particularly powerful for discriminating and classifying carbonaceous materials. Raman spectroscopy is basically a resonant process when applied to carbonaceous materials, as they absorb visible and UV photons. The Raman cross-sections of the G- and D-bands are therefore enhanced by several orders of magnitude with respect to those of free molecules and minerals. As a result, Raman measurements are less sensitive to minerals and SOM and are most useful for probing the IOM polyaromatic network. The first Raman spectroscopic studies of meteorites were published in the 1980s (Michel-Lévy and Lautie 1981), but this technique only became more widely used in the 2000s. Measurements on primitive (unmetamorphosed) chondrites have revealed a polyaromatic structure with a high degree of disorder, pointing to a kerogen-like structure in accordance with the results of NMR and degradative techniques (Quirico et al. 2003, 2005, 2009, 2014; Matrajt et al. 2005; Busemann et al. 2007; Starkey et al. 2013). The use of multi-wavelength Raman spectroscopy suggested that the IOM in primitive chondrites experienced some heating in the protosolar disk prior to accretion, although irradiation processes could not be fully excluded (Quirico et al. 2014).

Stable H, C and N isotopic compositions have been extensively investigated since the 1980s. Robert and Epstein (1982) reported large D-enrichments in several primitive carbonaceous chondrites and a large ^{15}N-excess in Renazzo (CR2). These values ruled out IOM synthesis from Fischer-Tropsch or Miller-Urey reactions. The authors suggested synthesis in cold conditions, presumably in the local ISM. A systematic survey of stable isotopic compositions in chondrites from the main groups later confirmed large D-enrichments in the most primitive chondrites, both D- and ^{15}N-enrichments in CR chondrites, and a solar C isotopic composition (Alexander et al. 1998, 2007, 2010). Large variations among chondrites were also observed and were partly attributed to the effects of parent processes. Apart from high enrichments of D- and ^{15}N, a major characteristic of IOM is the heterogeneity of its isotopic composition. Busemann et al. (2006) showed that very high D- and ^{15}N-enrichments are present in micrometric/sub-micrometric areas referred to as hot spots. Hydrous pyrolysis showed that part of the IOM is extractable with water and characterized by ^{13}C- and ^{15}N-enrichments, while the non-extractable, more refractory component is ^{13}C- and ^{15}N-depleted (Sephton et al. 2003). Kerridge et al. (1987) proposed the presence of three isotopically distinct components: aliphatic ^{13}C- and D-enriched, ^{13}C-depleted, D-enriched and ^{13}C-depleted, D-depleted aromatic components. Okumura and Mumura (2011) also pointed out a ^{13}C- and D-enriched aliphatic component, but a single ^{13}C- and D-depleted aromatic component. Wang et al. (2005) reported individual hydrogen isotopic compositions of pyrolysates of several chondrites and confirmed isotopic heterogeneity at the molecular level. Their study also evidenced similar patterns across several primitive chondrites although there were some differences particularly for the C2 Tagish Lake chrondrite, which experienced hydrothermalism at higher temperatures than the more primitive C1 and C2 chondrites. Remusat et al. (2006) combined pyrolysis and ruthenium oxide combustion on IOM extracted from the Orgueil chondrite and attempted to reconstruct the isotopic heterogeneity in the IOM molecular structure prior to degradation. Their study reports a selective deuteration of 1250, 550 and 150‰ in benzylic, aliphatic and aromatic carbons, respectively. The lower deuteration in aromatics with respect to aliphatics is consistent with the results from Okumura and Mimura (2011). The origin of this isotopic heterogeneity is, however, unclear and different mechanisms have been suggested: ion-reactions with H_3^+ in the upper layers of the protosolar disk (Remusat et al. 2006), interstellar heritage (Okumura and Mimura 2011) or radiolysis-induced fractionation (Le Guillou et al. 2012; Laurent et al. 2014). Note however that this radiolytic process is inhibited at low temperature and that these conclusions are derived from the study of only one chondrite (Orgueil). The selective deuteration in other chondrites is known to be different (Wang et al. 2005). Finally, note also that nitrogen fractionation remains unexplained by these interpretations.

IOM paramagnetic centers have been investigated with Electron Paramagnetic Spectroscopy (EPR). Binet et al. (2002, 2004) showed that Orgueil IOM, unlike type III kerogens and coals, contained paramagnetic radicals heterogeneously distributed in the material. The concentration of radicals also displays dramatic change with temperature, pointing to the occurrence of diradicaloids hosted in ~10-ring polyaromatic units at temperatures below 150 K. Gourier et al. (2008) showed that these paramagnetic centers host large D-enrichments (up to 100,000‰), just one order of magnitude below the highest fractionation reported in dense cores (Parise et al. 2006). Based on these results, Remusat et al. (2009) suggested that the deuterium *hot spots* in chondritic IOMs were coincident with paramagnetic centers. Delpoux et al. (2011) investigated their electronic structures and showed that they were dominated by biradicals and biradicaloids. They developed a model that suggested that this electronic structure was directly connected to the IOM structure and concluded that C–H breaking and D-enrichments were subsequent to IOM formation. This model, however, does not account for the C and N isotopic compositions.

A major finding in the last twenty years is the dramatic difference in the composition and structure of chondrite groups and between chondrites within the same group. Post-accretion processes have a major impact on IOM, which (1) urges us to select the most primitive objects in order to obtain insights into the protosolar disk and/or presolar cloud chemistry and (2) offers the opportunity to characterize the nature and extent of parent body processes. The effect of thermal metamorphism is very important and has been evidenced through variations in elemental composition (Alexander et al. 1998, 2007; Naraoka et al. 2004; Oba and Naraoka 2009), chemical composition (Kitajima et al. 2002; Cody and Alexander 2005; Wang et al. 2005; Yabuta et al. 2005, 2010; Cody et al. 2008; Alexander et al. 2014) and polyaromatic structure (Vis et al. 2002; Quirico et al. 2003, 2009, 2011, 2014; Bonal et al. 2006, 2007, 2016; Le Guillou et al. 2012; Busemann et al. 2007; Cody et al. 2008). The most common evolution results in the loss of H and heteroatoms as O, N and S (carbonization), accompanied by aromatization and an increase of structural order. The highest-order stage is a metastable O-rich carbonaceous material and, ultimately, the whole IOM is destroyed, at least in carbonaceous and ordinary chondrites (Alexander et al. 2007; Cody et al. 2008; Quirico et al. 2009). The degree of advancement of this transformation was found to be useful in rating the intensity of thermal metamorphism and to separate petrologic types. The determination of peak temperature has been proposed (Busemann et al. 2007; Cody et al. 2008), but these thermometric approaches have been considered as unreliable by other authors (see discussion in Bonal et al. 2016).

The impact of aqueous alteration on IOM is unclear even though it is a major issue given that it could blur valuable pre-accretion molecular or isotopic features for investigating the origin of IOM. Cody and Alexander (2005) proposed an oxidizing process that would lead to the loss of aliphatic groups in type 1 and 2 chondrites, based on four chondrites (Orgueil, Murchison, EET 92042, Tagish Lake). These effects were not confirmed by the subsequent studies of Orthous-Daunay et al. (2013) and Quirico et al. (2014), who observed alteration only in chondrites having experienced short duration thermal metamorphism. Herd et al. (2011) reported significant differences among different lithologies of the aqueously altered Tagish Lake, although it appears that this material experienced hydrothermalism at various elevated temperatures (Alexander et al. 2014). The alteration of the isotopic composition by fluid circulation is the subject of divergent conclusions (Alexander et al. 2010; Bonal et al. 2013; Piani et al. 2015; Remusat et al. 2016), although similar issues investigated in the field of terrestrial organic matter indicate H–D transfer between water and kerogens (Schimmelmann et al. 2006). To conclude, there is, to date, no consensus on the effect of low-temperature aqueous alteration and further studies, including laboratory experiments, will be required to reach firmer conclusions.

2.4 Organic Matter in Stratospheric IDPs and AMMs

Organics in IDPs and AMMs are less well-known than those in primitive chondrites. The very low mass of these particles (~1–100 ng) severely restricts the number of analytical techniques. In addition, the separation of organics from minerals and into soluble and insoluble fractions is extremely difficult to perform on these small particles and has been done only for a few cases. So far, most studies have focused on bulk raw grains.

The bulk carbon abundance in IDPs is higher than that of carbonaceous chondrites, reaching 45 wt.% and a mean value of ~15 wt.% (Thomas et al. 1993; Rietmeijer 1998). The carbon abundance in AMMs can reach even higher values (up to ~90 wt.% in UCAMMs) (Dartois et al. 2013). XANES spectroscopy at the C and, less frequently, O and N K-edges has provided valuable insights into the compositions of such organic material. This technique operates at a sub-micrometric spatial resolution (down to ~50 nm). Moreover, it can be combined with Transmission Electron Microscopy (TEM) and thus preserves the petrologic context. Electron Energy Loss Spectroscopy (EELS) is also reported in some studies, providing similar information, however sample stability under the electron beam is a very critical issue. Flynn et al. (2003) and Keller et al. (2004) report the characterization of 21 stratospheric IDPs, including 7 hydrated IDPs. They show that almost all particles contain aromatic rings and ketone groups (C=O). A reappraisal of the interpretation of their XANES spectra based the approach of Le Guillou et al. (2014) suggests that carboxylic groups, COOH, are also present. These XANES spectra share similarities with the spectra of chondritic IOM extracted from the Murchison chondrite, as they contain the same three peaks at 285, 286 and 289 eV. However, they display a broader range of variations that can be interpreted as (1) variations in the composition and structure of the IOM component and/or (2) the combined contribution of both IOM and SOM components. The second point is supported by XANES data collected on bulk matrices from Orgueil (CI) and Murchison (CM), which point to large spectral variations (Le Guillou et al. 2014), and by the infrared spectra of CI, CM and CR matrices (Beck et al. 2010; Bonal et al. 2013).

Micro-infrared spectroscopy measurements have been reported in several studies. Spectra show a broad peak around 2800–2900 cm^{-1}, with components that are due to the symmetric and anti-symmetric stretching modes of CH_2 and CH_3 functional groups (Flynn et al. 2003; Keller et al. 2004; Matrajt et al. 2005; Munoz Caro et al. 2006; Merouane et al. 2014). The CH_2/CH_3 ratio derived from these features is higher than that in the 3.4 μm band observed in the Diffuse Interstellar Medium, putting into questioning at which step of stellar evolution the IDP organics were formed (molecular cloud, protostar envelope or protosolar disk). The CH_2/CH_3 ratio is also found to be different from that of IOM extracted from primitive chondrites (which fits the 3.4 μm band observed in the Diffuse Interstellar Medium) and is, in fact, more consistent with thermally processed chondrites (Flynn et al. 2003;

Kebukawa et al. 2011; Orthous-Daunay et al. 2013). As mentioned above for XANES data, the interpretation of such similarities and differences is potentially biased by the fact that SOM and IOM are measured together in the case of IDPs, and by the fact that organics in IDPs could be modified during atmospheric entry. Matrajt et al. (2005) report measurements on carbon-enriched IDPs in which HF treatment was used to dissolve the silicates. The infrared spectra of these residues point to the presence of ketone C=O groups and likely carboxylic groups. The spectra also present differences with IOM extracted from primitive chondrites regarding the CH_2/CH_3 ratio, as well as the position and intensities of the bands in the range 1800–1000 cm^{-1}. Note that this procedure does not fully remove soluble molecules and that these spectra do not represent the insoluble fraction only of the studied IDPs.

The first Raman study on stratospheric IDPs focused on 20 particles, both anhydrous and hydrated, which were classified into six groups (Wopenka 1988). At the time, the low sensitivity of spectrometers produced spectra with low signal-to-noise ratios, which could not be fitted. The spectra were therefore analysed by eye. In addition, the acquisition conditions probably led to sample annealing (compared to current studies, the power applied to the sample was 30× higher and the collection times were 10× higher). This six-group classification scheme was not confirmed by subsequent studies based on 60 IDPs and 40 AMMs, which pointed to (1) the large internal structural heterogeneity within particles, sometimes encompassing the inter-particle heterogeneity, (2) the low degree of structural order (except for 3 of the 60 particles, whose provenance was likely thermally metamorphosed chondrites) and (3) structural signatures dissimilar to that of type 1 and 2 carbonaceous chondrites of CI, CM and CR types (Quirico et al. 2005; Bonal et al. 2006; Busemann et al. 2009; Dobrica et al. 2011; Starkey et al. 2013; Merouane et al. 2014). The structural differences between IDPs/AMMs and carbonaceous chondrites may point to different organics accreted by the parent body, but the effect of heating/oxidation during atmospheric entry and irradiation in space cannot be ruled out. Firm conclusions would require experimental simulations in order to determine the consequences of atmospheric heating.

Raman data collected with a single wavelength excitation (generally 514 or 532 nm) can be enhanced using spectra collected with other wavelengths. Due to the resonance effect, different excitation wavelengths probe different fractions of the samples and improve discrimination. This Multi-Wavelength Raman has been applied to IDPs and resulted in low dispersion of the G-band, i.e., 0.08 cm^{-1}/nm (Starkey et al. 2013). This low dispersion is consistent with values measured in carbonaceous chondrites, but is inconsistent with most amorphous carbons formed in the laboratory and might point to a heating event experienced by insoluble organics trapped in these IDPs (Quirico et al. 2014). Raman spectra collected with a 244 nm excitation have shown clear structural differences between an IDP, a UCAMM and primitive carbonaceous chondrites and have made it possible to identify the –CN cyanide chemical group (Dobrica et al. 2011). Overall, these data suggest that the insoluble component of organics in IDPs/AMMs might be inherited from a heating event. As organics are very similar in anhydrous and hydrated IDPs (Flynn et al. 2003; Merouane et al. 2014), this heating may have occurred prior to accretion in the inner region of the protosolar disk and hydrothermalism on the parent body may not have been involved in their formation.

The isotopic and elemental compositions of bulk organics (H, C and N) in IDPs and AMMs at the micrometric and sub-micrometric scales have been determined by SIMS. The C/H ration was found to range between 1 and 3 in IDPs (Aléon et al. 2001) and between 2 and 6 in UCAMMs (Duprat et al. 2010). Note however that bias due to instrumental fractionation might be serious in the case of the H/C ratio. The C/N ratio also shows variations at the micrometric scale, with areas characterized by N abundances as high as 10–20 wt.% (Aléon et al. 2003), while in UCAMM a high N/C ratio is observed across a large area (Dartois et al. 2013). In IDPs, as in IOM extracted from chondrites, the isotopic composition is heterogeneous at the micrometer and sub-micrometer scales, with *hot spots* that show very high D- and ^{15}N-enrichments (Messenger 2000, 2002; Aléon et al. 2001, 2003; Keller et al. 2004; Floss et al. 2006; Busemann et al. 2006, 2009). UCAMMs also show very high D-enrichments across larger areas and not restricted to tiny *hot spots* (Duprat et al. 2010). The D-enrichments have been interpreted as a fingerprint of low-temperature chemistry and ion-molecule reactions, which occurred either in the local ISM or in the protosolar disk. Due to the small size of IDPs and AMMs, several techniques—such as EPR—could not be applied and it is therefore unclear whether these high D-enrichments correspond to radicals, diradicals and diradicaloids as observed in the IOM extracted from the CI Orgueil chondrite. Lastly, AIB amino acids were identified in AMMs by Brinton et al. (1998) and Matrajt et al. (2004). These amino acids have been proved to be endogenous.

2.5 Origin and Formation of Organics in Chondrites and Dust

The origin of IOM in chondrites remains a debated issue. In the 1970s, formation in the protosolar disk through Fischer-Tropsch reactions was the favoured mechanism of production due to the similarities with laboratory analogs (Hayatsu et al. 1977) and to the general context of a fully gaseous solar nebula with a full reset of interstellar conditions (e.g., Grossman and Larimer 1974). This view was severely challenged with the discovery of large D-enrichments that pointed to a cold chemistry, presumably in the ISM (Robert and Epstein 1982). Fischer-Tropsch reactions were not able to provide the large H isotopic fractionations and phyllosilicates that were advocated to catalyse those reactions were formed in the parent body and not in the protosolar disk (Alexander et al. 1998). Concurrently, the spectral match of the aliphatic peak at 3.4 μm in the infrared spectra of Orgueil IOM and the diffuse ISM (collected towards the Galactic center) supported an interstellar origin (Ehrenfreund et al. 1991). This view was challenged around the mid-2000s. The 3.4 μm feature of diffuse ISM has since been assigned to hydrogenated amorphous carbon that is not similar to IOM (Pendleton and Allamandola 2002; Dartois et al. 2004, 2007). The cometary grains brought back by the Stardust mission confirmed observations made by the satellite ISO in the 1990s, i.e., crystalline minerals formed in the inner hot region of the protosolar disk are present in comets (Crovisier et al. 1997; Zolensky et al. 2006b). To date, estimating the abundance of amorphous *versus* crystalline minerals in comets remains difficult and the proportions of presolar *versus* protosolar materials is unclear.

The interpretation of the elevated D/H ratio in IOM in terms of ISM heritage has also been challenged. A debate has emerged about whether the conditions in the protosolar disk were favourable to ion-molecule reactions (Henning and Semenov 2013). The selective deuteration of molecular groups of IOM has, for instance, been interpreted by ion-molecule reactions in cold regions of the protosolar disk (Remusat et al. 2006, 2009; Delpoux et al. 2011), but other studies argue that the penetration of Galactic Cosmic Rays within a protoplanetary disk is shielded against by stellar winds (Cleeves et al. 2014). Recent publications have also experimentally demonstrated that radiolytic processes can produce moderate deuterium enrichment in organic solids, but these processes seem to be inhibited at low temperature and cannot account for the high values reported for *hot spots* (Laurent et al. 2014, 2015). On the other hand, the main carbon reservoirs in diffuse ISM are polycyclic aromatic hydrocarbons (PAHs) and hydrogenated amorphous carbons (HACs), which both formed in hot circumstellar environments. Therefore, they are not expected to bear deuterium enrichments. In conclusion, deuteration does not appear to be a clear and unambiguous tracer of IOM origin.

The distinction between solar *versus* interstellar is also somewhat simplistic and artificial. The evolution of diffuse ISM towards planets is actually a continuous succession of varying environments with overlaps of conditions between subsequent stages. PAHs and HACs are not recovered directly in meteorites. They may have been transformed to various extents either in molecular clouds through the action of GCRs or possibly in the inner hot region of the protosolar disk (Alexander et al. 2010; Quirico et al. 2014). To date, the respective effects of ion irradiation and thermal processing on these solids have not been clearly established (Bernstein et al. 2003; Brunetto et al. 2009) and require more observational and experimental investigations. Note that an alternate view proposes the formation of IOM from chondrites and IDPs through thermal processes within the parent body through the carbonization of H_2CO precursors in aqueous conditions (Cody et al. 2011; Kebukawa et al. 2013). This scenario is however inconsistent with the lack of liquid water in comets in the case of IDPs and it does not satisfactorily account for the IOM isotopic heterogeneity.

Organics in IDPs and AMMs are much less known, as outlined above, but comparison with chondritic organic matter sheds light on their origin. First, they share a number of similarities: (1) functional groups such as aliphatic, carboxyl, ketone and aromatic groups; (2) a disordered polyaromatic structure (with a weakly dispersive G-band position); (3) large and heterogeneous D- and ^{15}N-enrichments and (4) soluble and insoluble components. These similarities support the view that organics in meteorites and dusts were synthesized under fairly similar conditions. In this respect, solving the puzzle of the origin of chondritic organics might help resolve that of organics in dusts as well. However, this interpretation may not hold for the N-rich UCAMMs that might have been formed at the surface of TNOs or cometary nuclei in the Oort belt (Dartois et al. 2013).

In the near future, several complementary approaches should provide new clues and push the debate forward. Firstly, new techniques can simultaneously characterize the SOM and IOM at the sub-micrometer scale within their petrologic context, for instance XANES microscopy combining measurements at the C, N and O K-edge. The development of ultra-spectroscopy (infrared and Raman spectrometers coupled to an Atomic Force Microscope) should even provide insights into the composition at a spatial scale <10 nm (Dominguez et al. 2014). The development of protocols for characterizing SOM from micrometer-sized particles with High Resolution Mass Spectrometry and isolation of IOM from chondritic (small) clasts and IDPs and AMMs should soon provide very valuable data. Lastly, the ALMA and JWST instruments will provide insights into the composition of protoplanetary disks with an unprecedented angular resolution. This will provide important information on the chemical and isotopic heterogeneity in protoplanetary disks and thus on the extent to which an ISM-like chemistry can be induced.

2.6 Dust and Meteorites at the Surface of Mars

2.6.1 Flux of Exogenous Matter on Mars

The amount of extraterrestrial input to the surface of Mars has been constrained by several studies. In particular, Flynn and McKay (1990) concluded that fine-grained meteoritic material (60–1200 μm) is a major contributor to the exogenous input on Mars. They advocated the return of Mars soil samples in which the sampling of micrometeorites might reveal a complementary picture to micrometeorites on Earth. Most particles larger than 100 μm are melted on entry into the Earth's atmosphere, but given the lower density and lower gravity of the Martian atmosphere, 90% of these particles would survive Martian atmospheric entry without significant melting. For particles of 1000 μm diameter, the estimated 30% survival percentage is much higher than that on Earth. Flynn and McKay (1990) estimated the planet-wide meteoritic mass influx on Mars to be between 2700 and 59,000 t/year; the large range being related to the uncertainty in the terrestrial flux. With a production rate of Martian soil of 1 m per billion years, they estimated the concentration of micrometeorites ranging from 2% to 29% by mass (Flynn and McKay 1990).

In a complementary study, Bland and Smith (2000) focused on meteorite accumulation on Mars. Entry speed (from 5 to 30 km s^{-1}, mean value of 10.2 km s^{-1}), initial mass at the top of the atmosphere of Mars (10–130 g) and ablation rate are some of the main input parameters. Based on experimental studies, it is assumed that 1.6 km s^{-1} is the upper limit of survivability for stony meteorites impacting Mars. To estimate the impact rate of meteorites on Mars, a scaling factor of 2.6 compared to the present flux of meteoroids reaching Earth (based on crater statistics and reflecting for example the closer proximity of Mars to the asteroid belt) was considered. Bland and Smith (2000) then computed the proportions of meteoroids of a given entry mass surviving impact on Mars. They showed that a narrow range of small masses (10–50 g) should survive. Bland and Smith (2000) further calculated (in what they deemed a conservative estimation) a flux of 44–176 meteorites in the mass range 10–50 g at the surface of Mars, per 10^6 km^2 per year. Chemical and physical weathering operating on Mars is rather limited: oxidation would indeed occur on a timescale of 10^9 years, rather than 10^5 to 10^6 years as observed on Earth and based on analyses of meteorites from Antarctica. This absence of significant chemical alteration is confirmed by the remote study of iron meteorites found on Mars, in which iron oxides do not dominate (Schröder et al. 2008; Ashley et al. 2011). Bland and Smith (2000) thus concluded that meteorite accumulation on Mars could possibly be quite significant.

Finally, Davis (1993) demonstrated that the number of large impacts (>10 kg) should be similar to that on the Moon, with ~100 events per year.

This non-negligible contribution of meteoritic material to the surface of Mars is in agreement with analytical data. For example, based on X-Ray fluorescence data, Boslough (1988) suggested that the soils analysed by Viking are consistent with a mixture of 60% basaltic rock fragments and 40% meteoritic material. More recently, elemental analysis of siderophile elements (Ni, Ir) in the Martian meteorite NWA 7533 were explained by admixing up to 5% of meteoritic material in Martian regolith (Humayun et al. 2013). The composition of the Moon is also informative. Without an atmosphere, there is no deceleration to free-fall velocity of the impacting material. As a consequence, only a few projectiles survive as un-melted fragments on the Moon. This is consistent with only a few rare millimeter-diameter fragments of meteorites recovered in lunar soil samples (e.g., Rubin 1997). Nevertheless, extra-terrestrial matter contributes to the lunar regolith as indicated, for example, by the abundance of iridium in lunar soils (Anders et al. 1973).

To conclude, there might be a substantial input and preservation of extraterrestrial matter in Martian soils. Considering that this implies the presence of significant exogenous organic matter, several questions are raised concerning the preservation of organics once on Mars given the absence of clear detection of relatively abundant organics by the various rovers on Mars.

2.6.2 Exogenous Organics on Mars: Detection and Preservation

Mars' proximity to the asteroid belt may lead to sampling of different proportions of asteroidal material when compared with Earth. Moreover, meteorites arriving at the surface of Mars impact at a higher speed: their survival will thus depend on sample consistency and cohesive strength and whether the timescales of weathering are longer than on Earth. If this is the case, meteorites on Mars may reveal a slightly different history (older) to Earth (Flynn and McKay 1990). However, in the absence of reliable constraints, we here consider similar compositions for the material accreted by Mars and Earth.

In what follows, we discuss only meteoritic organics and do not consider carbonaceous molecules that could derive from Martian reservoirs (e.g., lithospheric magmatic carbon, weathering of carbonates in the hydrosphere, etc.). Since the meteoritic mass accreted by Mars is largely dominated by the flux of dust (see Sect. 2.6), we will neglect the flux of meteorites (Bland and Smith 2000) and large objects (Davis 1993) for the evaluation of exogenous organic matter. Taking into account (1) the present flux of dust at Mars (Flynn and McKay 1990), (2) the fraction of particles not heated above the pyrolysis temperature of carbonaceous matter (Flynn 1996) and (3) and assuming an average carbon content of 10 wt.% in IDPs (Thomas et al. 1993), the accretion rate of unaltered meteoritic carbon by Mars is estimated to be 2.4×10^5 kg/year (Flynn 1996).

Assessing past habitability and detecting potential preserved biosignatures are some of the major scientific goals of current and future missions on/to Mars. Models of dust flux on Mars clearly indicate that there is a non-negligible input of organic-containing extraterrestrial matter. Thus the detection, up to now, of only of a few organic molecules at low abundance is quite puzzling. This can either be interpreted as being due to (1) measurement biases and/or a low instrumental sensitivity or (2) the absence of organics in the analysed soil samples, explained for example by environmental conditions not in favour of the preservation of organics.

Our current knowledge concerning the organic compounds that are present in Martian soils has been derived only from molecular analysis from three experiments on board Martian landers: (1) Gas Chromatograph and Mass Spectrometer (GC-MS) experiments on board the Viking Lander (Biemann et al. 1977) and (2) on board MSL with the SAM instrument (Mahaffy et al. 2012) and (3) the Thermal and Evolved Gas Analyser (TEGA) on board Phoenix (Ming et al. 2009). These instruments require pyrolysis of the samples as a first step for analysis. In the case of Mars, clear difficulties were revealed in comparison with experiments conducted in laboratories on Earth, in particular for the interpretation of the data as the heating procedures induce an oxidation of the organics by the soils (Navarro-González et al. 2006; Steininger et al. 2012; Freissinet et al. 2015). The presence of perchlorates at the Martian surface (Hecht et al. 2009; Glavin et al. 2013) does not necessarily preclude the detection of organics by gas chromatography, but it clearly complicates data interpretation (e.g., Freissinet et al. 2015; Millan et al. 2016), for instance by the formation of new organic molecules through the decomposition of perchlorates (e.g., chlorohydrocarbons; Millan et al. 2016; Steininger et al. 2012) or by the combustion of organics through the release of O_2. Moreover, according to Benner et al. (2000), upon pyrolysis, meteoritic organic compounds are most likely to be converted to carboxylic acid derivatives under Martian conditions and these would be generated with variable yields, in addition to the generation of non-volatile components, carbon dioxide, carbon monoxide and water. All but the non-volatile material would be detectable by GC-MS, but since they are also components of the Martian atmosphere, data interpretation is more difficult (Benner et al. 2000). The identification of chlorinated organic compounds required numerous laboratory analogue studies, but it has finally been concluded that they originate from Martian organic carbon (as opposed to terrestrial contamination by the instrument) (Freissinet et al. 2015). Viking data have been reconsidered and may reflect a few ppm of organic material in Martian soils (Navarro-González et al. 2006, 2010; Navarro-González and McKay 2011), although even this is debated (Biemann and Bada 2011).

In addition to analytical issues, the environmental conditions on Mars may induce some modification of the exogenous organic input through photochemical (UV irradiation by sunlight and by galactic and solar energetic particles) and oxidation processes, which are characterized by distinct kinetics and scales of interaction. In particular, oxidation may be effective from a few meters to possibly hundreds of meters in depth (Davila et al. 2008), whereas UV radiation can interact only with the surface (a few mm) of the regolith. Solar energetic particles and galactic cosmic rays have much higher energy than UV photons, but with a much lower flux. These energetic particles may degrade organics on a timescale of hundreds of millions of years (e.g., Kminek and Bada 2006), but UV photons act much more rapidly, typically in a few days to months (e.g., Poch et al. 2014).

Experimental simulations of Martian soils processes have focused on UV irradiation and oxidative processes. Although indispensable for improving our understanding of the on-going processes operating in Martian soils and their consequences for organics, laboratory experiments simulate relatively simple systems for which the results are not easily directly applicable to Martian conditions. For example, several studies have focused only on simple organic molecules, while the most common organic matter in micrometeorites and IDPs a complex macromolecule. The regeneration of Martian soils is rarely taken into account. Mineral grains that potentially play a shielding role are present in only a few recent studies (e.g., Poch et al. 2015). Nevertheless, even though they might not be directly applicable to natural Martian conditions, these experiments are informative and imply that there is modification but not total destruction of the organics (e.g., Burns and Fisher, 1993; Stoker and Bullock 1997; Moores and Schuerger, 2012; Poch et al. 2013, 2014, 2015—a non-exhaustive list of references). The absence of detection of high abundances of organic compounds in Martian soils does not necessarily imply the absence of exogenous organics on Mars. The use of different analytical techniques that do not requiring pyrolysis as a first step will undoubtedly shed new light on the question of organics on Mars. This is the hope placed in the SuperCam instrument on board the Mars2020 rover, which combines Laser Induced Breakdown Spectroscopy, Raman spectroscopy, time-resolved fluorescence and visible and near-infrared spectroscopy (Maurice et al. 2015).

References

Aléon J, Engrand C, Robert F et al (2001) Clues to the origin of interplanetary dust particles from the isotopic study of their hydrogen-bearing phases. Geochim Cosmochim Acta 65:4399–4412

Aléon J, Robert F, Chaussidon M et al (2003) Nitrogen isotopic composition of macromolecular organic matter in interplanetary dust particles. Geochim Cosmochim Acta 67:3773–3783

Alexander CMO'D, Russell SS, Arden JW et al (1998) The origin of chondritic macromolecular organic matter: a carbon and nitrogen isotope study. Meteorit Planet Sci 33:603–622

Alexander CMO'D, Fogel M, Yabuta H et al (2007) The origin and evolution of chondrites recorded in the elemental and isotopic compositions of their macromolecular organic matter. Geochim Cosmochim Acta 71:4380–4403

Alexander CMO'D, Newsome SN, Fogel ML et al (2010) Deuterium enrichments in chondritic macromolecular material – implications for the origin and evolution of organics, water and asteroids. Geochim Cosmochim Acta 74:4417–4437

Alexander CMO'D, Cody GD, Kebukawa Y et al (2014) Elemental, isotopic and structural changes in Tagish Lake insoluble organic matter produced by parent body processes. Meteorit Planet Sci 49:503–525

Alexander CMO'D, Cody GD, De Gregorio BT et al (2017) The nature, origin and modification of insoluble organic matter in chondrites, the possibly interstellar source of Earth's C and N. Chem Erde 77:227–256

Altwegg K, Balsiger H, Bar Nun A et al (2016) Prebiotic chemicals amino acid and phosphorus in the coma of comet 67P/Churyumov-Gerasimenko. Sci Adv 2:e1600285

Anders E, Ganapathy R, Krähenbühl U et al (1973) Meteoritic material on the Moon. The Moon 8:3–24

Ash RD, Pillinger CT (1995) Carbon, nitrogen and hydrogen in Saharan chondrites: the importance of weathering. Meteoritics 30:85–92

Ashley JW, Golombek MP, Christensen PR et al (2011) Evidence for mechanical and chemical alteration of iron-nickel meteorites on Mars: process insights for Meridiani Planum. J Geophys Res 116:E00F20

Bandurski EL, Nagy B (1976) The polymer-like organic material in the Orgueil meteorite. Geochim Cosmochim Acta 40:1397–1406

Beck P, Quirico E, Montes-Hernandez G (2010) Hydrous mineralogy of CM and CI chondrites from infrared spectroscopy and their relationship with low albedo asteroids. Geochim Cosmochim Acta 74:4881–4892

Benner SA, Devine KG, Matveeva LN et al (2000) The missing organic molecules on Mars. Proc Natl Acad Sci USA 97:2425–2430

Bernstein MP, Moore MH, Elsila JE et al (2003) Side group addition to the polycyclic aromatic hydrocarbon coronene by proton irradiation in cosmic ice analogs. Astrophys J 582:L25–L29

Berthelot P (1868) Cosmologie – Sur la matière charbonneuse des météorites. C R Hebd Seances Acad Sci 67:849

Biemann K, Bada JL (2011) Comment on "Reanalysis of the Viking results suggests perchrlorate and organics at midlatitudes on Mars" by Rafael Navarro-Gonzáles et al. J Geophys Res 116: E12001

Biemann K, Oro J, Toulmin P III et al (1977) The search for organic substances and inorganic volatile compounds in the surface of Mars. J Geophys Res 82:4641–4658

Binet L, Gourier D, Derenne S et al (2002) Heterogeneous distribution of paramagnetic radicals in insoluble organic matter from the Orgueil and Murchison meteorites. Geochim Cosmochim Acta 66:4177–4186

Binet L, Gourier D, Derenne S et al (2004) Occurence of abundant diradicaloid moieties in the insoluble organic matter from the Orgueil and Murchison meteorites: a fingerprint of its extraterrestrial origin? Geochim Cosmochim Acta 68:881–891

Bland PA, Smith TB (2000) Meteorite accumulation on Mars. Icarus 144:21–26
Bland PA, Smith TB, Jull AJT et al (1996) The flux of meteorites to the Earth over the last 50000 years. Mon Not R Astron Soc 283:551–565
Bonal L, Quirico E, Bourot-Denise M (2006) Determination of the petrologic type of CV3 chondrites by Raman spectroscopy of included organic matter. Geochim Cosmochim Acta 70:1849–1863
Bonal L, Bourot-Denise M, Quirico E et al (2007) Organic matter and metamorphic history in CO chondrites. Geochim Cosmochim Acta 71:1605–1623
Bonal L, Alexander CMO'D, Huss GR et al (2013) Hydrogen isotopic composition of the water in CR chondrites. Geochim Cosmochim Acta 106:111–113
Bonal L, Quirico E, Flandinet L et al (2016) Thermal history of type 3 chondrites from the Antarctic meteorite collection determined by Raman spectroscopy of their polyaromatic carbonaceous matter. Geochim Cosmochim Acta 189:312–337
Boslough MD (1988) Meteoritic enrichment of Martian regolith. Abstracts of the Lunar and Planetary Science Conference 19:120
Bradley JP (2014) Early Solar Nebula grains – interplanetary dust particles. In: Davis AM (ed) Meteorites and cosmochemical processes. Volume 1 of Treatise on geochemistry, 2nd edn. Elsevier, Amsterdam, pp 287–308
Briani G, Gounelle M, Bourot-Denise M et al (2012) Xenoliths and microxenoliths in H chondrites: sampling the zodiacal cloud in the asteroid Main Belt. Meteorit Planet Sci 47:880–902
Brinton KLF, Engrand C, Glavin DP et al (1998) A search for extraterrestrial amino acids in carbonaceous Antarctic micrometeorites. Orig Life Evol Biosph 28:413–424
Brunetto R, Pino T, Dartois E, Cao A-T, d'Hendecourt L, Strazzulla G, Bréchignac P (2009) Comparison of the Raman spectra of ion irradiated soot and collected extraterrestrial carbon. Icarus 200:323–337
Burns RG, Fisher DS (1993) Rates of oxidative weathering on the surface of Mars. J Geophys Res 98(E2):3365–3372
Burton AS, Stern JC, Elsila JE et al (2012) Understanding prebiotic chemistry through the analysis of extraterrestrial amino acids and nucleobases in meteorites. Chem Soc Rev 41:5459–5472
Busemann H, Young A, Alexander CMO'D et al (2006) Interstellar chemistry recorded in organic matter from primitive meteorites. Science 312:727–730
Busemann H, Alexander CMO'D, Nittler LR (2007) Characterization of insoluble organic matter in primitive meteorites by microRaman spectroscopy. Meteorit Planet Sci 42:1387–1416
Busemann H, Nguyen AN, Cody GD et al (2009) Ultra-primitive interplanetary dust particles from the comet 26P/Grigg-Skjellerup dust stream collection. Earth Planet Sci Lett 288:44–57
Capaccioni F, Coradini A, Filacchione G et al (2015) The organic-rich surface of comet 67P/Churyumov-Gerasimenko as sees by VIRTIS/Rosetta. Science 347:aaa0628
Cleeves LI, Bergin EA, Alexander CMO'D et al (2014) The ancient heritage of water ice in the solar system. Science 345:1590–1593
Cody GD, Alexander CMO'D (2005) NMR studies of chemical structural variation of insoluble organic matter from different carbonaceous chondrite groups. Geochim Cosmochim Acta 69:1085–1097
Cody GD, Alexander CMO'D, Tera F (2002) Solid state (^{1}H and ^{13}C) NMR spectroscopy of the insoluble organic residue in the Murchison meteorite: a self-consistent quantitative analysis. Geochim Cosmochim Acta 66:1851–1865
Cody GD, Alexander CMO'D, Yabuta H et al (2008) Organic thermometry for chondritic parent bodies. Earth Planet Sci Lett 272:446–455
Cody GD, Heying E, Alexander CMO'D et al (2011) Establishing a molecular relationship between chondritic and cometary organic solids. Proc Natl Acad Sci USA 108:19171–19176
Cronin JR, Pizzarello S, Frye JS (1987) ^{13}C NMR spectroscopy of the insoluble carbon of carbonaceous chondrites. Geochim Cosmochim Acta 51:299–303

Crovisier J, Leech K, Bockelée-Morvan D et al (1997) The spectrum of comet Hale-Bopp (C/199501) observed with the Infrared Space Observatory at 2.9 astronomical units from the Sun. Science 275:1904–1907
Dartois E, Marco O, Munoz-Caro GM et al (2004) Diffuse interstellar medium organic polymers. Astron Astrophys 423:L33–L36
Dartois E, Geballe TR, Pino T et al (2007) IRAS 08572+3915: constraining the aromatic versus aliphatic content of interstellar HACs. Astron Astrophys 46:635–640
Dartois E, Engrand C, Brunetto R et al (2013) Ultracarbonaceous Antarctic micrometeorites, probing the solar System beyond the nitrogen snow-line. Icarus 224:243–252
Davila AF, Fairen AG, Gago-Duport L et al (2008) Subsurface formation of oxidants on Mars and implications for the preservation of organic biosignatures. Earth Planet Sci Lett 272:456–463
Davis PM (1993) Meteoroid impacts as seismic sources on Mars. Icarus 105:469–478
Delpoux O, Gourier D, Vezin H et al (2011) Biradical character of D-rich carriers in the insoluble organic matter of carbonaceous chondrites: a relic of the protoplanetary disk chemistry. Geochim Cosmochim Acta 75:326–336
DeMeo FE, Carry B (2013) The taxonomic distribution of asteroids from multi-filter all-sky photometric surveys. Icarus 226:723–741
Dermott SF, Durda DD, Grogan K et al (2002) Asteroidal dust. In: Bottke WF Jr, Cellino A, Paolicchi P et al (eds) Asteroids III. University of Arizona Press, Tucson, pp 423–442
Dobrica E, Engrand C, Duprat J et al (2009) Connection between micrometeorites and Wild 2 particles from Antarctic snow to cometary ices. Meteorit Planet Sci 44:1643–1661
Dobrica E, Engrand C, Quirico E et al (2011) Raman characterization of carbonaceous matter in CONCORDIA Antarctic micrometeorites. Meteorit Planet Sci 46:1363–1375
Dominguez G, McLeod AS, Gainsforth Z et al (2014) Nanoscale infrared spectroscopy as a non-destructive probe of extraterrestrial samples. Nat Commun 5:5445
Duprat J, Engrand C, Maurette M et al (2007) Micrometeorites from Central Antarctic snow: the CONCORDIA collection. Adv Space Res 39:605–611
Duprat J, Dobrica E, Engrand C et al (2010) Extreme deuterium excesses in ultracarbonaceous micrometeorites from central Antarctic snow. Science 328:742–745
Ehrenfreund P, Robert F, d'Hendecourt L et al (1991) Comparison of interstellar and meteoritic organic-matter at 3.4 μm. Astron Astrophys 252:712–717
Elsila JE, Dworkin JP, Berstein MP et al (2007) Mechanisms of amino acid formation in interstellar ice analogs. Astrophys J 660:911–918
Elsila JE, Glavin DP, Dworkin JP (2009) Cometary glycine detected in samples returned by Stardust. Meteorit Planet Sci 44:1323–1330
Engrand C, Maurette M (1998) Carbonaceous micrometeorites from Antarctica. Meteorit Planet Sci 33:565–580
Farley KA (1995) Cenozoic variations in the flux of interplanetary dust recorded by 3He in a deep-sea sediment. Nature 376:153–156
Floss C, Stadermann FJ, Bradle JP et al (2006) Identification of isotopically primitive interplanetary dust particles: a NanoSIMS isotopic imaging study. Geochim Cosmochim Acta 70:2371–2399
Flynn GJ (1996) The delivery of organic matter from asteroids and comets to the early surface of Mars. Earth Moon Planets 72:469–474
Flynn GJ, McKay DS (1990) An assessment of the meteorite contribution to the Martian soil. J Geophys Res 95:14497–14509
Flynn GJ, Keller LP, Feser M et al (2003) The origin of organic matter in the solar system: evidence from the interplanetary dust particles. Geochim Cosmochim Acta 67:4791–4806
Freissinet C et al (2015) Organic molecules in the Sheepbed Mudstone, Gale Crater, Mars. J Geophys Res Planets 120:495–514
Gabrielli P, Barbante C, Plane JMC et al (2004) Meteorite smoke fallout over the Halocene epoch revealed by iridium and plantinum in Greenland ice. Nature 432:1011–1014

Gardinier A, Derenne S, Robert F et al (2000) Solid state CP/MAS C-13 NMR of the insoluble organic matter of the Orgueil and Murchison meteorites: quantitative study. Earth Planet Sci Lett 184:9–21
Gattacceca J, Valenzuela M, Uehara M et al (2011) The densest meteorite collection area in hot deserts: the San Juan meteorite field (Atacama Desert, Chile). Meteorit Planet Sci 46:1276–1287
Genge MJ, Engrand C, Gounelle M et al (2008) The classification of micrometeorites. Meteorit Planet Sci 43:497–515
Glavin DP, Freissinet C, Miller KE et al (2013) Evidence for perchlorates and the origin of chlorinated hydrocarbons detected by SAM at the Rocknest aeolian deposit in Gale Crater. J Geophys Res Planets 118:1955–1973
Gounelle M, Zolensky ME, Liou J-C et al (2003) Mineralogy of carbonaceous chondritic microsclasts in Howardites: identification of C2 fossil micrometeorites. Geochim Cosmochim Acta 67:507–527
Gounelle M, Spurny P, Bland PA (2006) The orbit and atmospheric trajectory of the Orgueil meteorite from historical records. Meteorit Planet Sci 41:135–150
Gourier D, Robert F, Delpoux O et al (2008) Extreme deuterium enrichment of organic radicals in the Orgueil meteorite: revisiting the interstellar interpretation? Geochim Cosmochim Acta 72:1914–1923
Grady MM (2000) Catalogue of meteorites. Cambridge University Press, Cambridge
Grossman L, Larimer JW (1974) Early chemical of the solar system. Rev Geophys Space Phys 12:71–101
Halliday I, Blackwell AT, Griffin AA (1984) The frequency of meteorite falls on the Earth. Science 223:1405–1407
Halliday I, Blackwell AT, Griffin AA (1989) The flux of meteorites on the Earth's surface. Meteoritics 24:173–178
Harvey RP, Cassidy WA (1989) A statistical comparison of Antarctic finds and modern falls: mass frequency distributions and relative abundance by type. Meteoritics 24:9–14
Hayes JM (1967) Organic consistuents of meteorites – a review. Geochim Cosmochim Acta 31:1395–1440
Hayatsu R, Matsuoka S, Scott RG et al (1977) Origin of organic matter in earlt solar system. VII. The organic polymer in carbonaceous chondrites. Geochim Cosmochim Acta 41:1325–1339
Hayatsu R, Winans RE, Scott RG et al (1980) Phenolic esters in the organic polymer of the Murchison meteorite. Science 207:1202–1204
Hayatsu R, Anders E (1981) Organic compounds in meteorites and their origins. Top Curr Chem 99:3–37
Hecht MH, Kounaves SP, Quinn RC et al (2009) Detection of perchlorate and the soluble chemistry of Martian soil at the Phoenix Lander Site. Science 325:64–67
Henning T, Semenov D (2013) Chemistry in protoplanetary disks. Chem Rev 113:9016–9042
Herd CDK, Blinova A, Simku DN et al (2011) Origin and evolution of prebiotic organic matter as inferred from the Tagish Lake meteorite. Science 332:1304–1307
Hertkorn N, Harir M, Cawley KM et al (2015) Molecular characterization of dissolved organic matter from subtropical wetlands: a comparative study through the analysis of optical properties, NMR and FTICR/MS. Biogeosciences 13:2257–2277
Horst SM, Yelle RV, Buch A et al (2012) Formation of amino acids and nucleotides bases in a Titan atmosphere simulation experiment. Astrobiology 12:809–817
Humayun M, Nemchin A, Zanda B et al (2013) Origin and age of the earliest Martian crust from meteorite NWA 7533. Nature 503:513–516
Hunten DM, Turco RP, Toon OB (1980) Smoke and dust particles of meteoric origin in the mesosphere and stratosphere. J Atmos Sci 37:1342–1357
Huss GR, Rubin A, Grossman J (2006) Thermal metamorphism in chondrites. In: Lauretta DS, McSween HY Jr (eds) Meteorites and the early solar system II. University of Arizona Press, Tucson, pp 567–586

Jull AJT (2006) Terrestrial ages of meteorites. In: Lauretta DS, McSween HY Jr (eds) Meteorites and the early solar system II. University of Arizona Press, Tucson, pp 889–905
Kebukawa Y, Alexander CMO'D, Cody GD (2011) Compositional diversity in insoluble organic matter in type 1, 2 and 3 chondrites as detected by infrared spectroscopy. Geochim Cosmochim Acta 75:3530–3541
Kebukawa Y, Kilcoyne ALD, Cody GD (2013) Exploring the potential formation of organic solids in chondrites and comets through polymerization of interstellar formaldehyde. Astrophys J 771:19
Keller LP, Messenger S, Flynn GJ et al (2004) The nature of molecular cloud material in interplanetary dust. Geochim Cosmochim Acta 68:2577–2589
Kerridge J, Chang S, Shipp R (1987) Isotopic characterization of kerogen-like material in the Murchison carbonaceous chondrite. Geochim Cosmochim Acta 51:2527–2540
Kitajima F, Nakamura T, Takaoka N et al (2002) Evaluating the thermal metamorphism of CM chondrites by using the pyrolitic behavior of carbonaceous macromolecular matter. Geochim Cosmochim Acta 66:163–172
Kminek G, Bada JL (2006) The effect of ionizing radiation on the preservation of amino acids on Mars. Earth Planet Sci Lett 245:1–5
Komiya M, Shimoyama A (1996) Organic compounds from insoluble organic matter isolated from the Murchison carbonaceous chondrite by heating experiments. Bull Chem Soc Jpn 69:53–58
Krot AN, Keil K, Scott ERD et al (2014) Classification of Meteorites and their genetic relationships. In: Davis AM (ed) Meteorites and cosmochemical processes. Volume 1 of Treatise on geochemistry, 2nd edn. Elsevier, Oxford, pp 1–63
Kvenvolden K, Lawless J, Pering K et al (1970) Evidence for extraterrestrial amino-acids and hydrocarbons in the Murchison meteorite. Nature 228:923–926
Lanci L, Kent DV (2006) Meteoritic smoke fallout revealed by superparamagnetism in Greenland ice. Geophys Res Lett 33:L13308
Laurent B, Roskosz M, Remusat L et al (2014) Isotopic and structural signature of experimentally irradiated organic matter. Geochim Cosmochim Acta 142:522–534
Laurent B, Roskosz M, Remusat L et al (2015) The deuterium/hydrogen distribution in chondritic organic matter attests to early ionizing irradiation. Nat Commun 6:8567
Le Guillou C, Rouzaud J-N, Bonal L et al (2012) High resolution TEM of chondritic carbonaceous matter: metamorphic evolution and heterogeneity. Meteorit Planet Sci 47:345–362
Le Guillou C, Bernard S, Brearley AJ et al (2014) Evolution of organic matter in Orgueil Murchison and Renazzo during parent body aqueous alteration: In situ investigations. Geochim Cosmochim Acta 131:368–392
Love SG, Brownlee DD (1991) Heating and thermal transformation of micrometeoroids entering the Earth's atmosphere. Icarus 89:26–43
Love SG, Brownlee DE (1993) A direct measurement of the terrestrial mass accretion rate of cosmic dust. Science 262:550–553
Mahaffy PR, Webster CR, Cabane M et al (2012) The sample analysis at Mars investigation and instrument suite. Space Sci Rev 170:401–478
Matrajt G, Pizzarello S, Taylor S et al (2004) Concentration and variability of the AIB amino acid in polar micrometeorites: implications for the exogenous delivery of amino acids to the primitive Earth. Meteorit Planet Sci 39:1849–1858
Matrajt G, Munoz-Caro GM, Dartois E et al (2005) FTIR analysis of the organics in IDPs: comparison with the IR spectra of the diffuse interstellar medium. Astron Astrophys 433:979–995
Maurice S, Wiens RC, Anderson R et al and the SuperCam team (2015) Science objectives of the SuperCam instrument for the Mars2020 rover. LPSC #2818
Merouane S, Djouadi Z, Le Sergeant d'Hendecourt L (2014) Relations between aliphatics and silicate components in 12 stratospheric particles deduced from vibrational spectroscopy. Astrophys J 780:174–186

Messenger S (2000) Identification of molecular-cloud material in interplanetary dut particles. Nature 404:968–971
Messenger S (2002) Deuterium enrichments in interplanetary dust. Planet Space Sci 50:1221–1225
Michel-Lévy MC, Lautie A (1981) Microanalysis by Raman spectroscopy of carbon in the Tieschitz chondrite. Nature 292:321–322
Millan M, Szopa C, Buch A et al (2016) In situ analysis of martian regolith with the SAM experiment during the first mars year of the MSL mission: identification of organic molecules by gas chromatography from laboratory measurements. Planet Space Sci 129:88–102
Ming DW, Lauer HV Jr, Archer PD Jr et al (2009) Combustion of organic molecules by the thermal decomposition of perchlorate salts: implications for organics at the Mars Phoenix scout landing site. LPSC#2241
Moores JE, Schuerger AC (2012) UV degradation of accreted organics on Mars: IDP longevity, surface reservoir of organics, and relevance to the detection of methane in the atmosphere. J Geophys Res 117:E08008
Munoz Caro GM, Matrajt G, Dartois E et al (2006) Nature and evolution of the dominant carbonaceous matter in interplanetary dust particles: effects of irradiation and identification with a type of amorphous carbon. Astron Astrophys 459:147–159
Naraoka H, Mita H, Komiya M et al (2004) A chemical sequence of macromolecular organic matter in the CM chondrites. Meteorit Planet Sci 39:401–406
Navarro-González R, McKay CP (2011) Reply to comment by Biemann and Bada on "Reanalysis of the Viking results suggests perchlorate and organics at midlatitudes on Mars". J Geophys Res 116:E12002
Navarro-González R, Navarro KF, de la Rosa J et al (2006) The limitations on organic detection in Mars-like soils by thermal volatilization–gas chromatography-MS and their implications for the Viking results. Proc Natl Acad Sci USA 103:16089–16094
Navarro-González R, Vargas E, de la Rosa J (2010) Reanalysis of the Viking results suggests perchlorate and organics at midlatitudes on Mars. J Geophys Res 115:E12010
Nesvorný D, Jenniskens P, Levison HF et al (2010) Cometary origin of the Zodiacal Cloud and carbonaceous micrometeorites. Implications for hot debris disks. Astrophys J 713:816–836
Oba Y, Naraoka H (2009) Elemental and isotopic behavior of macromolecular organic matter from CM chondrites during hydrous pyrolysis. Meteorit Planet Sci 44:943–954
Okumura F, Mumura K (2011) Gradual and stepwise pyrolyses of insoluble organic matter from the Murchison meteorite revealing chemical structure and isotopic distribution. Geochim Cosmochim Acta 75:7063–7080
Orthous-Daunay FR, Quirico E, Lemelle L et al (2010) Speciation of sulfur in the insoluble organic matter from carbonaceous chondrites by XANES spectroscopy. Earth Planet Sci Lett 300:321–328
Orthous-Daunay FR, Quirico E, Beck P et al (2013) Mid-infrared study of the molecular structure variability of insoluble organic matter from primitive chondrites. Icarus 223:534–543
Parise B, Ceccarelli C, Tielens AGGM et al (2006) Testing grain surface chemistry: a survey of deuterated formaldehyde and methanol in low-mass class 0 protostars. Astron Astrophys 453:949–958
Peltzer ET, Bada J (1978) Alpha-hydroxycarboxylic acids in Murchinson meteorite. Nature 272:443
Pendleton YJ, Allamandola LJ (2002) The organic refractory material in the diffuse interstellar medium: mid-infrared spectroscopic constraints. Astrophys J Suppl 138:75–98
Piani L, Robert F, Remusat L (2015) Micron-scale D/H heterogeneity in chondrite matrices: a signature of the pristine solar system water? Earth Planet Sci Lett 415:154–164
Poch O, Noblet A, Stalport F et al (2013) Chemical evolution of organic molecules under Mars-like UV radiation conditions simulated in the laboratory with the "Mars organic molecule irradiation and evolution" (MOMIE) setup. Planet Space Sci 85:188–197

Poch O, Kaci S, Stalport F et al (2014) Laboratory insights into the chemical and kinetic evolution of several organic molecules under simulated Mars surface UV radiation conditions. Icarus 242:50–63
Poch O, Jaber M, Stalport F et al (2015) Effects of nontronite smectite clay on chemical evolution of several organic molecules under simulated Martian surface ultraviolet radiation conditions. Astrobiology 15:1–17
Quirico E, Raynal PI, Bourot-Denise M (2003) Metamorphic grade of organic matter in six unequilibrated ordinary chondrites. Meteorit Planet Sci 38:795–811
Quirico E, Borg J, Raynal P-I et al (2005) A micro-Raman survey of 10 IDPs and 6 carbonaceous chondrites 2005. Planet Space Sci 53:1443–1448
Quirico E, Montagnac G, Rouzaud J-N et al (2009) Precursor and metamorphic conditions effects on Raman spectra of poorly-ordered carbonaceous matter in chondrites and coals. Earth Planet Sci Lett 287:185–193
Quirico E, Bourot-Denise M, Robin C et al (2011) A reappraisal of the thermal history of EH3 and EL3 enstatite chondrites. Geochim Cosmochim Acta 75:3088–3102
Quirico E, Orthous-Daunay FR, Beck P et al (2014) Origin of insoluble organic matter in type 1 and 2 chondrites: new clues, new questions. Geochim Cosmochim Acta 136:80–99
Remusat L, Derenne S, Robert F et al (2005a) New pyrolytic and spectroscopic data on Orgueil and Murchison insoluble organic: a different origin than soluble? Geochim Cosmochim Acta 69:3919–3932
Remusat L, Derenne S, Robert F (2005b) New insight on aliphatic linkages in the macromolecular organic fraction of Orgueil and Murchison meteorites through ruthenium tetroxide oxidation. Geochim Cosmochim Acta 69:4377–4386
Remusat L, Palhol F, Robert F et al (2006) Enrichment of deuterium in insoluble organic matter from primitive meteorites: a solar system origin? Earth Planet Sci Lett 243:15–25
Remusat L, Robert F, Meibom A et al (2009) Protoplanetary chemistry recorded by D-rich organic radicals in carbonaceous chondrites. Astrophys J 698:2087–2092
Remusat L, Piani L, Bernard S (2016) Thermal recalcitrance of the organic D-rich component of ordinary chondrites. Earth Planet Sci Lett 435:36–44
Rietmeijer FJM (1998) Interplanetary dust particles. In: Papike JJ (ed) Planetary materials. Revs mineral, vol 36. Mineralogical Society of America, Washington, DC, pp 1–95
Robert F, Epstein S (1982) The concentration and isotopic composition of hydrogen, carbon and nitrogen in carbonaceous meteorites. Geochim Cosmochim Acta 46:81–95
Rubin AE (1997) The Hadley Rille enstatite chondrite and its agglutinate-like rim: impact melting during accretion to the Moon. Meteorit Planet Sci 32:135–141
Rudraswami NG, Shyam Prasad M, Jones RH et al (2016) In situ oxygen isotope compositions in olivines of different types of cosmic spherules: an assessment of relationships to chondritic particles. Geochim Cosmochim Acta 194:1–14
Schimmelmann A, Sessions A, Mastalerz M (2006) Hydrogen isotopic (D/H) composition of organic diagenesis and thermal maturation. Annu Rev Earth Planet Sci 34:501–533
Schmitt-Koplin P, Gabelica Z, Gougeon RD et al (2010) High molecular diversity of extraterrestrial organic matter in Murchison meteorite revealed 40 years after its fall. Proc Natl Acad Sci USA 107:2763–2768
Schmitz B, Tassinari M, Peucker-Ehrenbrink B (2001) A rain of ordinary chondrite meteorites in the early Ordovician. Earth Planet Sci Lett 194:1–15
Schröder C, Rodionov DS, McCoy TJ et al (2008) Meteorites on Mars observed with the Mars Exploration Rovers. J Geophys Res 113:E06S22
Sephton MA (2002) Organic compounds in carbonaceous meteorites. Nat Prod Rep 19:292–311
Sephton MA, Gilmour I (2001) Pyrolysis-gas chromatography-isotope ratio mass spectrometry of macromolecular material in meteorites. Planet Space Sci 49:465–471
Sephton MA, Pillinger CT, Gilmour I (2000) Aromatic moieties in meteoritic macromolecular materials: analyses by hydrous pyrolysis and $\delta^{13}C$ of individual compounds. Geochim Cosmochim Acta 64:321–328

Sephton MA, Verchovsky AB, Bland PA et al (2003) Investigating the variations in carbon and nitrogen isotopes in carbonaceous chondrites. Geochim Cosmochim Acta 67:2093–2108
Sephton MA, Love GD, Watson JS et al (2004) Hydropyrolysis of insoluble carbonaceous matter in the Murchison meteorite: new insights into its macromolecular structure. Geochim Cosmochim Acta 68:1385–1393
Starkey NA, Franchi IA, Alexander CMO'D (2013) A Raman spectroscopic study of organic matter in interplanetary dust particles and meteorites using multiple wavelength laser excitation. Meteorit Planet Sci 48:1800–1822
Steininger H, Goesmann F, Goetz W (2012) Influence of magnesium perchlorate on the pyrolysis of organic compounds in Mars analogue soils. Planet Space Sci 71:9–17
Stoker CR, Bullock MA (1997) Organic degradation under simulated martian conditions. J Geophys Res 102:10881–10888
Suavet C, Cordier C, Folco L et al (2011) Non carbonaceaous chondrite-related large cosmic spherules from the Transantarctic Mountains. Geochim Cosmochim Acta 75:6200–6210
Taylor S, Lever JH, Harvey RP (1998) Accretion rate of cosmic spherules measure at the South Pole. Nature 393:899–903
Thomas KL, Blanford GE, Keller LP et al (1993) Carbon abundance and silicate mineralogy of anhydrous interplanetary dust particles. Geochim Cosmochim Acta 57:1551–1566
Tonui E, Zolensky M, Hiroi T et al (2014) Petrographic, chemical and spectroscopic evidence for thermal metamorphism in carbonaceous chondrites I: CI and CM chondrite. Geochim Cosmochim Acta 126:284–306
Vernazza P, Delbo M, King PL et al (2012) High surface porosity as the origin of emissivity features in asteroid spectra. Icarus 221:1162–1171
Vis RD, Mrowiec A, Kooyman PJ et al (2002) Microscopic search for the carrier phase Q of the trapped planetary noble gases in Allende, Leoville and Vigarano. Meteorit Planet Sci 37:1391–1399
Wang Y, Huang Y, Alexander CMO'D et al (2005) Molecular and compound-specific hydrogen isotope analyses of insoluble organic matter from 80 different carbonaceous chondrites groups. Geochim Cosmochim Acta 69:3711–3721
Wopenka B (1988) Raman observations on individual interplanetary dust particules. Earth Planet Sci Lett 88:221–231
Yabuta H, Naraoka H, Sakanishi K et al (2005) Solid-state ^{13}C NMR characterization of insoluble organic matter from Antarctic CM2 chondrites: evaluation of the meteoritic alteration level. Meteorit Planet Sci 40:779–787
Yabuta H, Alexander CMO'D, Fogel ML et al (2010) A molecular and isotopic study of the macromolecular organic matter of the ungrouped C2 WIS 91600 and its relationship to Tagish Lake and PCA 91008. Meteorit Planet Sci 45:1446–1460
Yuen G, Blair N, DesMarias DJ et al (1984) Carbon isotope composition of low molecular weight hydrocarbons and monocarboxylic acids from Murchison meteorite. Nature 307:252
Zolensky ME, Bland PA, Brown P et al (2006a) Flux of extraterrestrial materials. In: Lauretta DS, McSween HY Jr (eds) Meteorites and the early solar system II. University of Arizona Press, Tucson, pp 869–888
Zolensky ME, Zega TJ, Yano H et al (2006b) Mineralogy and petrology of comet 81P/Wild 2 nucleus samples. Science 314:1735–1739
Zuckerman B, Ball JA, Gottlieb CA (1971) Microwave detection of interstellar formic acid. Astrophys J 163:L41–L45

Chapter 3
Biosignatures of Cellular Components and Metabolic Activity

David J. Des Marais and Linda L. Jahnke

Abstract The astrobiological search for biosignatures requires a working concept of the fundamental attributes of life. Life's basic capabilities of energy harvesting, metabolism, and self-replication can create objects, substances and patterns—biosignatures—that indicate their biological origins. High relative abundances of certain lipids, hydrocarbons, amino acids and polysaccharides are diagnostic products of billions of years of evolution. Lipid assemblages having narrow molecular weight ranges are key constituents of cellular membranes. The molecular structures of lipids provide details of their biosynthetic pathways. Some lipid biosignatures are diagnostic for particular groups of microorganisms. The biosynthesis of organic matter and biochemical oxidation-reduction reactions can discriminate against the heavier isotopes of carbon and sulfur and thereby create molecular isotopic patterns that indicate not only their biological origins, but also key details about biosynthetic pathways. Sulfur isotopic patterns can indicate biological redox reactions. Because microorganisms can greatly enhance, at relatively low to moderate temperatures, reaction rates between oxidized and reduced sulfur compounds, they can create a range in the stable isotopic compositions of these compounds that is substantially larger than nonbiological reactions can achieve under similar conditions. The simultaneous presence of multiple biosignature objects, substances and patterns in a demonstrably habitable earlier environment constitutes the most compelling evidence of past life.

3.1 Introduction

3.1.1 Biosignatures

The venue of exploration fundamentally determines the perspective taken for research on signatures of life, or "biosignatures". For example, because Mars and the early Earth are prominent targets of exploration for astrobiology, much research

D. J. Des Marais (✉) · L. L. Jahnke
NASA-Ames Research Center, Mountain View, CA, USA
e-mail: David.J.DesMarais@nasa.gov; Linda.L.Jahnke@nasa.gov

B. Cavalazzi, F. Westall (eds.), *Biosignatures for Astrobiology*, Advances in Astrobiology and Biogeophysics, https://doi.org/10.1007/978-3-319-96175-0_3

focuses on detecting signs of past life sequestered in the rock record of these planets. Earth's early geological record provides an opportunity to investigate the antiquity of life on Earth; it is also an instructive analogue that guides our search for biosignatures on other rocky bodies. The significance of Mars arises both from the relative similarity of Mars and Earth among the planets of the Solar System, and from the relative accessibility of the Martian geological record. It is possible to search in situ for preserved remnants of cellular components and evidence of past metabolic activity. Accordingly this chapter focuses on such features.

A biosignature is an object, substance, and/or pattern created by a biological source. The value of a biosignature is determined not only by the probability of life producing it, but also by the improbability of non-biological processes making it. The search for life beyond Earth rests upon the premise that biosignatures will be recognizable within the contexts of their planetary environments. Astrobiological investigations must therefore also consider how biosignatures might be generated, preserved, and/or detected within those contexts.

3.1.2 Concepts of Life

The search for biosignatures requires a working concept of the fundamental attributes of life. This concept helps to identify the life sustaining "services" that an environment must provide. This in turn helps to identify the most promising past or present environmental candidates for exploration. Such a working concept also helps to identify and interpret biosignatures.

Any working concept of life must admit the possibility that life elsewhere differs in fundamental ways from life on Earth. Without a second known example of life it is probably not possible to determine which characteristics are unique to terrestrial life and which are common or required for life elsewhere. Indeed some have proposed that our current perspectives might even preclude a definitive, universal concept of life (Cleland and Chyba 2007). But in order to achieve a better understanding we must start taking steps, however uncertain, along that path. Accordingly, it is productive to propose attributes of life as we know it that might be universal, as opposed to those that might represent local solutions that have been specific to survival on Earth.

National Research Council (2007) identified the following potentially universal attributes of life:

1. life must exploit thermodynamic disequilibrium in the environment in order to perpetuate its own disequilibrium state;
2. life most probably consists of interacting sets of covalently bonded molecules that include a diversity of heteroatoms (e.g., N, O, P, S, etc. as in Earth-based life) that promote chemical reactivity;
3. life requires a liquid solvent that supports these molecular interactions;
4. life employs a molecular system capable of Darwinian evolution.

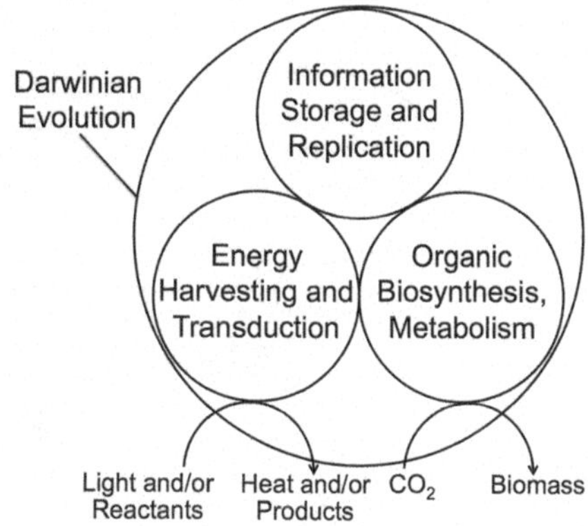

Fig. 3.1 The basic functions of life. Source: Adapted from Des Marais (2013)

These attributes implicate the following basic universal functions (Fig. 3.1):

1. life harvests energy from its environment and converts it to forms of chemical energy that directly sustain its other functions;
2. life sustains "metabolism"—a network of chemical reactions that synthesize the key chemical compounds required for maintenance, growth and self-replication;
3. life requires an "automaton," a multi-component system that is essential for self-replication and self-perpetuation (Von Neumann 1966).

Perhaps the most distinctive attribute of life is its capacity for self-replication to create populations upon which natural selection can act to maintain Darwinian evolution. Von Neumann's theory of the automaton (e.g., Von Neumann 1966) predicted the components that are essential for biological self-replication (Fig. 3.2). He had envisioned these essential components of the automaton before molecular biologists identified the actual molecular machinery of the DNA-RNA-protein system. The components required for self-replication consist of relatively complex molecules that must persist in environments so that they can function more rapidly than they are degraded. This perspective has implications regarding the unique virtues of carbon-based chemistry and it also imposes requirements on the attributes that an environment must possess in order to remain habitable.

Accordingly, life can be envisioned as a self-sustaining system that is capable of Darwinian evolution and that utilizes free energy to sustain and propagate an automaton, a metabolic network and functionally related larger structures. These functions specify a level of molecular complexity that in turn defines requirements for chemical ingredients, energy and environmental conditions that are essential to sustain life.

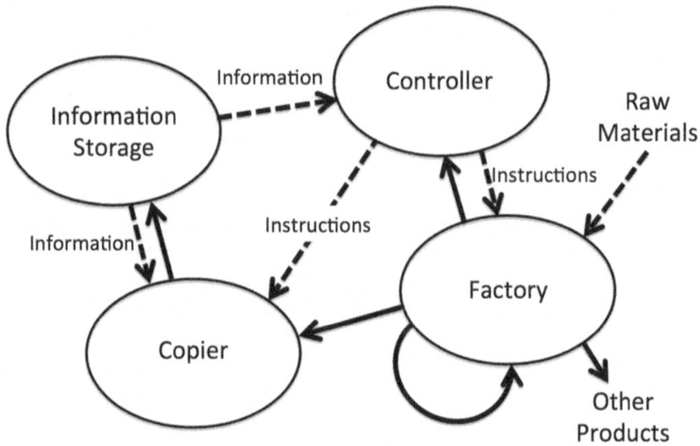

Fig. 3.2 An automaton is a system capable of self-replication and has several complex components. For example, in biological systems DNA stores and provides information (dashed arrows) and ribosome "factories" replicate other components, including proteins involved in other cellular processes. Source: Adapted from Des Marais (2013)

3.1.3 *Attributes of Life*

Because flight missions must make specific chemical observations to search for life, our working concept must provide more than just a short list of generic universal attributes. We can start by enumerating attributes that are universal among life as we currently understand it.

In his book "Beginnings of Cellular Life," Morowitz (1992) articulated several attributes that are shared by all life on Earth. Below are listed some key examples of these attributes and their applicability.

1. *"The chemistry of life is carried out in aqueous solutions or at water interfaces. Cells can survive the removal and restoration of cellular water, but water is essential to cellular function."* Liquid water is therefore essential to maintain the long-term survival of life.
2. *"The major atomic components in the covalently bonded portions of all functioning biological systems are C, H, N, O, P and S"*. Particular chemical compounds of these elements engage in genetic, metabolic and energy-harvesting functions and thus merit particular attention.
3. *"A cell is the most elementary unit that can sustain life."* Single-celled organisms are by far the most diverse, ancient and environmentally tolerant forms of life known. But we remain uncertain whether the earliest life was necessarily compartmentalized in cells.
4. *"There is a universal set of small organic molecules that constitutes the total mass of all cellular systems."* The set of key organic biomolecules on Earth represents an extremely small subset of all possible organic molecules. Perhaps this array is small due to the evolutionary selection for speed and efficiency of

biochemical processes. Living systems must efficiently maintain high information contents at molecular and structural levels.
5. *"Those reactions that proceed at appreciable rates in living cells are catalyzed by enzymes."* Catalysis enables cells to maintain their metabolism and to repair damage caused by environmental challenges such as radiation. Enzymes help to acquire essential energy by harvesting sunlight and by accelerating reactions between oxidized and reduced chemical species in the environment.
6. *"Sustained life is a property of an ecological system rather than a single organism or species."* For example, biofilms and microbial mats are highly successful ecological systems (e.g., Des Marais 2010). Fossilized biofilms have been identified as stromatolites in rocks more than 3.4 billion years old (Walter 1983).
7. *"All populations of replicating biological systems give rise to altered phenotypes that are the result of mutated genotypes."* All known living systems evolve by Darwinian-like natural selection. "Phenotype" is a cell's molecular and structural "machinery." Genotype is the information preserved in genomic molecules (DNA and RNA are examples in our biosphere). This interplay between phenotype and genotype is the automaton that is capable of Darwinian evolution.

This list does not necessarily preclude the possibility of radical alternatives to life as we know it. However, given the remarkable chemical and physical attributes of water and carbon compounds, together with their substantial cosmic abundances, the case can be made that at least a substantial and pervasive fraction of life elsewhere is based upon organic chemical reactions occurring in liquid water.

Basic attributes of life such as those enumerated above have allowed biochemists, paleontologists and astrobiologists to identify and interpret several categories of biosignatures in ancient geological deposits on Earth.

The following sections provide key examples of biosignatures that arise from carbon compounds and from biological redox reactions. Other categories of biosignatures include cellular microfossils (see Chap. 7), biominerals (Chap. 6) and biogenic sedimentary fabrics (Chap. 7).

3.2 Organic Molecular Biosignatures

Anomalously high relative abundances of specific organic molecules in a rock might constitute evidence of a biological origin (Sect. 3.1.3, example 4, above; Morowitz 1992). Molecules such as porphyrins, fatty acids, amino acids and polysaccharides help to sustain key biochemical functions and therefore are relatively abundant in living organisms. Certain key constituents of cellular membranes are relatively robust and can survive burial and storage in sedimentary rocks for long periods of time (more than a billion years, e.g. Brocks et al. 2005). Some of these compounds are diagnostic for particular groups of microorganisms (Sects. 3.2.1 and 3.2.2). Patterns of carbon isotopic abundances in organic compounds offer insights into metabolic processes (Sect. 3.3).

3.2.1 Membrane Polar Lipids

Membrane lipids are extremely valuable molecules for identifying living biota or its fossilized remains and have been equated to a "universal biomarker for life" (Georgiou and Deamer 2014). Indeed, membranes composed of amphiphilic lipids are the major structural components of all three domains of life—Eukarya, Archaea and Bacteria. While even relatively short chain fatty acids are good amphiphiles and readily self-assemble into micelle or monolayer structures in an aqueous environment, formation of a bilayer structure requires a hydrocarbon chain length of at least twelve carbon atoms (Deamer 1986). A biological membrane requires a more complex molecule composed of a polar head group and two long chain hydrocarbons with a C_{14} to C_{18} equivalent length capable of forming a bilayer ~40 nm thick. The hydrocarbon chains make up the membrane inner core and provide its unique hydrophobic barrier, while the polar moiety interacts with the aqueous environment and maintains cellular integrity. Such membrane amphiphiles are often referred to as intact polar lipids (IPL). They are the major component of the cytoplasmic membrane that plays a central role in energy storage and processing, both universal characteristics of life (Deamer 1997; Sojo et al. 2014).

IPL are composed of a core glycerol molecule. One hydroxyl group is attached to the polar moiety (e.g. choline, galactose) referred to as the 'head group' and the remaining two to long-chain hydrophobic moieties. A broad variety of polar head groups are universally shared throughout the three domains, however, glycerol stereochemistry, glycerol-hydrocarbon linkage and hydrocarbon chain structure vary and provide degrees of taxonomic attribution among the domains of life.

The hydrocarbon chains of Bacteria and Eukarya are acetogenic lipids composed of C2 units that derive from acetyl-CoA. These molecules are linked *via* ester-linkage to the *sn*-1 and *sn*-2 positions of the glycerol backbone. The polar head group is linked by either a phosphatide, sulfatide or glycoside linkage to the *sn*-3. Archaeal membrane lipids, aside from similar head-group architecture and fundamental amphiphilic properties, differ substantially. Archaeal IPL are composed of isoprenoid hydrocarbon chains linked by ether bonds to the *sn*-2 and *sn*-3 glycerol positions with the head group at *sn*-1. While some bacteria are known to have glycerol ether bonds, the hydrocarbon moieties are acetogenic with some limited methyl branching. Archaeal lipids are composed of C5 isoprene units that result in C_{20}, C_{25} or C_{40} chains. Biphytanyl tetraether lipids are composed of two C_{40} chains attached to a glycerol molecule at both ends that span the entire 40 nm width of the membrane. Until recently, this structural motif was thought to be unique to Archaea, but bacterial membrane-spanning lipids have now also been identified (Sinninghe Damsté et al. 2011).

Isoprenoid biosynthesis is essential for all domains of life and creates a diverse structural array of isoprene-based biomolecules—acyclic, exemplified by archaeal IPL and carotenoids, and cyclic, exemplified by the two types of triterpenoids discussed further below. There are two evolutionarily distinct pathways for the synthesis of C5 isoprene units, the mevalonate (MVA) and methylerythritol-phosphate (MEP) pathways (see Sect. 3.2.2). Archaea use the MVA pathway to synthesize C5 isoprene units (Lange et al. 2000).

Maintaining an appropriate membrane structure is crucial to cellular integrity and function. Biological membranes are exquisitely sensitive to changes in temperature, a process termed homeoviscous adaptation (Sinensky 1974). Modification of hydrocarbon moieties is of particular importance for maintaining a gel-like state associated with cell viability. In bacteria and microeukaryotes, the hydrophobic chains are continually modified or replaced in response to environmental conditions (López-Lara and Geiger 2017). Fatty acyl chains are readily modified to prevent crystallization at low environmental temperature by shortening the acyl chains and introducing methyl branching (e.g. iso-/anteiso-) or unsaturated double bonds. At higher temperatures survival depends on the synthesis of longer acyl chains or methylation of double bonds to form cyclopropyl rings (Ernst et al. 2016). Less is known about the maintenance of archaeal membrane fluidity but it appears to occur by a somewhat different process (Oger and Cario 2013). Phase transition of highly branched isoprenoid chains occurs at lower temperatures and over a broader temperature range than in bacteria (Koga 2012). Modifications of chain structure, such as unsaturation and synthesis of C_{40} chains with cyclopentane and cyclohexane rings, also occur (Schouten et al. 2013) but they are not well understood. Environmentally triggered modifications have been described for specific archaeal genera, however, intra-genera rules do not broadly apply. Thus, while membrane-spanning C_{40} and cyclic products have been associated with survival at high temperatures, similar compounds are found in colder environments (Oger 2015).

In studies of analogue ecosystems, IPL analyses provide important information related to community composition in an environmental context and, when combined with analyses of compound-specific isotope compositions (see Sect. 3.3.2.3), they can also reveal trophic structures. Generally, The IPL structure readily degrades upon cell death (enzymatic cleavage of acyl and polar head group linkages). Thus, IPL are excellent biomarkers for extant life. However, little is known about the enzymatic degradation of the ether bond (White et al. 1996). Archaeal ether lipids are known to survive longer than bacterial ester-based lipids (Harvey et al. 1986) and have been shown to have greater potential for geological preservation (Brassell et al. 1981).

The methodology of many geochemical studies of microbial mat lipids (i.e. acid and base hydrolysis) focuses almost exclusively on free lipids (fatty acids, alcohols, hydrocarbons). Free lipids are often the products of the initial stages of diagenesis, which can partially or totally obscure the identities of the source organisms. In order to achieve a holistic perspective on biomarker diagenesis, an investigation must examine biomarkers both within living biomass and at various stages of diagenesis and preservation.

Modern microbial mats have been studied as analogs of ancient stromatolites for many years. At Exportadora de Sal, Guerrero Negro, Baja California Sur, Mexico, a wide variety of mats grow across a wide salinity gradient in an extensive network of evaporation ponds (Des Marais 1995, 2010). These ponds provide relatively stable ecosystems for the study of production, degradation and survival of biomarker lipids in microbial mats. In Pond 4 for example, over a salinity range of ~60‰ to 110‰,

Microcoleus cyanobacteria form extensive cohesive, laminated microbial mats. In order to establish the microbial community structure of this Guerrero Negro ecosystem, we initiated analysis of a full depth profile for the abundance of the acyl and alkyl IPL using Pond 4 mat cores (Orphan et al. 2008; Jahnke et al. 2008). Intervals of surface mat and underlying sediment were analysed to document changes in the bacterial and archaeal population. Archaeal lipids are not abundant in these hypersaline mats. Even in the organic-rich sediments of the lower salinity ponds, methanogenic archaea are relatively limited. IPL abundances near the mat surface indicate an almost exclusive bacterial population. Archaea are only more abundant than bacteria at the lowest terminal trophic level of the mat core system (Jahnke et al. 2008). However phylogenetic analyses revealed that archaeal diversity varied substantially at depth. Most importantly, microcosm experiments revealed that growth and concomitant changes in diversity of the archaea had occurred and could be detected by monitoring changes in archaeal biomarkers (Orphan et al. 2008).

3.2.2 Cyclic Triterpenoids

Sterols and hopanoids are two classes of cyclic triterpenoids that are both important membrane lipids, however, sterols are almost exclusively found in eukaryotes and hopanoids in bacteria. As mentioned earlier, biological membranes play a crucial role in cellular function. One of the unifying principles of thermodynamically stable membranes is the fluid mosaic model in which amphipathic globular proteins alternate among sections of phospholipid bilayer to form a viscous solution that supports permeability of the cellular boundary layer (Singer and Nicolson 1972). The physical advantage associated with the inclusion of sterols in the eukaryotic membrane is well known. Essentially any higher sterol with a planar geometry (e.g. cholesterol, ergosterol, sitosterol) will enhance lipid order and provide a more mechanically robust, flexible membrane (Lingwood and Simons 2010). Although it is still unclear whether some of the initial enzymatic steps involved for sterol biosynthesis are of bacterial origin, phylogenomics indicates that sterol synthesis was well established in the earliest of the eukaryotic lineages (Pearson et al. 2003; Desmond and Gribaldo 2009; Wei et al. 2016).

The synthesis of one molecule of cholesterol requires 11 molecules of O_2 (Summons et al. 2006). However, relatively low levels of O_2 are required (Jahnke and Klein 1983; Waldbauer et al. 2011), which is consistent with co-evolution of sterol synthesis with cyanobacteria prior to the Great Oxidation Event. Indeed, it has been suggested that the initial role of sterols in eukaryotes was protection against oxidative stress (Galea and Brown 2009). Squalene is the last common intermediate in the biosynthetic pathways for biosynthesis of sterol and hopane carbon skeletons. For sterol synthesis, the first dedicated enzymatic step is the oxygen-dependent epoxidation of squalene to 2,3-oxidosqualene (squalene monoxygenase), followed by enzymatic cyclization by oxidosqualene cyclase (OCS) to either of two C30 protosterols, lanosterol or cycloartenol. Further modifications of these molecules result in synthesis of the structurally broad class of sterols representative of the eukaryotic kingdom (Summons et al. 2006; Nes 2011).

As noted above, hopanoids are also the cyclization products of squalene. By contrast, bacterial hopanoid synthesis is an anaerobic process. The squalene-hopene cyclase (SHC) is an ortholog of OCS with some unique properties (see Welander et al. 2012 and references therein). In contrast to the strict stereo specificity of eukaryotic OCS, bacterial SHC have very low substrate specificity (Ourisson and Rohmer 1992). It has been demonstrated in cell-free systems that SHC will accept squalene epoxide as a substrate and is not selective in stereochemistry in that both enantiomers are cyclized (Rohmer et al. 1980). SHC cyclization in bacteria results in the production of both diploptene and diplopterol and, in some bacteria, the product is tetrahymenol, a pentacyclic of the gammacerane series.

The primary hopanoid identified in natural environments is a class of complex molecules referred to collectively as bacteriohopanepolyol (BHP). BHP is synthesized by linking a polar side-chain to the hopane skeleton. A wide range of structural variations is observed in the side chain (Rohmer 1993; Talbot et al. 2007, 2008). Methylation at C-2 or C-3 of the hopane skeleton also occurs in a limited number of bacterial groups. The molecular dimensions and shape of hopanoid molecules, such as diplopterol and BHP, are similar to those of sterol molecules; this observation led Ourisson and colleagues to suggest that hopanoids in bacterial membranes were the structural equivalent of, and phylogenetic precursor to, eukaryotic sterols (Ourisson and Rohmer 1982; Ourisson et al. 1987 and references therein). The main difference in BHP or diplopterol is that the polar moiety is attached to the E ring of the hopane nucleus at C-21, whereas with sterol the hydroxyl group is located in the A ring at C-3. In vitro studies demonstrate that both diplopterol and a BHP with a tetrol side chain can order synthetic membranes in a similar manner to the order achieved by sterols (Sáenz et al. 2012; Poger and Mark 2013). In vivo physiological studies have shown that different hopanoids have distinct cellular roles, some of which clearly involve outer membrane integrity and some with yet undetermined functional roles (Welander et al. 2012). Indeed, more recent in vivo studies directly support the role that hopanoids play in outer membrane integrity by interaction with lipid A in a fashion similar to that shown for eukaryotic microdomains (Sáenz et al. 2015). Thus, similar functional roles span the bacterial-eukarya domains in meeting the demands for environmental adaptation over evolutionary time.

3.2.3 *Modern Microbial Mat Ecosystems as Analogs for Life on the Early Earth*

Within the context of this discussion, microbial mats are defined as organo-sedimentary structures formed by vertical, upward growth in response to burial by sediment trapping and binding, and/or encrustation by mineral precipitation and the requirement for light to fuel photosynthetic primary production. Stromatolites are their fossil equivalents and can occur as diverse forms of layered mounds, columns, and laminated sedimentary rock fabrics throughout the Archaean and Proterozoic (Awramik 1984; Bosak et al. 2013; Pierson et al. 1992; Walter 1983). One of the earliest *bona fide* examples has been described in the 3.43 Ga Strelley Pool Chert, Pilbara Craton, Australia, within what is thought to have been a hypersaline lagoonal environment in a region of active volcanism (Walter 1994; Allwood et al. 2006). While there is no definitive evidence that these earliest Archaean mats were not formed by anoxygenic phototrophs, the massive accumulation of free oxygen likely beginning in the Neoarchaean points to an early evolution of oxygenic photosynthesis and cyanobacteria.

Early mats were widespread and abundant throughout shallow water continental shelves, lagoons and lakes, and their fossil counterparts, stromatolites, remain one of the most important indications of early life. Modern microbial mats are widely distributed globally but are more restricted environmentally. Conditions must be extreme to limit eukaryotic grazers but relatively benign for benthic microbes. Such conditions occur in numerous tidal flats, hypersaline sabkhas, polar and high latitude lakes, hydrothermal springs and various evaporative environments (e.g., Gerdes et al. 1985; Franks and Stolz 2009; Jahnert and Collins 2011; Javor 1989; Stal 2012; Vincent and Quesada 2012; Ward et al. 2012). The primary phototrophs in modern mats are cyanobacteria and the broad morphological diversity seen throughout cyanobacterial taxa are also apparent in extant mats (Allen et al. 2009; Golubic and Abed 2010; Nübel et al. 2000). In marine mats, salinity, UV tolerance and desiccation resistance are major determinants. Such environmental factors broadly define cyanobacterial diversity and, thus, mat morphology.

Highly laminated phototrophic mat communities often develop in protected marine environments with elevated salinities due, in part, to their characteristically higher evaporation rates. *Microcoleus chthonoplastes* is a cosmopolitan builder of flat, laminated marine mats (Garcia-Pichel et al. 1996) in such environments. The filaments of this cyanobacterium are composed of multiple trichomes within a common sheath. This morphological characteristic (intertwined filaments) imparts a robust, cohesive surface and the development of distinct fine laminae to the mat. Excellent examples of such cyanobacterial mats develop along a salinity gradient in evaporating ponds of the Exportadora de Sal, S.A., Guerrero Negro saltern (Des Marais et al. 1989; Des Marais 1995, 2010 and references therein). The optimal salinity range for *Microcoleus*-dominant mats lies between 75‰ and 100‰ (salinity expressed as parts per thousand). As salinity increases above 110‰, unicellular and other filamentous cyanobacteria become increasingly dominant and the mat texture becomes more gelatinous and the laminations less well-defined. At the higher salinities, endoevaporatic mats develop in gypsum crusts (Jahnke et al. 2014b). These mats are exposed to high solar irradiance throughout the day and darkness at night, which results in dramatic shifts in physicochemical gradients over mat depth and the development of distinct trophic layers, each accompanied by a complex assemblage of organic molecules.

It is generally considered that Precambrian microbial mats are the prototypical source of early petroleum reserves, whose hydrocarbon molecules form the base of our lipid biomarker record (Peters et al. 2005). From this perspective, our knowledge of structure, function and physiological dynamics within microbial mats, particularly as they relate to development of biosignatures, such as organic molecules and stable carbon isotopic compositions, is fundamental to identifying definitively the biologic origins of stromatolites. However, organic analyses of a microbial mat entail distinct challenges. Molecules synthesized in some horizons are indeed lipid biomarkers associated specifically with various microbial groups, while others more broadly define microbes at the kingdom level. Some compounds are freely extractable (e.g. saturated hydrocarbons) and available over the full mat depth for analysis, while others that are highly functionalized (e.g. bacteriohopanepolyol) are readily entrained within the sulfide-rich matrix and form a non-extractable fraction. Many organic geochemical studies only focus on surface or near surface horizons yet, below the surface photic zone of primary production and even intermingled within, diverse groups of respiring aerobes, sulfate-reducers, fermentative bacteria and methanogenic archaea carry out continuous degradation and heterotrophic mineralization of primary organics and, in the process, contribute their own biosignatures to the mix.

Cyanobacteria in the surface layer are characterized by numerous biomarker lipids (Jahnke et al. 2004, 2008, 2014b; Palmisano et al. 1989 and references therein): IPL galactosyl-polyunsaturated fatty acid, normal chain alkanes (primarily *n*-17:1 and *n*-17:0), mid-chain branched alkanes (e.g. 7-methylhexadecane), chlorophyll *a* and numerous carotenoids (myxoxanthophyll, zeaxanthin and canthaxanthin). The presence of chlorophyll *c*, fucoxanthin, the highly branched isoprenoid C_{20} and C_{25} alkanes (HBI) and sterols indicate the presence of diatoms or other microeukaryotes. Polyunsaturated fatty acids, *n*-17:1 alkene and Chl *a*, while excellent biomarkers for vital populations, do not preserve well and decrease rapidly with depth. However, many isoprenoid lipids (HBI, sterols, carotenoids, hopanoids) survive well below the photic zone. Below the cyanobacteria a secondary layer of anoxygenic photosynthetic bacteria (*Chlorothrix, Chromatium, Rhodovulum*) harvest the residual light that penetrates into the sulfide rich zone (Jørgensen and Des Marais 1986). Biomarker lipids for these phototrophs include phospholipid-derived fatty acids $\Delta 11$-$C_{18:1}$ and cyclopropyl-C_{19}. Distinct lipophilic pigments (γ-carotene, spirilloxanthin, bacteriochlorophyll *a*, and Bchl *c* represent the purple and *Chloroflexus*-like bacteria in this horizon (Klappenbach and Pierson 2004; Oren et al. 1995; Palmisano et al. 1989; Pierson et al. 1994).

Many lipid biomarker molecules have been documented in various organic geochemical mat studies (Abed et al. 2008; Allen et al. 2010; Boon et al. 1983; Boudou et al. 1986; Bühring et al. 2009; Grimalt et al. 1992; Jahnke et al. 2008, 2014a; Pagés et al. 2014a, b; Rontani and Volkman 2005), but generally their fates have not. For instance, in the Guerrero Negro *Microcoleus* mats, the carotenoid/chlorophyll ratio increased with depth (Palmisano et al. 1989), indicating that Chl *a* had decomposed relative to carotenoid substantially over a timescale of months. For Camargue mats, the cyanobacterial biomarkers *n*-$C_{17:1}$, *n*-C_{17} and monomethylalkane (MMA) all decreased rapidly below the surface layer with only slightly greater survival potential documented for the saturated C_{17} and MMA within the deeper anoxic horizon

(Wieland et al. 2008). A similar pattern has been observed for the Guerrero Negro *Microcoleus* mat (Jahnke et al. 2014a). Hopanoids, particularly BHP, appear to be ubiquitous components of microbial mats. As noted above, numerous bacteria synthesize hopanoids. BHP is often most abundant in the surface layer and generally decreases with depth (Jahnke et al. 2014b; Pagés et al. 2014a). However, in other mats BHP appears to be synthesized primarily at depth by the anaerobic community (Blumenberg et al. 2013). Archaeal isoprenoid IPL were present in much lower abundances compared with bacterial fatty acids in the GN *Microcoleus* mat (Jahnke et al. 2008). Archaeol was the primary IPL over the entire core depth of 130 mm although a GDGT-C_{40} isoprenoid represented a relatively greater proportion in the anoxic layers.

Studies of lipid biosignatures in modern microbial mats help to establish ecological relationships, identify the diversity of biosignatures that might be preserved, and also understand the processes of taphonomy that have shaped the organic molecular fossil record.

3.2.4 Linking Geological and Biological Records

Cyclic triterpenoid molecules are valuable for understanding the evolution of functional adaptation in microorganisms. Accordingly, they help to elucidate the organic geochemical record in rocks. The cyclic skeletons of these molecules are resistant to diagenesis and catagenesis. Hopanes and steranes, defunctionalized products of catagenesis, are perhaps the two most important categories of organic biomarkers, and both have substantial records dating back to the 1.64 Ga Barney Creek Formation in northern Australia (Brocks et al. 2005; Brocks and Schaeffer 2008). Beyond this age, attempts to identify *bona fide* syngenetic structures have been plagued by contamination with younger organics and the highly mature nature of available samples (French et al. 2015). Major advances in technologies for analysis and the protocols for recovery of kerogen-bound syngenetic biomarkers have set the stage for a more comprehensive exploration of ancient rocks for evidence of early life. In this respect, open-system catalytic hydropyrolysis (HyPy) has been essential (Love et al. 1995). On the biological front, great progress has also been made in answering questions related to source organisms, environmental distribution and physiological functions. This is attributable largely to new molecular tools that allow for identification of the genes responsible for the biosynthesis of cyclic triterpenoids (Welander et al. 2010, 2012).

A case in point is the evolved history of hopanoid biomarkers. The first discoveries of hopanoid source organisms exclusively involved bacterial genera that were facilitated by some form of aerobiosis (obligate, microaerobic, faculative). All of the strict anaerobes and archaea initially analzyed had no hopanoids (Rohmer et al. 1984). As a consequence, hopanoids became popular indicators for environmentally available O_2. However, hopanoid biosynthesis does not require O_2 and we are now aware that some strict anaerobes indeed synthesize hopanoids in anoxic environments

(Blumenberg et al. 2006; Fischer et al. 2005; Härtner et al. 2005; Sinninghe Damsté et al. 2004). At the time of this writing, hopanoids have still not been found in archaea. The identification of the 2-methyl homologue of BHP in numerous cyanobacteria further spurred the proposal that the 2-methylhopane prevalent in the geological organic record was a biomarker for cyanobacteria (Summons et al. 1999). However, the subsequent identification of 2-methyl-BHP in an anoxygenic phototroph, *Rhodopseudomonas palustris*, somewhat diminished the usefulness of 2-methyl-BHP as a cyanobacterial biomarker (Rashby et al. 2007). With the identification of the 2-methylase gene (HpnP) in *R. palustris* by Welander et al. (2010), the phylogenomic interrogation for HpnP identified both cyanobacteria and two groups within the Alphaproteobacteria (the *Rhizobiales* including *R. palustris* and *Methylobacterium*) as the likely evolutionary source. However, based on the limited number of genomes then available, the precise branching order could not be definitively established (Welander et al. 2010). More recently, increased availability of HpnP-containing genomes has allowed Ricci et al. (2015) to establish that C-2 methylation arose within a group of *obligately aerobic* alpha-proteobacteria. Thus, 2-methyl-hopanes identified in contemporary environments or sedimentary rocks are still biomarkers for the presence of O_2, but cannot be considered specific to cyanobacteria.

The complexities of interpretation of the organic sedimentary rock record are numerous and, in many cases, require multiple lines of evidence. The recovery of abundant quantities of hopanes from the geological record emphasizes the importance of BHP as a bacterial biomarker and has stimulated investigation of the molecular variability of the side chain moiety. In particular, the analysis of BHP structures from numerous bacteria and recent sediments by the groups of Rohmer (2008 and references therein) and Talbot et al. (2007, 2008) have established the detailed structures and molecular distributions of many BHP molecules, and in some cases, provided insights into the sources and environmental contexts of geohopanes (Farrimond et al. 2000, 2003). As noted above, some hopanoids are methylated at C-2 and others at C-3 of ring A, but these are limited in distribution. Molecular structures in combination with the carbon isotopic compositions of individual biomarkers can further refine attribution. Diplopterol and/or diploptene are common in all hopanoid-producing bacteria and are the easiest to analyze, however, quantitatively they are not generally significant (Rohmer et al. 1984; Eickhoff et al. 2014). This is not the case for the 2-methyldiplopterol-producing alphaproteobacteria. Diplopterol and its 2-methyl homologue can be quantitatively significant in this group, while the synthesis of the 2-methyl homologue of BHP generally is not (Knani et al. 1994; Vilcheze et al. 1994; Renoux and Rohmer 1985; Bravo et al. 2001; Welander et al. 2010). Additionally, in *R. palustris* and *Bradyrhizobium* spp., tetrahymenol, its 2-methyl homologue and their hydrocarbon derivatives, gammaceranes, are also abundant (Bravo et al. 2001). In the geological record, gammacerane is considered the diagenetic product of tetrahymenol (Ten Haven et al. 1989) and thus the identification of co-occurring methylgammacerane and methylhopane would provide taxonomic attribution to this group.

As noted above, sterol biosynthesis can occur at very low O_2 levels (Jahnke and Klein 1979, 1983), the O_2 K_m for the squalene monooxygenase is 70 nM (Jahnke and Klein 1983), and synthesis has been demonstrated at levels as low as 7 nM (Waldbauer et al. 2011). Such O_2 requirements are consistent with levels near the base of the oxic-anoxic chemocline of a cyanobacterial mat, or roughly 10^{-2} to 10^{-3} present atmospheric levels (PAL). Thus cyanobacterial mats and the ecosystem that they support provide a viable scenario for the evolution of this biosynthetic pathway. As witnessed by the stromatolite record, these microbial mats were plausibly the first environments with available biogenic O_2.

Organic biomarker molecules are primarily derived from cellular lipids. Such lipid biomarkers have been valuable for characterizing source rocks in petroleum science (Peters et al. 2005), which has until recently formed the basis of our knowledge for understanding the organic geological record. Indeed the assertion that biohopanoids are 'more abundant globally than any other group of natural products' with an estimated abundance of 10^{12} tons (Ourisson and Albrecht 1992) has led to a fruitful exploration for these molecules in the biological record. Investigating the generation and preservation of geohopanoids is now central to understanding Earth's organic record and its potential for detection of past or present life on Mars (Eigenbrode 2008). Biohopanoids are subject to a wide range of early diagenetic reactions, including loss or alteration of functional groups, structural modification and rearrangements. C_{31} to C_{35} geohopanes are the most abundant in sedimentary organics and record BHP molecules (Peters et al. 2005). C_{30} geohopanes that represent simple hopanoids (e.g. diploptene, diplopterol) are much less abundant. Major degradation of organics is carried out by aerobic microbes, however, under such conditions, modification of the pentacyclic skeleton of BHP and C_{30} molecules is limited and no alteration to the BHP linear side-chain occurs (Tritz et al. 1999). In recent anoxic environments, sulfurization is thought to play a major role in the preservation of organic biomarkers. Inorganic sulfur species (e.g. H_2S) react with functionalized biolipids such as the polyol side chain of BHP during the early stages of diagenesis and form organic sulfur compounds, namely thiophenes (Sinninghe Damsté and de Leeuw 1990) that remove these molecules from further bacterial degradation. Relatively strong chemical degradation is required to release bound organosulfur compounds (OSC) from the macromolecular structure and not always with satisfactory results. However the development of open-system catalytic hydropyrolysis (HyPy) resolves many of these issues and allows release of covalently bound biomarker molecules as their defunctionalized homologues with minimal structural rearrangement (Love et al. 1995). Indeed, HyPy is also well-suited to the analysis of older, catagenically matured rocks, and has become an essential tool for characterizing syngenetic organic material bound in the overmature kerogens in Archaean rocks (Bishop et al. 1998; Brocks et al. 2003; Marshall et al. 2007).

3.3 Stable Isotope Abundance Patterns as Biosignatures

3.3.1 Stable Isotope Basics

Several elements that are abundant in biological materials have two or more stable isotopes. The stable isotopes of an element differ in the number of neutrons in their nuclei. This difference in nuclear mass has minimal effects on the atoms' electron orbitals but it creates differences in the vibrational energy states of their polyatomic molecules. Molecules having heavier isotopes have lower vibrational energy states than those having lighter isotopes, creating differences in the molecules' physical properties and in the isotopic compositions of suites of molecules at mutual chemical equilibrium. Molecules having heavier isotopes have slightly more stable bonds and therefore tend to react more slowly. Accordingly, biochemical processes can affect the stable isotopic compositions of reactants and products in ways that differ from those caused by non-biological processes. Such differences form a basis for distinguishing between biosignatures and the products of other processes.

In the natural sciences, the stable isotopic composition of a sample is typically expressed as the ratio of the abundance of the lighter isotope over that of the heavier isotope, relative to a standard. For example, carbon isotopic abundances are represented as follows:

$$\delta^{13}C_{PDB} = \left(\left({}^{13}C/{}^{12}C\right)_{sample} / \left({}^{13}C/{}^{12}C\right)_{PDB} - 1\right) 1000 \tag{3.1}$$

$\delta^{13}C_{PDB}$ is the difference in permil (parts per thousand) between a sample and carbon isotopic standard (the Peedee Belemnite carbonate—PDB). Herein we will use the terms $\delta^{13}C_{org}$ and $\delta^{13}C_{carb}$ to indicate $\delta^{13}C$ values of organic matter and carbonates, respectively. Sulfur isotopic abundances are represented as follows:

$$\delta^{34}S_{CD} = \left(\left({}^{34}S/{}^{32}S\right)_{sample} / \left({}^{34}S/{}^{32}S\right)_{CDT} - 1\right) 1000 \tag{3.2}$$

$\delta^{34}S_{CD}$ is the difference in permil between a sample and the Canyon Diablo Troilite meteorite sulfur standard. Analogous equations express isotopic compositions for nitrogen (${}^{15}N/{}^{14}N$, air nitrogen standard), hydrogen (D/H of 'standard mean ocean water'—SMOW), and oxygen (e.g., ${}^{18}O/{}^{16}O$, SMOW).

3.3.2 Carbon Compounds and Microbial Metabolism

The objective is to determine whether and how patterns of isotopic compositions among suites of carbon compounds might indicate biological activity and, perhaps, even identify particular microbial populations. The approach is first to understand

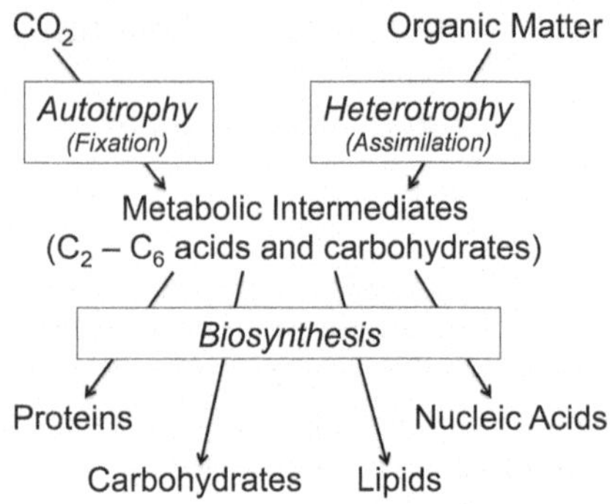

Fig. 3.3 Biosynthesis in autotrophic and heterotrophic organisms. Source: Adapted from Hayes (2001)

how enzymatic reactions that are centrally important for life can play major roles in determining isotopic compositions of the major classes of biochemicals, e.g., carbohydrates, proteins, nucleic acids and lipids. Figure 3.3 provides an overview of the relevant processes.

Enzymes create reaction products whose isotopic compositions can serve as isotopic signatures of those enzymes. In the absence of life, rates of organic reactions are typically very sluggish under ambient conditions. Enzymes generally accelerate organic reaction rates substantially, as noted above (*"Those reactions that proceed at appreciable rates in living cells are catalyzed by enzymes"* Sect. 3.1.3 example 5). In kinetically controlled chemical reactions such as these, the isotopic compositions of reaction products reflect the compositions and molecular configurations of particular intermediate chemical species in the reaction pathways. Because at least some of these intermediate species have elevated vibrational energy states, they are thereby enriched in the lighter isotope(s) relative to the reactants. The magnitude of the differences in isotopic compositions between reactants and products is related to the isotopic discrimination that occurs during the reaction. The magnitude of this discrimination can, in turn, help to identify particular reaction intermediates and thereby serve as biosignatures of particular enzymatic reaction mechanisms. Ultimately, the reaction networks that synthesize and recycle these compounds can create patterns of molecular isotopic abundances that reflect these metabolic pathways. Such isotopic patterns can thereby become biosignatures.

Biologically-mediated molecular isotopic patterns among biomolecules can reflect certain basic attributes of life (Morowitz 1992): *"There is a universal set of small organic molecules that constitutes the total mass of all cellular systems,"* and *"Those reactions that proceed at appreciable rates in living cells are catalyzed by enzymes."* Such patterns are useful biosignatures to the extent that they can be distinguished from patterns created by any non-biological processes.

The following discussion explores the following examples that illustrate how carbon isotopic patterns can be interpreted as biosignatures: (1) isotopic discrimination during the assimilation of carbon (autotrophy), (2) isotopes in intermediary metabolism, (3) lipid isotopic compositions, (4) isotopic patterns within molecules, and (5) isotopic contrasts between reduced and carbonate carbon in ancient sedimentary rocks.

Table 3.1 Isotopic discrimination by major carbon fixation pathways (Hayes 2001; House et al. 2003)

Enzyme/pathway	E, ‰
RuBisCO, C3 pathway (algae)	30
RuBisCO, C3 pathway (bacteria, cyanobacteria)	22
Acetyl CoA pathway	15–36
Reductive TCA cycle	4–13
3-Hydroxypropionate cycle	0–4

E (epsilon) is the difference, in permil, between the instantaneous isotopic compositions of products and reactants in the carbon fixation reactions

3.3.2.1 Isotopic Discrimination by Autotrophic Carbon Fixation

The biological fixation of small molecules like CO_2 is typically accompanied by significant isotopic discrimination of the fixed carbon. The magnitude of this fractionation establishes the isotopic relationships between autotrophs and dissolved inorganic carbon (DIC) and atmospheric CO_2.

The following carbon fixation pathways are particularly prominent among microorganisms: (1) reductive pentose phosphate cycle (Calvin-Benson-Bassham cycle), (2) reductive tricarboxylic acid (rTCA) cycle, (3) 3-hydroxypropionate cycle, and (4) reductive acetyl-CoA pathway (Preuss et al. 1989; House et al. 2003). These pathways exhibit systematic differences in the magnitude of the isotopic discrimination that accompanies carbon fixation (e.g., Table 3.1).

Ribulose Bisphosphate Carboxylase-Oxygenase (RuBisCO) is the principal enzyme utilized for carbon fixation by cyanobacteria, algae and plants. Today, RuBisCO is by far the quantitatively most important enzyme for fixing CO_2 (e.g., Dhingra et al. 2004; Raven 2013). In geologically recent sedimentary rocks, isotopic discrimination by primary producers is principally responsible for creating a 20 to 30 permil difference (e.g. Hayes et al. 1999) between the $\delta^{13}C_{PDB}$ values of organic and carbonate carbon. The $\delta^{13}C_{PDB}$ values of recent marine carbonates lie typically in the range between -2 and +3 permil. In the modern biosphere this isotopic dichotomy between reduced carbon and carbonates is a biosignature for biological carbon fixation.

In Earth's early biosphere isotopic discrimination during carbon fixation has probably contributed substantially to maintaining the dichotomy in isotopic compositions between sedimentary organic matter and coeval carbonates (e.g., Hayes et al. 1983, Hayes 1993; Des Marais 2001). But it is interesting to inquire whether carbon fixation pathways other than the Calvin cycle were relatively more prominent earlier in Earth history. Later in this chapter, this topic will be addressed by examining the record of carbon isotopic compositions of reduced carbon and carbonates in Earth's ancient sedimentary rocks.

3.3.2.2 Isotopic Fractionation Within Intermediary Metabolism

The arrows depicted in Fig. 3.3 indicate the net carbon flows in a cell that is growing and building biomass. The isotopic compositions of the major components of cells (carbohydrates, proteins, nucleic acids, lipids) reflect the outcome of isotopic competition between the enzymatic reactions that lead to these components (Hayes 2001). Changes in the physiological state of a cell can, in some cases, shift the relative abundances of these cellular components. Hayes (2001, 2004) provides comprehensive reviews of carbon isotopic discrimination associated with biosynthetic processes.

3.3.2.3 Isotopic Compositions of Individual Lipids

The production of biomass is associated with numerous variables that depend upon the specific organism, its growth substrate and its environment. In this respect the use of stable isotopes provides a valuable tool for further characterization of organic material. In this section we will address the role of compound specific isotopic analyses (CSIA) of carbon in individual lipid molecules, in particular, in distinguishing source and determining metabolic status. Section 3.3.3 discusses further the broader topic of bulk stable isotopes for biogeochemical analysis.

Organisms preferentially use the lighter isotope of carbon (^{12}C), which results in fractionation between the substrate (heavier or ^{13}C enriched) and product (lighter or ^{13}C depleted biomass). This is generally the rule but as discussed below, bulk isotope effects for a few organisms can result in ^{13}C enriched biomass. Factors that determine the ^{13}C content of individual organic molecules follow similar kinetic processes within the biosynthetic networks that distribute ^{12}C and ^{13}C throughout cellular systems (see Hayes 1993, 2001 for a detailed discussion). The complementary use of structurally diagnostic molecules, together with their particular C-isotopic compositions, can refine biomarker attribution to specific physiologies and metabolic groups of organisms.

The lipid biomarkers discussed above are the products of two biosynthetic pathways that result in carbon skeletons that are either linear (acetogenic) or multi-branched (isoprenoid). The biosynthetic pathway for acetogenic lipids, such as fatty acids, derives from acetyl-CoA and is fundamentally similar in all organisms (Reilly 2016). Some exceptions in bacteria are the synthesis of branched fatty acids (primarily iso- and anteiso-) and anaerobic synthesis of unsaturated fatty acids (Cronan and Thomas 2009), whereas in eukaryotes, oxidative desaturation results in the synthesis of both monounsaturated and polyunsaturated fatty acids with a few bacterial exceptions, including cyanobacteria (Murata et al. 1992; Cook and McMaster 2002).

Isoprenoid biosynthesis is far more complex. The synthesis of the universal C5-isopentenyl diphosphate (IPP) subunit occurs by two different mechanisms (Fig. 3.4). In the well-known mevalonate (MVA) pathway, IPP carbon is derived

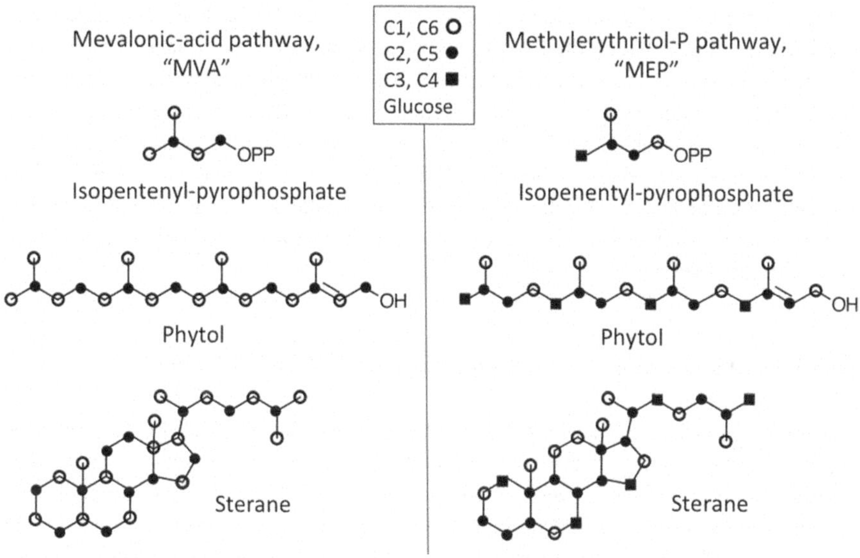

Fig. 3.4 Sources of carbon atoms in isopentyl-pyrophosphate (IPP), phytol and sterane for the mevalonic acid (MVA) and methylerythritol-P (MEP) pathways. In the MVA pathway all five IPP carbon atoms derive from C1 and C2 atoms of glucose. For MEP, one carbon atom in IPP derives from C3/C4 of glucose. Source: Adapted from Hayes (2001)

from acetyl-CoA, while in the more recently described methylerythritol phosphate (MEP) pathway, the substrates are pyruvate and glyceraldehyde-3-phosphate. From the initial discovery of the MEP pathway by Rohmer (2008) and his colleagues, extensive work has demonstrated that this new mechanism for IPP synthesis functions in bacteria and plant plastids. Most organisms only use one of these pathways for isoprene synthesis, however plant cells also use the MVA pathway for IPP biosynthesis in their cytoplasm. The MVA pathway appears to be the ancestral pathway for biosynthesis of IPP in archaea and the cytoplasm of non-photosyntheyic eukaryotes (Lange et al. 2000). The second stage of isoprenoid synthesis commences with condensation of the C5-isoprene subunits to generate linear polymers of defined chain lengths (C_{20}, C_{25}, C_{30}, C_{40}, i.e. C_{5n}). Cyclization results in a vast number of biomolecules of geochemical importance (Peters et al. 2005). From the perspective of CSIA networks, it is important to note that, in plant cells and particularly in algae, exchange of cytosolic- and plastid-isoprenoid precursors can and does occur (Vranová et al. 2012) providing a potentially additional level of complexity to the interpretation of algal biomarker lipids. The isotopic consequences of biosynthesis, carbon source and discrimination, and the added dimension of intracytoplasmic compartmentalization in algae result in significant differences among isoprenoids in phototrophic eukaryotes (see Hayes 2001; Pearson 2014 for synopsis and discussion).

Distinct differences in the carbon isotopic compositions of acetogenic and isoprenoid lipids, relative to bulk organic carbon, were first noted in the geological record by Logan and colleagues (Logan et al. 1995, 1997). They suggested that transition in biota between the Proterozoic and Phanerozoic was marked by a major change in the carbon isotopic relationships among bitumen, associated lipids and syngenetic kerogen. These observations have stimulated discussions related to the likely causes (Close et al. 2011; Pawlowska et al. 2013) that rely on available CSIA for lipid biomarkers, preservation of lipids and likely depositional environments, and point to the importance of continued efforts to understand ^{13}C discrimination within cellular products in environments of potential relevance to interpretation of the ancient organic record.

3.3.2.4 Carbon Isotopic Abundance Patterns Within Molecules

A distinctive characteristic of the biosynthesis of organic molecules is the intramolecular carbon isotopic patterns developed by the flow of precursor molecules through reaction networks. Individual carbon atoms of a carbohydrate molecule can be either enriched or depleted relative to the average molecular composition (see Hayes 2001 for full discussion related to all compound classes). Similar consequences are expected regarding the intramolecular isotopic compositions of lipid molecules and regarding their effects upon the $\delta^{13}C$ values of biomarkers.

For any class of compound, carbon isotope abundances will ultimately reflect initial fractionations associated with carbon assimilation (see 3.3.2.1), metabolic carbon flow (Figs. 3.3 and 3.4) and distinct fractionation events associated with enzymatic reaction mechanisms. As noted above, *n*-alkyl carbon skeletons are essentially acetate polymers derived from acetyl-coenzyme A. The general ^{13}C-depletion noted for fatty acids relative to bulk carbon in most organisms arises primarily due to a fractionation associated with the enzymatic decarboxylation of pyruvate to form acetyl-CoA (DeNiro and Epstein 1977; Monson and Hayes 1980). Fractionation is localized in the carboxyl position of acetyl-CoA and is carried through the isotopic order in biosynthesis of the fatty acid molecule. Additional discriminations may further be associated with the individual sources and sinks for acetyl-CoA in individual organisms.

The isoprene carbon skeleton is constructed from isopentenyl-pyrophosphate subunits. However, as noted above, IPP can be biosynthesized by two different mechanisms (Rohmer 2008). Characterization of intramolecular ordering depends heavily upon ^{13}C labeling experiments and NMR analysis (Eisenreich et al. 2004). As with *n*-alkyl synthesis, acetyl-CoA is the fundamental subunit for biosynthesis of IPP *via* the MVA pathway. Three molecules of acetyl-CoA are used to synthesize mevalonate; decarboxylation results in IPP. The consequence is incorporation of three carbons derived from the methyl position and two carbons from the carboxyl position (Fig. 3.4). Thus MVA isoprenoids are expected to be somewhat enriched relative to the accompanying acetogenic lipids.

For the MEP pathway, the IPP subunit is biosynthesized by condensation of one molecule of pyruvate and one of glyceraldehyde 3-phosphate to form 1-deoxyxylulose-5-phosphate. The mechanism involves transfer of an acyl anion synthon by thiamine

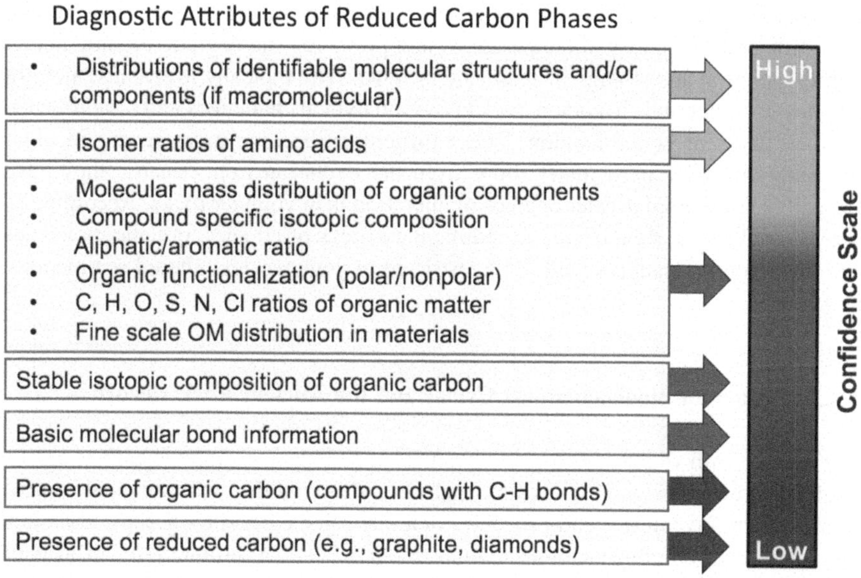

Fig. 3.5 The confidence in establishing the origins of reduced carbon and its compounds increases with the extent to which compound classes, specific molecular structures and stable isotopic compositions are characterized. Source: Adapted from Mustard et al. (2013)

pyrophosphate (activated acetaldehyde) to glyceraldehyde-phosphate. The reaction mechanism of the transketolase is very similar to other enzymes such as pyruvate dehydrogenase and pyruvate carboxylase to which specific fractionation is attributed (Eisenreich et al. 2004; Hayes 2001, 2004). The resulting IPP molecule reflects the contribution of glucose carbon atoms C1–C2 and C1–C2–C3, respectively. Intramolecular isotopic analyses indicate that IPP-C3 is depleted in ^{13}C compared to the other atoms (Schouten et al. 2008). The isotopic difference for the IPP-C3 *via* the MVA pathway was found to be significantly smaller. Accordingly different pathways for IPP biosynthesis produce IPP having different intramolecular isotopic distributions (Fig. 3.4), but ultimately these differences must be assessed within the context of metabolic sources for intermediary carbon and the reaction networks (Tang et al. 2017).

3.3.3 *Reduced Carbon and Carbonates in Sedimentary Rocks*

The discussions above indicate how our confidence in establishing the origins of reduced carbon and its compounds increases with the extent to which compound classes, specific molecular structures and stable isotopic compositions can be characterized (Fig. 3.5).

It should be noted, however, that processes such as microbial diagenesis, oxidation, radiation and elevated temperatures and pressures can degrade organic matter during its burial and storage in sedimentary rocks. The typically low abundances of diagnostic compounds in ancient rocks poses analytical challenges and is aggravated by the effects of contamination. These difficulties become more severe in older Precambrian rocks, particularly those from the Archaean Eon, because they have typically experienced greater degrees of alteration than younger rocks. Accordingly, whereas reduced carbon occurs in sedimentary rocks of all ages, organic molecules and their isotopic patterns tend to be restricted to younger, less altered sedimentary rocks.

3.3.3.1 Carbon Biogeochemical Cycles and the Ancient Rock Record

The fidelity with which isotopic composition of sedimentary organic carbon can be established as a biosignature requires that we understand the environmental context of the carbon inventories and processes that have shaped Earth's oceans, crust and interior. These components interact *via* the biogeochemical carbon cycle, an array of carbon reservoirs linked by a network of physical, chemical and biological processes. The overall carbon cycle actually consists of multiple nested cyclic pathways that differ with respect to some of their reservoirs and processes (Fig. 3.6; Des Marais 2001). However, all pathways ultimately pass through the hydrosphere and atmosphere, and it is this common course that unites the entire carbon cycle and allows even its most remote constituents to influence our environment and biosphere.

3.3.3.2 Early Isotopic Record in Reduced Carbon and Carbonates

Figure 3.6 illustrates schematically the biologically induced dichotomy in isotopic compositions between reduced carbon reservoirs *versus* carbonates. The near-identical ranges of $\delta^{13}C$ values of biomass and recent sedimentary organic matter indicate the biological origins of reduced carbon in sedimentary rocks. Thermal alteration of sedimentary organic matter tends to increase the $\delta^{13}C$ values of the residual kerogens (e.g., Hayes et al. 1983), therefore metamorphism likely accounts for the shift in the range of $\delta^{13}C_{org}$ values observed between the reservoirs of sedimentary organic matter *versus* the reduced carbon in metamorphic rocks (e.g., Des Marais 1997). The isotopic dichotomy persists in sedimentary rocks despite these effects, in part because the $\delta^{13}C_{org}$ values of kerogen, a highly refractory form of reduced carbon in ancient sedimentary rocks, are remarkably resistant to alteration (Des Marais 1997). This dichotomy spans more than 3.5 billion years (e.g., Schidlowski 1988; Fig. 3.7) and is perhaps the oldest, most widespread evidence of early life preserved in the geologic record.

As a cautionary note, carbonaceous chondrite meteorites also exhibit carbonates that have $\delta^{13}C$ values that are greater than those of co-occurring reduced carbon phases (Pearson et al. 2006). The similarity of these patterns to those in Earth's

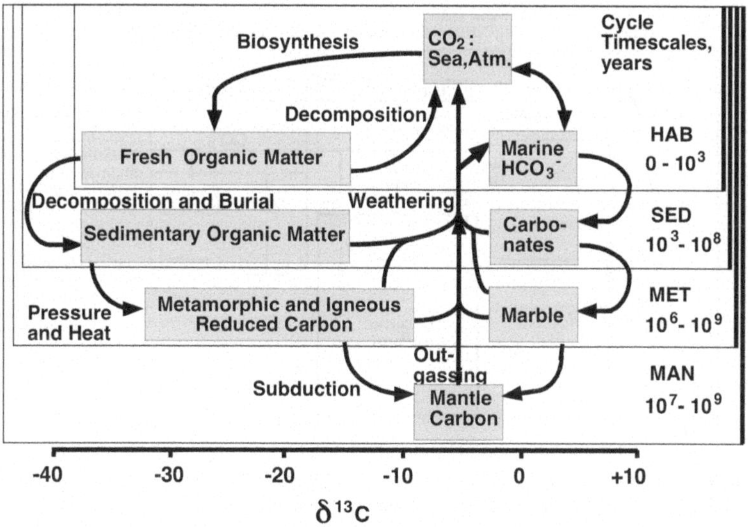

Fig. 3.6 Biogeochemical C cycle, showing principal C reservoirs (boxes) in the mantle, crust, oceans and atmosphere, and showing the processes (arrows) that unite these reservoirs. The range of each of these reservoir boxes along the horizontal axis gives a visual estimate of the $\delta^{13}C$ values most typical of each reservoir. The vertical bars at the right indicate the timeframes within which carbon typically completely traverses each of the four carbon sub-cycles (the "HABitable, SEDimentary, METamorphic and MANtle" sub-cycles). For example, C can traverse the hydrosphere-atmosphere-biosphere (HAB) sub-cycle typically within a time scale between 0 and 1000 years. Source: Adapted from Des Marais (1997)

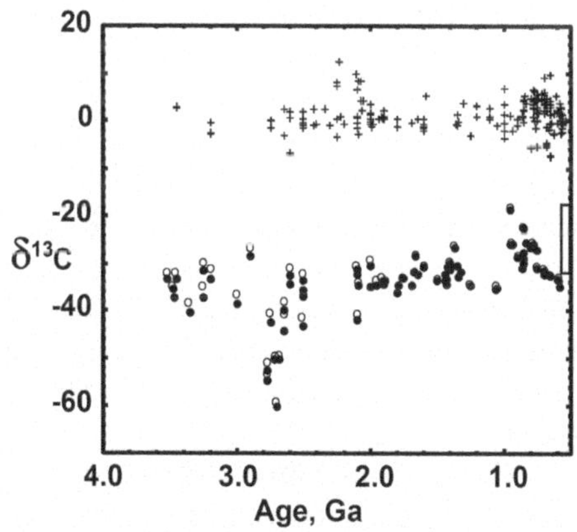

Fig. 3.7 Plot of age *versus* δ_{carb} (crosses) and δ_{org} (circles) for Archaean and Proterozoic kerogens. Kerogen data (filled circles) are corrected for the effects of thermal alteration (Des Marais 1997). Uncorrected data are shown as open circles. The $\delta^{13}C_{org}$ range for Phanerozoic kerogens is indicated by the narrow box along the right margin of the figure. Source: Adapted from Des Marais (1997)

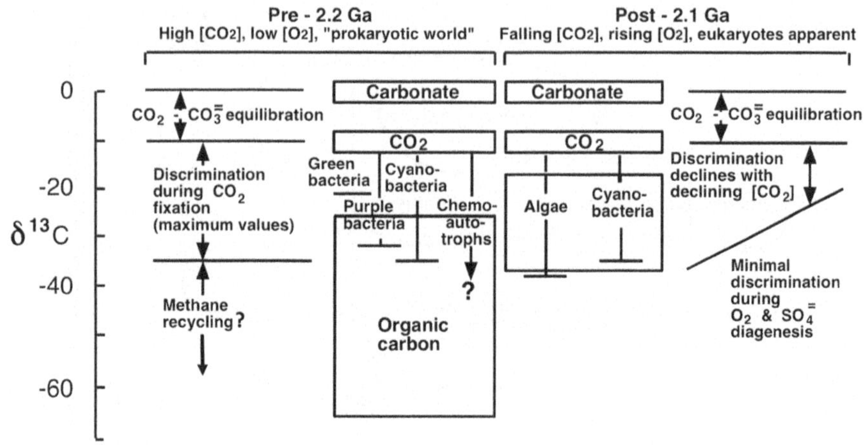

Fig. 3.8 Ranges of $\delta^{13}C$ values of carbonates, seawater CO_2 and sedimentary organic carbon, together with the processes proposed to explain their distribution prior to 2.2 Ga and subsequent to 2.1 Ga. The arrows associated with the various groups of autotrophic bacteria and algae illustrate the maximum isotopic discrimination expected for each group. The sloped line on the right depicts declining discrimination over time, perhaps in response to declining CO_2 levels in the environment. Source: Adapted from Des Marais (1997)

sedimentary record illustrates that interpretations of carbon isotopic patterns must be supported by other geochemical data and observations about paleoenvironments and associated processes. Such supporting evidence has been obtained for Earth's early geologic record (e.g., Hayes et al. 1983; Schopf and Klein 1992; Schidlowski 1993), however analogous data are still needed in order to interpret the carbon isotopic record on Mars.

Figure 3.8 represents schematically the long-term trends in the range of $\delta^{13}C_{org}$ and $\delta^{13}C_{carb}$ values observed through hundreds of analyses. In kerogens older than 2.2 Ga, $\delta^{13}C_{org}$ values are widely scattered, ranging from values around −25 to as low as −65, with many values more negative than −35. In contrast, $\delta^{13}C_{org}$ for kerogens younger than 2.1 Ga lie in the narrower range −25 to −35; virtually no values are more negative than −36. The range of $\delta^{13}C_{org}$ values typical of Phanerozoic kerogens is somewhat more positive (Fig. 3.7). The range of $\delta^{13}C_{carb}$ values typically lies within a few permil of 0, except ca. 2.2 billion years ago, and also during the Neoproterozoic.

For the ca. 3.2–3.5 Ga volcano-sedimentary sequences in the Kaapvaal (-South Africa) and Pilbara (Australia) cratons, $\delta^{13}C_{carb}$ averages 0(+/−2) ‰ (Veizer et al. 1999), and $\delta^{13}C_{org}$ ranges from −25 to −41‰ (Fig. 3.6; Strauss and Moore 1992; Des Marais 1997). Such $\delta^{13}C_{org}$ values are consistent with the notion that early Archaean ecosystems were driven by autotrophy. Such values are conventionally interpreted to indicate discrimination by the pentose phosphate (Calvin) cycle operating under conditions of high CO_2 that favor maximum isotopic discrimination (Fig. 3.8; e.g., Schidlowski 1993). However, the broad $\delta^{13}C_{org}$ range observed for these early Archaean sequences is reminiscent of the wide range of discrimination exhibited during autotrophic C assimilation by diverse microorganisms, anaerobes in particular (Table 3.1). Therefore, the carbon isotopic record of early Archaean carbonates and kerogens also might indicate that anoxygenic photoautotrophic bacteria, chemoautotrophic microorganisms and methanogens contributed substantially to global primary productivity.

Sedimentary carbon isotope abundances might also indicate the environmental consequences of biological activity. As noted above, in sedimentary rocks younger than 2.1 Ga, $\delta^{13}C_{org}$ values lie in the range −20 to −35 and virtually no values are more negative than −36 (Figs 3.7 and 3.8), consistent with contributions from various photoautotrophs. However, in rocks older than 2.2 Ga, $\delta^{13}C_{org}$ values range from −25 to as low as −65. $\delta^{13}C_{org}$ values <−37‰ require contributions from strongly ^{13}C-depleted biomass, which has been attributed to the consumption of methane (see Eigenbrode and Freeman 2006; Eigenbrode et al. 2008 and references and discussions therein). Extremely low $\delta^{13}C_{org}$ values ca. 2.7 Ga might indicate the prevalence of biogeochemical methane cycling in oxidant-deficient closed basin depositional environments (Flannery et al. 2016). Consumption of ^{13}C-depleted methane by either anaerobic archaeal consortia or aerobic methane oxidizers, or by both of these, provides mechanisms for the production and transport of highly ^{13}C-depleted organic matter to sedimentary environments (Hayes 1994; Hinrichs et al. 2000). The absence of very negative $\delta^{13}C_{org}$ values after 2.1 Ga might indicate that such environments had become more oxidized by either oxygen or sulfate. The proposal that sulfate levels in the deep ocean became substantial during the Mesoproterozoic (Canfield 1998) is consistent with this carbon isotopic evidence.

The $\delta^{13}C_{org}$ and $\delta^{13}C_{carb}$ record is also consistent with the oxidation of the Proterozoic environment between 2.3 and 2.0 Ga (Fig. 3.8). A large positive $\delta^{13}C_{carb}$ excursion between 2.2 and 2.06 Ga (Fig. 3.7; Baker and Fallick 1989; Karhu and Holland 1996) indicates that the relative rate of organic burial increased. Increased net organic burial would have led to increased $[O_2]$, $[SO_4^{=}]$ and sedimentary $[Fe^{3+}]$. Oxic respiration and bacterial sulfate reduction became, as they are today, globally dominant pathways for organic utilization and decomposition. Notably these pathways express minimal carbon isotopic discrimination (e.g., Blair et al. 1985; Kaplan and Rittenberg 1964), thus, they apparently created few opportunities to create sedimentary organic C having $\delta^{13}C_{org}$ values lower than −35, consistent with the post-2.1 Ga $\delta^{13}C_{org}$ record (Fig. 3.7). In the post-1.9 Ga world, isotopic discrimination associated with the CO_2-assimilation by aerobic photoautotrophs became the dominant mechanism controlling the dichotomy between $\delta^{13}C_{org}$ and $\delta^{13}C_{carb}$.

The search for carbon isotopic evidence of life in rocks older than 3.5 Ga is challenged both by their rarity and by metamorphism. Claims of potential biosignatures have been consistently challenged by counterclaims questioning their validity. Examples include the debate surrounding the origin and significance of isotopically light reduced carbon phases in Archaean quartzose rocks at Akilia, Greenland (Mojzsis et al. 2002; Friend et al. 2002; Fedo and Whitehouse 2002). Some postulate that mantle processes contribute isotopically light carbon phases to crustal rocks (e.g., Horita 2005; Horita and Polyakov 2015). These debates underscore the need to establish the actual age when any potential biosignatures were emplaced in a deposit and whether the environment was indeed habitable at that time (e.g., Rosing and Frei 2004). The report of potential biosignatures in early Archaean hydrothermal vent precipitates (Dodd et al. 2017) is a recent example of efforts to adopt this approach.

3.3.4 Isotopic Patterns Arising from Biological Redox Reactions

The relationships between patterns of stable isotopic composition and biological activity can also be illustrated by the energy requirements of life (Fig. 3.1): "Life must exploit thermodynamic disequilibrium in the environment in order to perpetuate its own disequilibrium state" (e.g., National Research Council 2007; Sect. 3.1.2). Enzymes harvest energy by accelerating reactions between oxidized and reduced chemical species ("redox reactions") in the environment (e.g., oxidation of reduced C, S, and N compounds). "Those reactions that proceed at appreciable rates in living cells are catalyzed by enzymes" (Morowitz 1992; Sect. 3.1.3). Life can capture energy by catalyzing redox reactions to exceed rates that would occur in the absence of life.

3.3.4.1 Sulfur

Sulfate reducing bacteria (SRB) obtain energy by coupling the reduction of sulfate to sulfide with the oxidation of organic matter or H_2 (Postgate 1984). Laboratory cultures of SRB incubated at ambient temperatures discriminate against ^{34}S, generating sulfides having $\delta^{34}S$ values that range typically between 10 to 40 permil lower than the $\delta^{34}S$ of sulfate when its abundance is in excess of the amount consumed (e.g., Canfield 2001). Theoretical calculations indicate that, under near-ambient conditions, the $\delta^{34}S$ values of sulfates and sulfides in isotopic equilibrium can differ by several 10s of permil. In the absence of biological activity, the sulfate oxyanion is strongly kinetically inhibited against achieving isotopic equilibrium within the temperature range required to achieve such large $\delta^{34}S$ values. Therefore, if the depositional environment can be constrained, the differences in $\delta^{34}S$ between sulfates and sulfides can become potential biosignatures and also record changes in SRB activity and environmental conditions over time.

Figure 3.9 illustrates $\delta^{34}S$ values for the 3.5 billion-year record of sedimentary sulfides. $\delta^{34}S$ values for sulfides older than 2.5 Ga typically lie within a few permil of 0, a range that is also typical of $\delta^{34}S$ values of mantle-derived peridotites and pyroxenites (Chaussidon and Lorand 1990). However, ~3.49 Ga sedimentary sulfides from the Dresser Formation in Western Australia exhibit $\delta^{34}S$ values that are

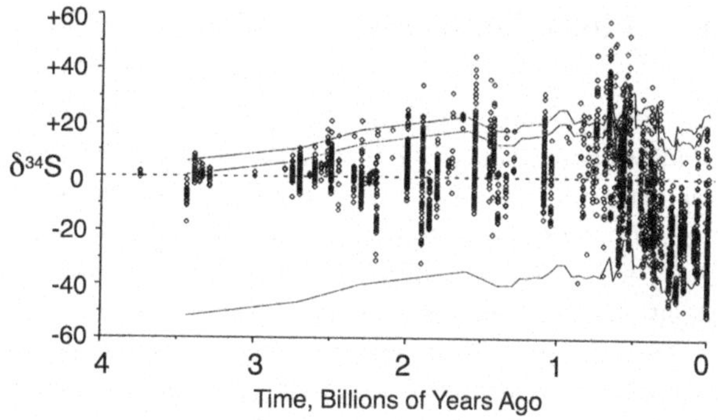

Fig. 3.9 Sulfur isotopic compositions of sulfides in ancient sedimentary rocks. Source: Adapted from Canfield (2001)

some 20 permil lower than values of coexisting sulfates (Shen et al. 2001). The authors interpreted these $\delta^{34}S$ values to indicate biologically driven sulfur transformations. The significant increase in the range of $\delta^{34}S$ values of sedimentary sulfides at ~2.3 Ga documents a substantial increase in the magnitude of isotopic fractionation by microorganisms (Canfield 2001). This change probably reflects an increase in seawater sulfate concentrations, consistent with an associated increase in the oxidation state of the global surface environment (Canfield 2001).

3.3.4.2 Metals

Stable isotope abundances of biologically important metals have also been explored as biosignatures. For example, iron isotopes have received considerable attention during the past fifteen years (e.g., Johnson et al. 2008a). The most substantial differences between various iron species are due to isotopic discrimination that occurs during changes in the redox state of iron or between iron-bearing species having different bonding states (e.g., between different iron complexes, etc.). In near-neutral pH waters where Fe^{2+} is the most mobile dissolved iron species, the $^{56}Fe/^{54}Fe$ value of Fe^{2+}_{aq} is low relative to the corresponding value of Fe^{3+} in iron-bearing minerals. Microbially mediated Fe^{3+} reduction produces the greatest abundances of isotopically distinct iron, compared to quantities produced by any non-biological processes. The quantities of iron cycled across redox boundaries were much larger in the Archaean than at present and were due to the much larger pools of Fe^{2+} and Fe^{3+} that interacted in these ancient environments. Isotopic evidence for the expansion of microbial Fe^{3+} reduction during the late Archaean is consistent with the increased production of large inventories of Fe^{3+} and organic matter (Johnson et al. 2008b). Very likely this increased production was driven by the expansion of photosynthetic communities. The magnitude of the isotopic signature of Fe^{3+} reduction in sediments decreased between ca. 2.4 to 2.2 Ga, perhaps due to the ascent in global prominence of microbial sulfate reduction which, in turn, decreased the availability of reactive iron for Fe^{3+} reduction. Research to interpret stable isotopes of iron and other metals offers a promising future both for environmental geochemistry (e.g., Wiederhold 2015) and for astrobiology.

3.4 Final Comments

The extraordinary complexity of organic chemistry can record an enormous volume of information that can be interpreted to reveal the origins and processes of organic deposits. Analyses of preserved organic molecular biosignatures can reveal not only the former presence of life but also key details about diverse physiologies and ecosystems. Stable isotopic biosignatures provide complementary information about processes and environmental conditions.

The relevance of organic biosignatures has now been firmly extended to Mars exploration, given the discoveries of organic matter in Martian meteorites (e.g., McKay et al. 1996) and in Gale Crater, Mars by the Curiosity rover (e.g., Freissinet et al. 2015). The ongoing challenge is to locate organic matter that has been sufficiently well preserved (see Summons et al. 2011) to retain key attributes (Fig. 3.5) that could indicate its origins. Recent plans for a proposed Europa lander mission include a search for potential organic biosignatures (Hand et al. 2017).

Other categories of biosignatures (metal isotopes, morphologies, minerals and biofabrics) are still vitally important. The simultaneous presence of multiple biosignature objects, substances and patterns in a formerly habitable environment would constitute the most compelling evidence of life.

Acknowledgements NASA Ames Research Center and the NASA Astrobiology program supported the preparation of this chapter. This work is dedicated to the memory of John M. Hayes—scholar, mentor and lifelong friend.

References

Abed RM, Kohls K, Schoon R et al (2008) Lipid biomarkers, pigments and cyanobacterial diversity of microbial mats across intertidal flats of the arid coast of the Arabian Gulf (Abu Dhabi, UAE). FEMS Microbiol Ecol 65:449–462

Allen MA, Goh F, Burns BP et al (2009) Bacterial, archaeal and eukaryotic diversity of smooth and pustular microbial mat communities in hypersaline lagoon of Shark Bay. Geobiology 7:82–89

Allen MA, Neilan BA, Burns BP et al (2010) Lipid biomarkers in Hamelin Pool microbial mats and stromatolites. Org Geochem 41:1207–1218

Allwood AC, Walter MR, Kamber BS et al (2006) Stromatolite reef from the Early Archaean era of Australia. Nature 414:714–718

Awramik SM (1984) Ancient stromatolites and microbial mats. In: Cohen Castenholz RW, Halvorson HO (eds) Microbial mats: stromatolites. Alan R. Liss, New York, pp 1–22

Baker AJ, Fallick AE (1989) Evidence from Lewisian limestones for isotopically heavy carbon in two-thousand-million-year-old-sea water. Nature 337:352–354

Bishop AN, Love GD, Mcaulay AD et al (1998) Release of kerogen-bound hopanoids by hydropyrolysis. Org Geochem 29:989–1001

Blair N, Leu A, Munoz E et al (1985) Carbon isotopic fractionation in heterotrophic microbial metabolism. Appl Environ Microbiol 50:996–1001

Blumenberg M, Krüger M, Nauhaus K et al (2006) Biosynthesis of hopanoids by sulfate-reducing bacteria (genus *Desulfovibrio*). Environ Microbiol 8:1220–1227

Blumenberg M, Arp G, Reitner J et al (2013) Bacteriohopanepolyols in a stratified cyanobacterial mat from Kiritimati (Chistmas Island, Kiribati). Org Geochem 55:55–62
Boon JJ, Hines H, Burlingame AL et al (1983) Organic geochemical studies of Solar Lake laminated cyanobacterial mats. Adv Org Geochem 1981:207–227
Bosak T, Knoll AH, Petroff AP (2013) The meaning of stromatolites. Annu Rev Earth Planet Sci 41:21–44
Boudou JP, Trichet J, Robinson N et al (1986) Lipid composition of a recent Polynesian microbial mat sequence. Org Geochem 10:705–709
Brassell SC, Wardroper AMK, Thomson ID et al (1981) Specific acyclic isoprenoids as biological markers of methanogenic bacteria in marine sediments. Nature 290:693–696
Bravo JM, Perzl M, Härtner T et al (2001) Novel methylated triterpenoids of the gammacerane series from the nitrogen-fixing bacterium *Bradyrhizobium japonicum* USDA 110. Eur J Biochem 268:1323–1331
Brocks JJ, Schaeffer P (2008) Okenane, a biomarker for purple sulfur bacteria (Chromatiaceae), and other new carotenoid derivatives from the 1640 Ma Barney Creek Formation. Geochim Cosmochim Acta 72:1396–1414
Brocks JJ, Love GD, Snape CE et al (2003) Release of bound aromatic hydrocarbons from late Archean and Mesoproterozoic kerogens via hydropyrolysis. Geochim Cosmochim Acta 67:1521–1530
Brocks JJ, Love GD, Summons RE et al (2005) Biomarker evidence for green and purple sulphur bacteria in a stratified Palaeoproterozoic sea. Nature 437:866–870
Bühring SI, Smittenberg RH, Sachse D et al (2009) A hypersaline microbial mat from the Pacific Atoll Kiritimati: insights into composition and carbon fixation using biomarker analyses and a ^{13}C-labeling approach. Geobiology 7:308–323
Canfield DE (1998) A new model for Proterozoic ocean chemistry. Nature 396:450–453
Canfield DE (2001) Biogeochemistry of Sulfur Isotopes. In: Valley JW, Cole DR (eds) Stable isotope geochemistry. Mineralogical Society of America, Washington, DC, pp 607–636
Chaussidon M, Lorand J-P (1990) Sulphur isotope composition of orogenic spinel lherzolite massifs from Ariege (north-eastern Pyrenees, France): an ion microprobe study. Geochim Cosmochim Acta 54:2835–2846
Cleland CE, Chyba CF (2007) Does 'Life' have a definition? In: Sullivan WT, Baross JA (eds) Astrobiology and life. Cambridge University Press, Cambridge, pp 119–131
Close HG, Bovee R, Pearson A (2011) Inverse carbon isotope patterns of lipids and kerogen record heterogeneous primary biomass. Geobiology 9:250–265
Cook HW, McMaster CR (2002) Fatty acid desaturation and chain elongation in eukaryotes. New Compr Biochem 36:181–204
Cronan JE, Thomas J (2009) Bacterial fatty acid synthesis and its relationships with polyketide synthetic pathways. Methods Enzymol 459:395–433
Deamer D (1986) Role of amphiphilic compounds in the evolution of membrane structure on the early Earth. Orig Life Evol Biosph 17:3–25
Deamer D (1997) The first living system: a bioenergetics perspective. Microbiol Mol Biol Rev 61: 239–261
DeNiro MJ, Epstein S (1977) Mechanism of carbon isotope fractionation associated with lipid synthesis. Science 197:261–263
Des Marais DJ (1995) The biogeochemistry of hypersaline microbial mats. In: Jones G (ed) Advances in microbial ecology. Springer, New York, pp 251–274
Des Marais DJ (1997) Isotopic evolution of the biogeochemical carbon cycle during the Proterozoic Eon. Org Geochem 27:185–193
Des Marais DJ (2001) Isotopic evolution of the biogeochemical carbon cycle during the Precambrian. In: Valley JW, Cole DR (eds) Stable isotope geochemistry (Reviews in mineralogy and geochemistry), vol 43. Mineralogy Society of America, Washington, pp 555–578
Des Marais DJ (2010) Marine hypersaline *Microcoleus*-dominated cyanobacterial mats in the Saltern at Guerrero Negro, Baja California Sur, Mexico. In: Seckbach J (ed) Microbial mats,

Cellular Origins, Life in Extreme habitats and astrobiology (COLE) series. Springer, Berlin, pp 401–420
Des Marais DJ (2013) Planetary climate and the search for life. In: Mackwell SJ, Simon-Miller A, Harder JW et al (eds) Comparative climatology of terrestrial planets. The University of Arizona Press, Tucson, pp 583–601
Des Marais DJ, Cohen Y, Nguyen H et al (1989) Carbon isotopic trends in the hypersaline ponds and microbial mats at Guerrero Negro, Baja California Sur, Mexico: implications for Precambrian stromatolites. In: Cohen Y, Rosenberg E (eds) Microbial mats: physiological ecology of benthic microbial communities. American Society for Microbiology, Washington, pp 191–205
Desmond E, Gribaldo S (2009) Phylogenomics of sterol synthesis: insights into the origin, evolution, and diversity of a key eukaryotic feature. Genome Biol Evol 1:364–381
Dhingra A, Portis AR, Daniell H (2004) Enhanced translation of a chloroplast-expressed RbcS gene restores small subunit levels and photosynthesis in nuclear RbcS antisense plants. Proc Natl Acad Sci USA 101:6315–6320
Dodd MS, Papineau D, Grenne T et al (2017) Envidence for early life in Earth's oldest hydrothermal vent precipitates. Nature 543:60–64
Eickhoff M, Birgel D, Talbot HM (2014) Oxidation of Fe(II) leads to increased C-2 methylation of pentacyclic *Rhodopseudomonas palustris* strain TIE-1. Geobiology 11:268–278
Eigenbrode JL (2008) Fossil lipids for life-detection: a case study from the early Earth record. Space Sci Rev 135:161–185
Eigenbrode JL, Freeman KH (2006) Late Archean rise of aerobic microbial ecosystems. Proc Natl Acad Sci USA 103:15759–15764
Eigenbrode JL, Freeman KH, Summons RE (2008) Methylhopane biomarker hydrocarbons in Hamersley Province sediments provide evidence of Neoarchean aerobiosis. Earth Planet Sci Lett 273:323–331
Eisenreich W, Bacher A, Arigoni D et al (2004) Biosynthesis of isoprenoids via the non-mevalonate pathway. Cell Mol Life Sci 61:1401–1426
Ernst R, Ejsing CS, Antonny B (2016) Homeoviscous adaptation and the regulation of membrane lipids. J Mol Biol 428:4776–4791
Farrimond P, Head IM, Innes HE (2000) Evironmental influence on the biohopanoids composition of recent sediments. Geochim Cosmochim Acta 64:2985–2992
Farrimond P, Love GD, Bishop AN et al (2003) Evidence for the rapid incorporation of hopanoids into kerogen. Geochim Cosmochim Acta 67:1383–1394
Fedo CM, Whitehouse MJ (2002) Origin and significance of Archaean quartzose rocks at Akilia, Greenland. Science 298:917a
Fischer WW, Summons RE, Pearson A (2005) Targeted genomic detection of biosynthetic pathways: anaerobic production of hopanoid biomarkers by a common sedimentary microbe. Geobiology 3:33–40
Flannery DT, Allwood AC, Van Kranendonk MJ (2016) Lacustrine facies dependence of highly ^{13}C-depleted organic matter during the global age of methanotrophy. Precambrian Res 285: 216–241
Franks J, Stolz JF (2009) Flat laminated microbial mat communities. Earth Sci Rev 96:163–172
Freissinet C, Glavin DP, Mahaffy PR et al (2015) Organic molecules in the Sheepbed Mudstone, Gale Crater, Mars. J Geophys Res 120:495–514
French KL, Hallmann C, Hope JM et al (2015) Reappraisal of hydrocarbon biomarkers in Archean rocks. Proc Natl Acad Sci USA 112:5915–5920
Friend CRL, Nutman AP, Bennett VC (2002) Origin and significance of Archaean quartzose rocks at Akilia, Greenland. Science 298:917a
Galea AM, Brown AJ (2009) The special relationship between sterols and oxygen: were sterols an adaptation to aerobic life? Free Radic Biol Med 47:880–889
Garcia-Pichel F, Prufert-Bebout L, Muyzer G (1996) Phenotypic and phylogenetic analyses show *Microcoleus chthonoplastes* to be a cosmopolitan cyanobacterium. Appl Environ Microbiol 62: 3284–3291

Georgiou CD, Deamer DW (2014) Lipids as universal biomarkers of extraterrestrial life. Astrobiology 14:541–549

Gerdes G, Krumbein WE, Hotkamp E (1985) Salinity and water activity related zonation of microbial communities and potential stromatolites of the Gavish Sabkha. In: Friedman GM, Krumbein WE (eds) Hypersaline ecosystems: the Gavish Sabkha. Springer, Berlin, pp 238–266

Golubic S, Abed RM (2010) Entophysalis mats as environmental regulators. In: Seckbach J, Oren A (eds) Microbial mats, modern and ancient microorganisms in stratified systems. Springer, Dordrecht, pp 237–251

Grimalt JO, DeWit R, Teixior P et al (1992) Lipid biogeochemistry of *Phormidium* and *Microcoleus* mats. Org Geochem 19:509–530

Hand KP, Murray AE, Garvin J et al (2017) Science goals, objectives and investigations of the 2016 Europa Lander Science Definition Team report. Lunar and Planetary Science XLVIII: 2492

Härtner T, Straub KL, Kannenberg E (2005) Occurrence of hopanoid lipids in anaerobic *Geobacter* species. FEMS Microbiol Lett 243:59–64

Harvey HR, Fallon RD, Patton JS (1986) The effect of organic matter and oxygen on the degradation of bacterial membrane lipids in marine sediments. Geochim Cosmochim Acta 50: 795–804

Hayes JM (1993) Factors controlling ^{13}C contents of sedimentary organic compounds: principles and evidence. Mar Geol 113:111–125

Hayes JM (1994) Global methanotrophy at the Archean-Proterozoic transition. In: Bengtson S (ed) Early life on Earth. Nobel symposium, vol 84. Columbia University Press, New York, pp 220–236

Hayes JM (2001) Fractionation of carbon and hydrogen isotopes in biosynthetic processes. Rev Mineral Geochem 43:225–277

Hayes JM (2004) Isotopic order, biogeochemical processes, and earth history. Geochim Cosmochim Acta 68:1691–1700

Hayes JM, Kaplan IR, Wedeking KW (1983) Precambrian organic geochemistry, preservation of the record. In: Schopf JW (ed) Earth's earliest biosphere. Princeton University Press, Princeton, pp 93–134

Hayes JM, Strauss H, Kaufman AJ (1999) The abundance of ^{13}C in marine organic matter and isotopic fractionation in the global biogeochemical cycle of carbon during the past 800 Ma. Chem Geol 161:103–125

Hinrichs KU, Summons RE, Orphan V et al (2000) Molecular and isotopic analysis of anaerobic methane-oxidizing communities in marine sediments. Org Geochem 31:1685–1701

Horita J (2005) Some perspectives on isotopic biosignatures for early life. Chem Geol 218:171–186

Horita J, Polyakov VB (2015) Carbon-bearing iron phases and the carbon isotopic composition of the deep Earth. Proc Natl Acad Sci USA 112:31–36

House C, Schopf JW, Stetter KO (2003) Carbon isotopic fractionation by Archeans and other thermophilic prokaryotes. Org Geochem 34:345–356

Jahnert RJ, Collins LB (2011) Significance of subtidal microbial deposits in Shark Bay, Australia. Mar Geol 286:106–111

Jahnke LL, Klein HP (1979) Oxygen as a factor in eukaryote evolution: some effects of low levels of oxygen on *Saccharomyces cerevisiae*. Orig Life 9:329–334

Jahnke LL, Klein HP (1983) Oxygen requirements for formation and activity of the squalene epoxidase in *Saccharomyces cerevisiae*. J Bacteriol 155:488–492

Jahnke LL, Embaye T, Hope J et al (2004) Lipid biomarker and carbon isotopic signatures for stromatolite-forming, microbial mat communities and Phormidium cultures from Yellowstone National Park. Geobiology 2:31–47

Jahnke LL, Orphan VJ, Embaye T et al (2008) Lipid biomarker and phylogenetic analyses to reveal archaeal biodiversity and distribution in hypersaline microbial mat and underlying sediment. Geobiology 6:394–410

Jahnke LL, Lee C, Parenteau MN et al (2014a) Transformations and fates of lipid biomarkers in microbial mat ecosystems. Goldschmidt Abstracts 2014:1112

Jahnke LL, Turk-Kubo KA, Parenteau MN et al (2014b) Molecular and lipid biomarker analysis of a gypsum-hosted endoevaporitic microbial community. Geobiology 12:62–82
Javor B (1989) Hypersaline environments: microbiology and biogeochemistry. Springer, Berlin
Johnson CM, Beard BL, Roden EE (2008a) The iron isotope fingerprints of redox and biogeochemical cycling in modern and ancient Earth. Annu Rev Earth Planet Sci 36:457–493
Johnson CM, Beard BL, Klein C et al (2008b) Iron isotopes constrain biologic and abiologic processes in banded iron formation genesis. Geochim Cosmochim Acta 72:151–169
Jørgensen BB, Des Marais DJ (1986) Competition for sulfide among colorless and purple sulfur bacteria in cyanobacterial mats. FEMS Microbiol Ecol 38:179–186
Kaplan IR, Rittenberg SC (1964) Carbon isotope fractionation during metabolism of lactate by *Desulfovibrio desulfuricans*. J Gen Microbiol 34:213–217
Karhu JA, Holland HD (1996) Carbon isotopes and the rise of atmospheric oxygen. Geology 24: 867–870
Klappenbach JA, Pierson BK (2004) Phylogenetic and physiological characterization of a filamentous anoxygenic photoautotrophic bacterium '*Candidatus* Chlorothrix halophila' gen. nov., sp. Nov., recovered from hypersaline microbial mats. Arch Microbiol 181:17–25
Knani MH, Corpe WA, Rohmer M (1994) Bacterial hopanoids from pink-pigmented facultative methylotrophs (PPFMs) and from green plant surfaces. Microbiology 140:2755–2759
Koga Y (2012) Thermal adaptation of the archaeal and bacterial lipid membranes. Archaea 2012: 1–6
Lange BM, Rujan T, Martin W et al (2000) Isoprenoid biosynthesis: the evolution of two ancient and distinct pathways across genomes. Proc Natl Acad Sci USA 97:13172–13177
Lingwood D, Simons K (2010) Lipid rafts as a membrane-organizing principle. Science 327:46–50
Logan GA, Hayes JM, Hieshima GB et al (1995) Terminal Proterozoic reorganization of biogeochemical cycles. Nature 376:53–56
Logan GA, Summons RE, Hayes JM (1997) An isotopic biogeochemical study of Neoproterozoic and Early Cambrian sediments from the Centralian Superbasin, Australia. Geochim Cosmochim Acta 61:5391–5409
López-Lara IM, Geiger O (2017) Bacterial lipid diversity. Biochim Biophys Acta 1862:1287–1299
Love GD, Snape CE, Carr AD et al (1995) Release of covalently bound alkane biomarkers in high yields from kerogen via catalytic hydropyrolysis. Org Geochem 23:981–986
Marshall CP, Love GD, Snape CE et al (2007) Structural characterization of kerogen in 3.4 Ga Archaean cherts from the Pilbara Craton, Western Australia. Precambrian Res 155:1–23
McKay DS, Gibson EK Jr, Thomas-Keprta KL et al (1996) Search for past life on Mars: possible relic biogenic activity in Martian meteorite ALH84001. Science 273:924–930
Mojzsis SJ, Harrison TM, Friend CRL et al (2002) Origin and significance of Archaean quartzose rocks at Akilia, Greenland. Science 298:917a
Monson KD, Hayes JM (1980) Biosynthetic control of the natural abundance of carbon 13 at specific positions within fatty acids in *Escherichia coli*. J Biol Chem 255:11435–11441
Morowitz H (1992) Beginnings of cellular life. Yale University Press, New Haven, CT
Murata N, Wada H, Gombos Z (1992) Modes of fatty-acid desaturation in cyanobacteria. Plant Cell Physiol 33:933–941
Mustard JF, Adler M, Allwood A et al (2013) Report of the Mars 2020 Science Definition Team, 154 p, posted July 2013, by the Mars Exploration Program Analysis Group (MEPAG) at http://mepag.jpl.nasa.gov/reports/MEP/Mars_2020_SDT_Report_Final.pdf
National Research Council (2007) The limits of organic life in planetary systems. The National Academies Press, Washington, DC. https://doi.org/10.17226/11919
Nes WD (2011) Biosynthesis of cholesterol and other sterols. Chem Rev 111:6423–6451
Nübel U, Garcia-Pichel F, Clavero E (2000) Matching molecular diversity and ecophysiology of benthic cyanobacteria and diatoms in communities along a salinity gradient. Environ Microbiol 2:217–226
Oger PM (2015) Homeoviscous adaptation of membranes in archaea. High Pressure Biosci 72: 383–403

Oger PM, Cario A (2013) Adaptation of the membrane in Archaea. Biophys Chem 183:42–56

Oren A, Kühl M, Karsten U (1995) An endoevaporitic microbial mat within a gypsum crust: zonation of phototrophic photopigments, and light penetration. Mar Ecol Prog Ser 128:151–159

Orphan VJ, Jahnke LL, Embaye T et al (2008) Characterization and spatial distribution of methanogens and methanogenic biosignatures in hypersaline microbial mats of Baja California. Geobiology 6:376–393

Ourisson G, Albrecht P (1992) Hopanoids. 1. Geohopanoids: the most abundant natural products on Earth? Acc Chem Res 25:398–402

Ourisson G, Rohmer M (1982) Prokaryotic polyterpenes: phylogenetic precursors of sterols. Curr Top Membr Transport 17:153–182

Ourisson G, Rohmer M (1992) Hopanoids. 2. Biohopanoids: a novel class of bacteria lipids. Acc Chem Res 25:403–408

Ourisson G, Rohmer M, Poralla K (1987) Prokaryotic hopanoids and other polyterpenoid sterol surrogates. Annu Rev Microbiol 41:301–333

Pagés A, Grice K, Ertefai T et al (2014a) Organic geochemical studies of modern microbial mats from Shark Bay: Part I: influence of depth and salinity on lipid biomarkers and their isotopic signatures. Geobiology 12:469–487

Pagés A, Grice K, Vacher M et al (2014b) Characterizing microbial communities and processes in a modern stromatolite (Shark Bay) using lipid biomarkers and two-dimensional distributions of porewater solutes. Environ Microbiol 16:2458–2474

Palmisano AC, Cronin SE, D'Amelio ED et al (1989) Distribution and survival of lipophilic pigments in a laminated microbial mat community near Guerrero Negro, Mexico. In: Cohen Y, Rosenberg E (eds) Microbial mats, physiological ecology of benthic microbial communities. American Society for Microbiology, Washington, DC, pp 138–152

Pawlowska MM, Butterfield NJ, Brocks JJ (2013) Lipid taphonomy in the Proterozoic and the effect of microbial mats on biomarker preservation. Geology 41:103–106

Pearson A (2014) Lipidomics for geochemistry. In: Holland HD, Turekian KK (eds) Treatise on geochemistry, vol 2, 2nd edn. Elsevier, Oxford, pp 291–336

Pearson A, Budin M, Brocks J (2003) Phylogenetic and biochemical evidence for sterol synthesis in the bacterium *Gemmata obscuriglobus*. Proc Natl Acad Sci USA 100:15352–15357

Pearson VK, Sephton MA, Franchi IA et al (2006) Carbon and nitrogen in carbonaceous chondrites: elemental abundances and stable isotopic compositions. Meteorit Planet Sci 41:1899–1918

Peters KE, Walters CC, Moldowan JM (2005) The biomarker guide. Cambridge University Press, Cambirdge

Pierson BK, Bauld J, Castenholz RW (1992) Modern mat-building microbial communities: a key to the interpretation of proterozoic stromatolitic communities. In: Schopf JW, Klein C (eds) The proterozoic biosphere. Cambridge University Press, Cambridge, pp 245–260

Pierson BK, Valdez D, Larsen M et al (1994) *Chloroflexus*-like organisms from marine and hypersaline environments: distribution and diversity. Photosynth Res 41:35–52

Poger D, Mark AE (2013) The relative effect of sterols and hopanoids on lipid bilayers: when comparable is not identical. J Phys Chem B 117:16129–16140

Postgate JR (1984) The sulphate-reducing bacteria, 2nd edn. Cambridge University Press, Cambridge

Preuss A, Schauder R, Fuchs G et al (1989) Carbon isotope fractionation by autotrophic bacteria with three different CO_2 fixation pathways. Z Naturforsch 44c:397–402

Raven JA (2013) Rubisco: still the most abundant protein of Earth? New Phytol 198:1–3

Rashby SE, Sessions AL, Summons RE, Newman DK (2007) Biosynthesis of 2-methylbacteriohopanepolyols by an anoxygenic phototroph. Proc Natl Acad Sci U S A 104: 15099–15104

Reilly P (2016) Biosynthesis of fatty acids. American Oil Chemists Society. http://lipidlibrary.aocs.org/Biochemistry

Renoux M, Rohmer M (1985) Prokaryotic triterpenoids. New bacteriohopanetetrol cyclitol ethers from the methylotrophic bacterium *Methylobacterium organophilum*. Eur J Biochem 151: 405–410

Ricci JN, Michel AJ, Newman DK (2015) Phylogenetic analysis of HpnP reveals the origin of 2-methylhopanoid production in Alphaproteobacteria. Geobiology 13:267–277

Rohmer M (1993) The biosynthesis of triterpenoids of the hopane series in the Eubacteria: a mine of new enzymatic reactions. Pure Appl Chem 65:1293–1298

Rohmer M (2008) From molecular fossils of bacterial hopanoids to the formation of isoprene units: discovery and elucidation of the methylerythritol phosphate pathway. Lipids 43:1095–1107

Rohmer M, Anding C, Ourisson G (1980) Non-specific biosynthesis of hopane triterpenes by a cell-free system from *Acetobacter pasteurianum*. Eur J Biochem 112:541–547

Rohmer M, Bouvier-Nave P, Ourisson G (1984) Distribution of hopanoid triterpenes in prokaryotes. J Gen Microbiol 130(5):1137–1150

Rontani J-F, Volkman JK (2005) Lipid characterization of coastal hypersaline cyanobacterial mats from the Camargue (France). Org Geochem 36:251–272

Rosing MT, Frei R (2004) U-rich Archaean sea-floor sediments from Greenland-indications of >3700 Ma oxygenic photosynthesis. Earth Planet Sci Lett 217:237–244

Sáenz JP, Sezgin E, Schwille P et al (2012) Functional convergence of hopanoids and sterols in membrane ordering. Proc Natl Acad Sci USA 109:14236–14240

Sáenz JP, Grosser D, Bradley AS et al (2015) Hopanoids as functional analogues of cholesterol in bacterial membranes. Proc Natl Acad Sci USA 112:11971–11976

Schidlowski M (1988) A 3,800 million year isotopic record of life from carbon in sedimentary rocks. Nature 333:313–318

Schidlowski M (1993) The initiation of biological processes on Earth; Summary of the empirical evidence. In: Engel MH, Macko SA (eds) Organic geochemistry. Plenum, New York, pp 639–655

Schopf JW, Klein C (1992) The proterozoic biosphere: a multidisciplinary study. Cambridge University Press, Cambridge

Schouten S, Özdirekcan S, van der Meer MT et al (2008) Evidence for substantial intramolecular heterogeneity in the stable carbon isotopic composition of phytol in photoautotrophic organisms. Org Geochem 39:135–146

Schouten S, Hopmans EC, Sinninghe Damsté JS (2013) The organic geochemistry of glycerol dialkyl glycerol tetraether lipids: a review. Org Geochem 54:19–61

Shen Y, Buick R, Canfield DE (2001) Isotopic evidence for microbial sulphate reduction in the early Archaean era. Nature 410:77–81

Sinensky M (1974) Homeoviscous adaptation—a homeostatic process that regulates the viscosity of membrane lipids in *Escherichia coli*. Proc Natl Acad Sci USA 7:522–525

Singer SJ, Nicolson GL (1972) The fluid mosaic model of the structure of cell membranes. Science 175:720–731

Sinninghe Damsté JS, de Leeuw JW (1990) Analysis, structure and geochemical significance of organically-bound sulphur in the geosphere: state of the art and future research. Org Geochem 16:1077–1101

Sinninghe Damsté JS, Rijpstra WIC, Schouten S et al (2004) The occurrence of hopanoids in planctomycetes: implications for the sedimentary biomarker record. Org Geochem 35:561–566

Sinninghe Damsté JS, Rijpstra WI, Hopmans EC et al (2011) 13,16-dimethyl octacosanedioic acid (iso-diabolic acid), a common membrane-spanning lipid of Acidobacteria Subdivisions 1 and 3. Appl Environ Microbiol 77:4147–4154

Sojo V, Pomiankowski A, Lane N (2014) A bioenergetics basis for membrane divergence in Archaea and Bacteria. PLoS Biol 12:e1001926

Stal LJ (2012) Cyanobacterial mats and stromatolites. In: Whitton BA (ed) Ecology of cyanobacteria II. Springer, New York, pp 65–125

Strauss H, Moore TB (1992) Abundances and isotopic compositions of carbon and sulphur species in Whole Rock and Kerogen. In: Schopf JW, Klein C (eds) The Proterozoic biosphere: a multidisciplinary study. Cambridge University Press, Cambridge, pp 709–797

Summons RE, Jahnke LL, Hope JM et al (1999) 2-Methylhopanoids as biomarkers for cyanobacterial oxygenic photosynthesis. Nature 400:554–557

Summons RE, Bradley AS, Jahnke LL et al (2006) Steroids, triterpenoids and molecular oxygen. Philos Trans R Soc B 361:951–968

Summons RE, Amend JP, Bish D et al (2011) Preservation of martian organic and environmental records. Astrobiology 11:157–181

Talbot HM, Rohmer M, Farrimond P (2007) Rapid structural elucidation of composite bacterial hopanoids by atmospheric pressure chemical ionization liquid chromatography/ion trap mass spectrometry. Rapid Commun Mass Spectrom 21:880–892

Talbot HM, Summons RE, Jahnke LL et al (2008) Cyanobacterial bacteriohopanepolyol signatures from cultures and natural environmental settings. Org Geochem 39:232–263

Tang T, Mohr W, Sattin SR et al (2017) Geochemically distinct carbon isotope distributions in *Allochromatium vinosum* DSM 180^T grown photoautotrophically and photheterotrophically. Geobiology 15:324–339

Ten Haven TL, Rohmer M, Rullkötter J et al (1989) Tetrahymanol, the most likely precursor of gammacerane occurs ubiquitously in marine sediments. Geochim Cosmochim Acta 53:3073–3079

Tritz J-P, Herrmann D, Bisseret P et al (1999) Abiotic and biological hopanoid transformation: towards the formation of molecular fossils of the hopane series. Org Geochem 30:499–514

Veizer J, Ala D, Amzy K et al (1999) $^{87}Sr/^{86}Sr$, $\delta\ ^{13}C$ and $\delta\ ^{18}O$ evolution of Phanerozoic seawater. Chem Geol 161:59–88

Vilcheze C, Llopiz P, Neunlist S et al (1994) Prokaryotic triterpenoids: new hopanoids from the nitrogen-fixing bacteria *Azotobacter vinelandii*, *Beijerinckia indica* and *Beijerinckia mobilis*. Microbiology 140:2749–2753

Vincent WF, Quesada A (2012) Cyanobacteria in high latitude lakes, rivers and seas. In: Whitton BA (ed) Ecology of cyanobacteria II. Springer, New York, pp 371–399

Von Neumann J (1966) Theory of self-reproducing automata. In Burks AW (ed) University of Illinois Press, Urbana

Vranová E, Coman D, Gruissem W (2012) Structure and dynamics of the isoprenoid pathway network. Mol Plant 5:318–333

Waldbauer JR, Newman DK, Summons RE (2011) Microaerobic steroid biosynthesis and the molecular fossil record of Archean life. Proc Natl Acad Sci USA 108:13409–13414

Walter MR (1983) Archean stromatolites – evidence of the earth's earliest benthos. In: Schopf JW (ed) Earth's earliest biosphere: its origin and evolution. Princeton University Press, Princeton, NJ, pp 187–213

Walter MR (1994) Stromatolites: the main geological source of information on the evolution of the early benthos. In: Bentson S (ed) Early life on earth. Nobel symposium, vol 84. Columbia University Press, New York, pp 270–286

Ward DM, Castenholz RW, Miller SR (2012) Cyanobacteria in geothermal habitats. In: Whitton BA (ed) Ecology of cynobacteria II. Springer, Rotterdam, pp 39–69

Wei JH, Yin X, Welander PV (2016) Sterol synthesis in diverse bacteria. Front Microbiol 7:990

Welander PV, Coleman ML, Sessions AL et al (2010) Identification of a methylase required for 2-methylhopanoid production and implications for the interpretation of sedimentary hopanes. Proc Natl Acad Sci USA 107:8537–8542

Welander PV, Doughty DM, Wu C-H et al (2012) Identification and characterization of *Rhodopseudomonas palustris* TIE-1 hopanoid biosynthesis mutants. Geobiology 10:163–177

White GF, Russell NJ, Tidswell EC (1996) Bacterial scission of ether bonds. Microbiol Rev 60: 216–232

Wiederhold JG (2015) Metal stable isotope signatures as tracers in environmental geochemistry. Environ Sci Technol 49:2606–2624

Wieland A, Pape T, Möbius J et al (2008) Carbon pools and isotopic trends in a hypersaline cyanobacterial mat. Geobiology 6:171–186

Chapter 4
The Deep Subseafloor and Biosignatures

Frédéric Gaboyer, Gaëtan Burgaud, and Virginia Edgcomb

Abstract A critical issue in astrobiology is "where to look for present or past life?" and which types of environments could be relevant, i.e. environments associated with high probabilities to (have) support(ed) life and preserve(d) biosignatures. Due both to the large reservoir it represents and to its protective effect against harmful surface conditions, for example radiation, oxidation, the subsurface is of considerable interest in astrobiology. On Earth, living microorganisms have been documented buried in the subsurface up to depths of several kilometers, demonstrating that the deep subsurface can be inhabited by complex microbial communities for millions of years and offering astrobiologists the possibility to better understand how life could be supported, and what kind of biosignatures could be expected, in the subsurface of other planetary bodies. In this chapter we present general trends in the microbial ecology of deep subsurface environments and their peculiar conditions, with a focus on sedimentary microbial ecosystems. We provide a case study of the Canterbury Basin subseafloor as an analogue, subsurface ecosystem on extraterrestrial planetary bodies, and discuss analytical methods for studying microbial lifestyles and preservation in that ecosystem.

F. Gaboyer (✉)
Centre de Biophysique Moléculaire, CNRS, Orléans, France
e-mail: frederic.gaboyer@cnrs-orleans.fr

G. Burgaud
Université de Brest, Laboratoire Universitaire de Biodiversité et Ecologie Microbienne (EA3882), ESIAB, Plouzané, France
e-mail: gaetan.burgaud@univ-brest.fr

V. Edgcomb
Woods Hole Oceanographic Institution, Woods Hole, MA, USA
e-mail: vedgcomb@whoi.edu

B. Cavalazzi, F. Westall (eds.), *Biosignatures for Astrobiology*, Advances in Astrobiology and Biogeophysics, https://doi.org/10.1007/978-3-319-96175-0_4

4.1 The Deep Biosphere: An Unseen World of Contrasting Habitats

4.1.1 Definition

The deep biosphere (DB) is represented by a wide range of terrestrial and aquatic environments (Fig. 4.1), including (1) the terrestrial DB: deep aquifers, mines and caves; (2) the aquatic DB: the deep ocean, its (sub)surface sediments and igneous crust, and lacustrine habitats (e.g., subglacial lakes).

The DB encompasses a huge variety of ecological niches characterised by site-specific physical and geochemical conditions that are difficult to unify in a single definition. Additionally, deep environments share characteristics, such as a relative disconnection from the surface, the presence of chemosynthetic-based food webs, and atypical environmental conditions. Since the depth of subsurface habitats varies between sites, it seems unreasonable to set a depth threshold defining the DB.

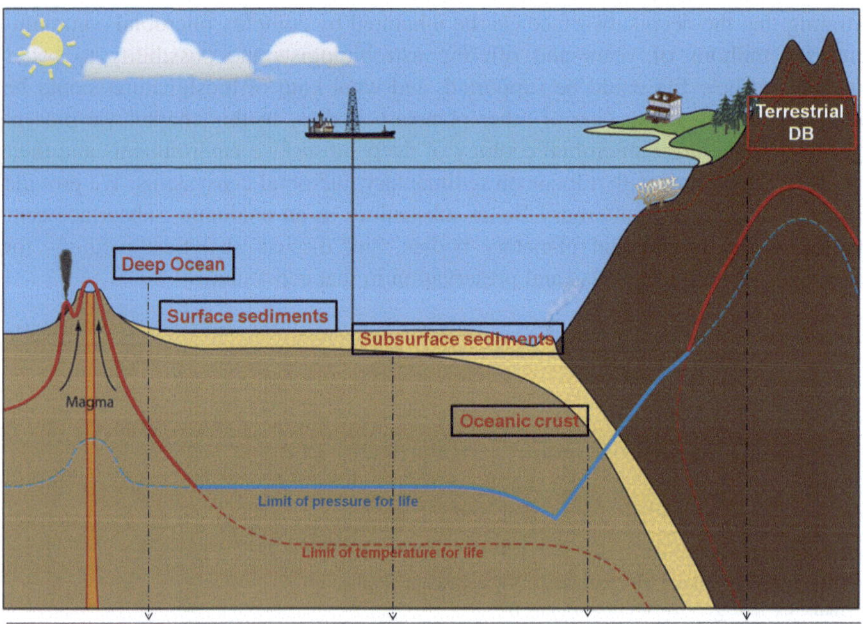

Cell abundance (cell.mL^{-1} or cell.gr,$^{-1}$)	10^4 - 10^5	10^7 - 10^9 (eutrophic sediments) 10^3 - 10^5 (oceanic gyres)	10^5 - 10^9	10^4 - 10^7
Major microbial taxa	*Bacteroidetes*, *Chloroflexi*, *Planctomycetes*, *Proteobacteria*, *Thaumarcheota* MG-I, Group I and II *Euyarcheota*, *Alveolata*, *Fungi*, *Rhizaria*	*Actinobacteria*, *Bacteroidetes*, *Chloroflexi*, *Planctomycetes*, *Proteobacteria*, *Thaumarcheota* MG-I, ANME-1, ANME-2, ANME-3, *Fungi*	*Bacteroidetes*, *Firmicutes*, *Proteobacteria*	*Actinobacteria*, *Firmicutes*, *Proteobacteria* SAGMEG

Fig. 4.1 Schematic representation of the DB and associated cell abundances and taxa per type of habitat. Source: Adapted from Oger and Jebbar (2010)

Considering the subseafloor, Teske and Sørensen (2008) proposed "*sediment layers with distinct microbial communities that lack a microbial imprint of water column communities should be considered deep subsurface*." Extending this idea to other environments, we suggest as a guideline for defining the DB that it represents *environments that are physically isolated from the surface with microbial communities that are distinct from those in shallow layers of the Earth and with food webs that can function independently from solar energy*. However, as mentioned above, the disconnection of the DB from the surface is relative, as highlighted by several lines of evidence, including that microbial cell numbers decrease in subsurface sediments in areas of lower sedimentation, or that archaeal $\delta^{13}C$ isotopic signatures can overlap with cells of photosynthetic origin (Biddle et al. 2006). The similarity between fungal DNA sequences from the deep subseafloor and DNA sequences of terrestrial Fungi (Rédou et al. 2015) also indicates that the surface biosphere contributes to shaping the subsurface biosphere.

4.1.2 Environmental Conditions of Subsurface Environments

The marine deep subsurface represents a vast reservoir for life on Earth, extending several hundred to more than a thousand meters into deeply buried sedimentary habitats and even further down into igneous crust hydrated by fluid flows. The environmental conditions of the DB are hostile to most forms of life, favouring the selection of so-called extremophiles. We note that physicochemical extremes interacting in a synergetic way (Harrison et al. 2013) should not be considered individually when characterising the DB.

4.1.2.1 Temperature and Pressure

An increasing temperature gradient with depth creates a progressive succession of thermal conditions for psychro/meso/thermophiles. This gradient is estimated to average 20 °C/km in the oceanic crust, 25 °C/km in the terrestrial subsurface and 30–60 °C/km in subsurface sediments (Turcotte and Schubert 2002). Near-surface sediments typically have a temperature range of 2–4 °C (the temperature of the deep ocean). The upper temperature limit for life is 122 °C under high pressure for *Methanopyrus kandleriir* (Takai et al. 2008) and 113 °C at atmospheric pressure for *Pyrolobus fumarii* (Blöchl et al. 1997), implying that life in the DB could persist up to 2–5 kilometers below the seafloor (kbsf). However, it should be kept in mind that the laboratory experiments in which these organisms grew were conducted under energy-replete conditions. Thus, considering the low amount of energy available for cells in the DB, microorganisms are unlikely to survive in the DB to their experimentally-determined upper temperature limits.

Large differences in temperature and pressure exist due to the heterogeneity of local environments, e.g. hydrothermal fluids emanating from vents near tectonic and volcanic boundaries can reach 450 °C and circulate through subseafloor fractures. Geothermal activity in terrestrial systems also provides local hot circulating groundwater, rendering some horizons ideal for (hyper-)thermophiles. Thermophiles have been obtained from Japanese gold mine samples at 350 m below land surface and at 80 °C (Inagaki et al. 2003), whereas the terrestrial subsurface at similar depth in crust not particularly influenced by hydrothermal circulation tends to be considerably cooler.

Pressure has various effects on microbial physiology (for a thorough review, see Oger and Jebbar 2010) but is unfortunately not routinely measured during sampling. The pressure gradient occurring with depth depends on the material properties of rocks (e.g., porosity, compaction degree). In oceanic systems, in addition to the pressure gradient of sediments (15–25 MPa/km) and oceanic crust (30 MPa/km), the water column imposes a pressure gradient of 10 MPa/km (Dziewonski and Anderson 1981). Since seafloor is located at an average depth of 3800 m, the basal pressure of marine surface sediments is ~38 MPa. Extreme piezophilic or piezotolerant organisms can survive and remain active in buried habitats of the DB. *Colwellia* BNL-1 and *Methanocaldococcus jannaschii* grow optimally at 93 MPa/10 °C (Deming et al. 1988) and 75 MPa/86 °C (Jones et al. 1983), respectively. The maximum pressure limit compatible with cell activity reported to date is 120 MPa for *Pyrococcus yayanossi* (Birrien et al. 2011), which corresponds to a depth limit of 3.3–5.5 kbsf in sediments (25–15 MPa/km) and of 2.7 kbsf (30 MPa/km) in the oceanic crust, considering a basal pressure of 38 MPa at the seafloor.

4.1.2.2 Activity of Water and Porosity

The low water activity (A_w) within the DB leads to physiological challenges for cells, such as osmotic pressure changes and difficulties in membrane transport. Enzymatic activities are also inhibited due to the catalytic roles played by H_2O. To cope with this, cells must respond with energy-consuming processes in environments with minimal energy sources (Jørgensen and Boetius 2007).

Evolution has also produced life forms adapted to extremely dry environments, thus life may remain active in the DB down to A_w ~0.64, based on reported microbial growth in laboratory experiments with A_w, as low as 0.635 and 0.64 for haloarcheal and fungal species, respectively (Stevenson et al. 2015). Note that, even though materials can be humid, the lack of liquid water available for biological activity can result in very low A_w. In deeper subsurface habitats, originally humid sediments give way to progressively drier and compacted rocks. Clay-rich sediments for instance are typically replaced with sedimentary rocks, such as marlstone and limestone.

Porosity is also a major parameter controlling life in the DB and is correlated to subseafloor cell abundance (Parkes et al. 2000). Indeed, pores are microniches in which fluid circulation produces chemical gradients, thus, potentially supplying microorganisms with nutrients, electron donors and acceptors and enabling microbial growth. The low porosity and the absence of macro-fractures limit the presence of water in the DB. Although porosity and water activity stresses are related, we note that in some places there is simply insufficient space for buried cells. Black shales are an excellent illustration of this point since, despite high organic content, they are barely inhabited by cells due to their compaction.

4.1.2.3 Energy Sources

It is estimated that less than 1% of organic matter (OM) in the oceans is accumulated in the subseafloor and is largely consumed in the first centimeters of sediments (Orcutt et al. 2011). Since most subseafloor sediments are located in the low-productivity abyssal regions of the open ocean, the OM content of sediments in those regions can be extremely low. The maturation of OM with depth during diagenesis produces refractory carbon unavailable to most microorganisms (Burdige 2007). The marine DB is thus an extreme oligotrophic environment, and very low metabolic rates and long generation times are expected for the cells inhabiting it (Jørgensen and Boetius 2007; Hoehler and Jørgensen 2013; Jørgensen and Marshall 2016).

As mentioned above, circulating water can provide chemical gradients that supply microbial metabolisms with electron donors (mainly organic matter, hydrogen, reduced sulphur compounds, reduced iron compounds, and ammonium) and acceptors (mainly oxygen, nitrate/nitrite, manganese and iron oxides and sulphite/sulphates). Since redox couples with the highest free energy are preferentially consumed in the top layers of sediments, this leads to zonation where oxygen, nitrate, manganese, iron and sulphate are consumed in a successive and sequential pattern (DeLong 2004). As a consequence, only traces of electron donors and acceptors are present in deeper horizons in the absence of fluid circulation.

Other natural processes can supply microbial metabolisms with alternative energy sources: serpentinisation produces methane and hydrogen during the interaction of water with olivine and pyroxenes, whereas radiolysis of water leads to hydrogen production from ^{236}U, ^{232}Th and ^{40}K radioactivity, a process that was shown to strongly support microbial life in subsurface sediments of the South Pacific Gyre (D'Hondt et al. 2009).

Specific lifestyles may be particularly adapted to the nutrient-poor conditions of the DB. For example, it has been reported that a single Firmicute species, *Candidatus Desulforudis audaxviator*, inhabits a gold mine at 2.8 kms depth with a lifestyle well-suited to long-term isolation from the photosphere, based on inorganic carbon and nitrogen fixation, sulphate reduction, or complete anabolism (Chivian et al. 2008). This kind of independent lifestyle is an elegant example of how life could autonomously persist in the subsurface of extraterrestrial, basaltic, planetary bodies, such as Mars.

4.2 Tools to Detect Subsurface Biosignatures

4.2.1 Contamination Issues

As discussed in Lever et al. (2006), there are significant challenges in eliminating contamination of core samples collected from below the seafloor. The huge amount of surface seawater injected into the borehole is a major potential source of contamination. Various protocols can be employed at sea to clean the exteriors of collected rock, and/or to remove exterior material from rock or sediment cores.

DNA sequencing approaches are required to evaluate the extent to which core material may still be contaminated, for instance sequencing different negative controls (kit and the drilling fluids). Signatures of organisms found to be identical in both the drilling fluid and rock/sediment samples may be a sign of contamination, and the most prudent approach is to eliminate those signatures from downstream analyses.

Chemical tracers and microsphere beads are also injected into the seawater drilling fluids to assist microbiologists in the evaluation of potential contamination but inconsistencies in the distribution of microbeads have limited their utility for this purpose. Perfluoromethylcyclohexane (PFMC) is the most commonly used perfluorocarbon tracer by microbiologists (Smith et al. 2000). The sample is flamed upon recovery to volatilize the PFMC, to release it into the air and thereupon quantify it using gas chromatography. Any PFMC detected on the interior of a sample means that drilling fluid could have penetrated into the sample, and thus it is likely contaminated. A problem with this approach is that PFMC is extremely volatile and, once a few samples have been run, it can generally be detected in most samples because of airborne contamination. A less volatile tracer, perfluoromethyldecalin (PFMD) was recently tested during the Expedition 360 (South West of the Indian ridge), of the International Ocean Discovery Program (IODP). PFMD was found to produce more reliable results (Edgcomb and Sylvan unpublished data). Another approach used a modified microsphere tracer dispersed in an aqueous fluorescent pigment (Friese et al. 2017).

4.2.2 Sensitive Analytical Methods

The most challenging aspect of investigating life in the subsurface is that, despite its apparent ubiquity in the DB, cell numbers can be extremely low (Kallmeyer et al. 2012; Parkes et al. 2014), particularly in crustal samples. This necessitates extremely sensitive analytical procedures for the detection of cells and cell activities (Lomstein et al. 2012).

Stable- and radio-isotope tracer-based techniques represent one possible approach based on sample incubation under conditions mimicking in situ conditions. This has been used to measure microbial activities such as sulphate reduction

(Jørgensen and Boetius 2007), methanogenesis, methane oxidation (Orcutt et al. 2010), and hydrogenase activity (Nunoura et al. 2009). The resulting data can be influenced by artefacts associated with physicochemical changes that occur during, and following, sample recovery. In this respect, the development of new in situ mass and laser absorption spectrometers hold great promise (Cowen et al. 2012).

To quantify active subsurface microbial populations using microscopy, dead cells must be differentiated from living cells and their biosignatures. Some of the first studies of DB sediments applied epifluorescence microscopy to quantify cell abundances but were restricted to organic-rich sediments with high cell abundance. Recent advances that allow cells to be separated from sediments (Kallmeyer et al. 2008) with counting automation (Morono et al. 2009) have resulted in a downward revision of estimated total cells in the DB sediments (Kallmeyer et al. 2012).

Most assessments of microbial diversity within the marine DB have utilized PCR amplification of target genes, most commonly small subunit ribosomal RNA (SSU rRNA) genes, from extracted DNA. Because these extracts include DNA from active, inactive but viable, and dead cells, and also extracellular DNA, they do not exclusively represent active cells. This has spurred researchers to pursue analyses targeting the more labile RNA molecule, with its reverse transcription into complementary DNA (cDNA), that can be then analyzed either as a metatranscriptome, capturing functional information from active communities (Orsi et al. 2013b; Pachiadaki et al. 2016), or as a template to amplify target genes (Orsi et al. 2013a; Rédou et al. 2014). Curiously, nucleic acids extracted from subsurface marine sediments exhibit signatures of taxa that are clearly not endogenous to this habitat, indicating that preservation of DNA and small subunit ribosomal RNA preservation of bacterial spores, fossilized diatoms with silica-rich frustules, cysts of protists and pollen or spores from plants up to 2.7 My ago is possible in sediments and rocks with reduced water content (Orsi et al. 2013a). The degree to which DNA is preserved in these cells is unknown, however the presence of this paleome challenges the interpretation of deep subsurface molecular data making the design of integrated approaches highly desirable.

Lipid biomarkers also provide good targets for the detection of deeply buried cells with high sensitivity. High-pressure liquid chromatography coupled to mass spectrometry has been applied to lipids extracted from subsurface hydrothermal samples (Sturt et al. 2004), and can also be used to identify particular microbial groups. Intact polar lipids are thought to derive from viable intact cells, as polar head groups are hydrolyzed after cell death in sedimentary environments within days, although their persistence as head groups for millions years may also be possible in immature sediments (Schouten et al. 2013).

Polar head groups devoid of core lipids reflect fossil remains and can, thus, provide information about past microbial communities. Ultra-sensitive triple quadrupole mass spectrometry may also be required for the analysis of low biomass samples, such as crustal rocks. Paired with confocal Raman spectroscopy analyses, a non-destructive method, it may be possible to visualize microfossils or living cells in such samples since confocal Raman spectroscopy allows the recognition of living and fossil microorganisms in rock material (Foucher et al. 2010, 2015).

4.3 Microbiology of Subsurface Sediments Using the Canterbury Basin (CB) as a Case Study

4.3.1 Drilling Expeditions and Environmental Context

The IODP Leg 317 (December 2009) expedition took place in the CB, offshore New Zealand. The *JOIDES Resolution* drilled four holes (U1352A to D) at a water depth of 344 m at site U1352, ~75 km from shore. A total core of 1927 m was recovered, spanning the Holocene to late Eocene (the deepest layers aged 32 Ma) and divided into three units of clay and calcareous sandy mud (unit I, from surface to 709 mbsf), sandy marlstone and limestone (unit II, 709 to 1852 mbsf) and limestone and cherts (unit III, 1852 to 1927 mbsf). Abundance in calcareous nannofossils and planktonic foraminifera provides good biostratigraphic age controls (Fulthorpe et al. 2011).

Investigation of environmental conditions throughout the core showed that temperatures at the bottom were ~60 °C. Measured pH varied from 7 to 8 and fluctuated with sulphate reduction, methanogenesis and possibly carbonate precipitation. Sediment porosity ranged from 40% near the surface to >10% in the deepest layers, and TOC varied from 0.25% to 0.75%. Gas contents ranged between 1 to 10 ppmv (ethane and propane) and to 10^4 to 10^2 ppmv (methane) throughout the core (Fulthorpe et al. 2011).

4.3.2 Cell Abundance

Although initial estimates showed that microorganisms of the DB represented 35% to 47% of Earth's total biomass (Whitman et al. 1998), these data were largely obtained in organic-rich sediments close to continents and associated with high sedimentation rates. Considering the sedimentation rates and distances from continents, Kallmeyer et al. (2012) offered more reserved and realistic estimates of the subseafloor biomass of 3.10^{29} cells and 4.10^{15}g C in total, representing 0.6% of Earth's total biomass. Taking into account recent studies that have reported intact cells down to 1922 mbsf (Ciobanu et al. 2014), this estimate can be further revised at $5.39.10^{29}$ cells (Parkes et al. 2014). Without doubt, estimates of cell abundances are likely to be amended in coming years, especially now that microbial life has been documented down to 2.5 kbsf with atypical cell concentrations along the core (Inagaki et al. 2015).

Although the presence of spores and of viral-like particles (VLPs) in the DB is poorly studied, it is of great importance since it is known that spores can persist under stressful conditions and can reactivate after several million years (Vreeland et al. 2000). Lomstein et al. (2012) quantified spores (dipicolinic acid) and bacterial (muramic acid and D-amino acids) biomarkers, finding that "*endospores are as abundant as vegetative cells and microbial activity is extremely low, leading to microbial biomass turnover times of hundreds to thousands of years*", highlighting the critical, but underestimated, role of spores in microbial ecology of the DB.

Viral-like particles (VLPs) strongly impact upon the carbon cycle and microbial populations. Bacteriophages were first revealed by induction of the viral cycle on strains isolated from subsurface sediments (Engelhardt et al. 2011), showing head-tailed morphologies, typical of *Myoviridae* and *Siphoviridae*. The presence of VLPs in subsurface sediments has been confirmed by in situ studies down to 320 mbsf, where VLP cell abundances in sediments are 10^9 cells cm^{-3} (Engelhardt et al. 2014) and virus/cell ratios from 1 to 10 suggest ongoing viral production (Engelhardt et al. 2012).

4.3.2.1 The Canterbury Basin Subseafloor Case Study

In the CB, mean cell numbers decrease with depth from about $1.5 \times 10^6 \pm 4.7 \times 10^4$ cells cm^{-3}at the surface to $2.5 \times 10^4 \pm 4.9 \times 10^3$ cells cm^{-3} within the deepest samples (1922 m.b.s.f.) (Fig. 4.2a). The detection limit was estimated to be 2.94×10^3 cells cm^{-3}. The detected cells were very small (0.3–0.8 μm). This depth profile is consistent with the general depth distribution of prokaryotic cells from other subsurface sediments, showing a logarithmic decrease in cell number with depth. No spores or viruses were observed, despite the fact that genes for sporulation and prophages were detected using metagenomic analyses (Gaboyer et al. 2015).

4.3.3 Evolution of Diversity with Depth

Complex microbial communities have been revealed in subsurface sedimentary habitats, including representatives from all three domains of life (Bacteria, Archaea and mostly Fungi among the micro-Eukarya). As the taxonomic composition appears to reflect oceanic regions and site-specific conditions (Parkes et al. 2014), it is difficult to highlight any diversity pattern. Nevertheless, for Bacteria the dominant phyla are Chloroflexi, Proteobacteria, Planctomyces, the candidate phylum Atribacteria, Firmicutes and Actinobacteria; for Archaea, the dominant members are Crenarchaeota with the Lokiarchaeota (previously MBG-B), Euryarchaeota with the Hadesarchaea (previously SAGMEG) and Marine Benthic Group-D (MBG-D, an archaeal group within the Thermoplasmatales) and Thaumarchaeota with the Bathyarchaoeta (previously MCG) and the Marine Group I (MG-1). These acronyms reflect that, to date, most subsurface microbial groups have largely escaped cultivation attempts, and their phylogeny remains uncertain at the phylum-level (Brochier-Armanet et al. 2008).

There is certainly a pattern in the diversity of subsurface microbial communities, controlled by organic matter supply and redox status (Durbin and Teske 2011). Whereas the patterns of distribution in Bacteria tend to be relatively similar between

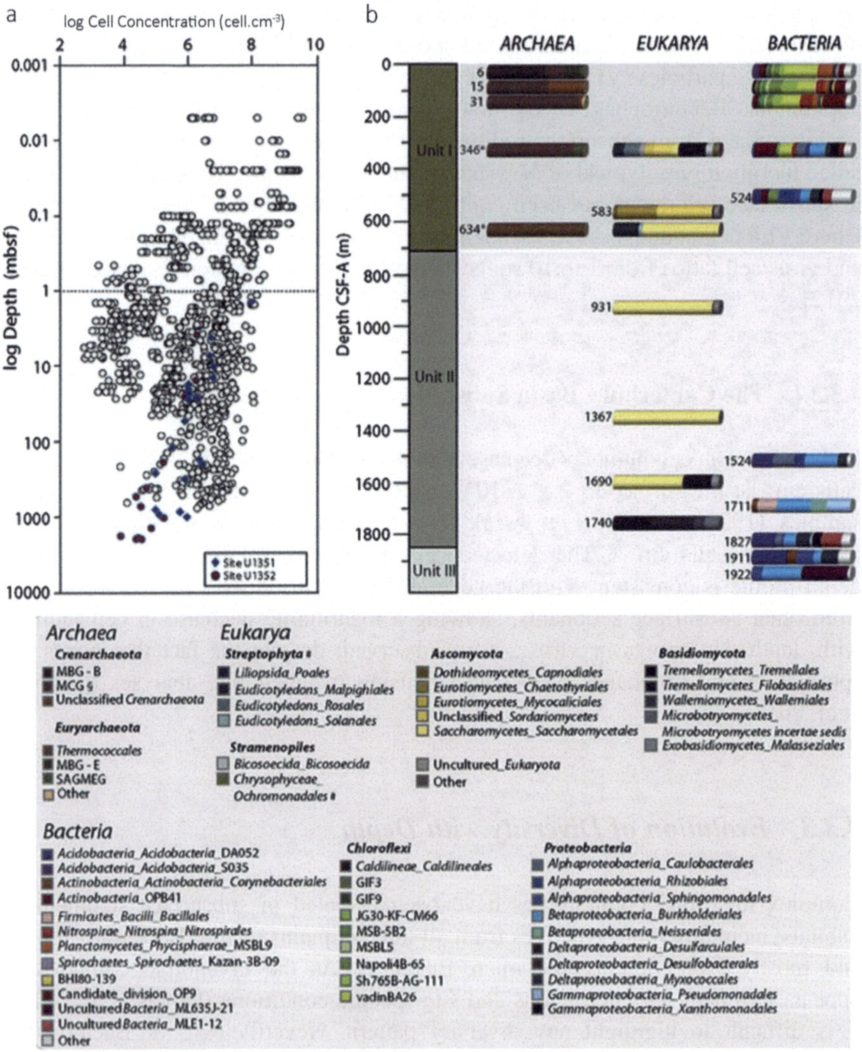

Fig. 4.2 (**a**) Cell counts at two sites in the Canterbury Basin (red and blue circles) compared to general cell counts (white circles) in sub-seafloor sediments. Source: Adapted from Kallmeyer et al. (2012). (**b**) Phylum Class/Order distribution of archaeal, eukaryotic and bacterial 16S/18S rRNA gene-tag sequences (based on SILVA111 classification) from OTUs containing >100 sequences. Source: Adapted from Ciobanu et al. (2014)

sites, contrasting results on archaeal diversity strongly suggest that each subsurface site, characterised by its own environmental factors, has its own archaeal fingerprint.

Nevertheless, without standardized methodologies (extraction methods, targeted genes, primers, sequencing technologies and bioinformatics pipelines), it is difficult to draw any firm conclusions about the presence of particular prokaryotic groups and their relative abundance in different subsurface habitats.

While the focus of debate has mainly centered on bacterial or archaeal dominance, less is known about the eukaryotes. Culture-independent approaches targeting microeukaryotes suggest their presence and activity in deep subsurface sediments on the basis of DNA and SSU rRNA (Edgcomb et al. 2011; Orsi et al. 2013a; Ciobanu et al. 2014; Rédou et al. 2014). Fungal sequences appear to dominate the fraction of microeukaryote signatures, with some fungal representatives having been successfully isolated from Northwestern Pacific margin samples up to 2457 mbsf (Liu et al. 2016).

Irrespective of the domain of life concerned, patterns in microbial diversity with depth are certainly not random but governed by the selection of cells that can survive starvation, such that a progressive selection of adapted survivors occurs from the surface to the depth, as proposed recently on the basis of DNA sequencing and genome analysis (Starnawski et al. 2017). Both the abundance of cells and their diversity within microbial communities decreased down the CB core (Fig. 4.2b). Multidimensional statistical analyses highlighted that, among environmental factors, changes in microbial taxa were mainly defined by depth (Ciobanu et al. 2014). The number of OTUs (Operational Taxonomic Units) per domain was very low with 198, 16 and 40 unique bacterial, archaeal and eukaryotic OTUs being detected, most of them retrieved in the first hundred meters of the CB core. Both iTAG and qPCR data indicated that recovered archaeal SSU rRNA genes decreased between 6 and ~600 mbsf. Archaeal sequences could not be amplified and sequenced below 634 mbsf. Sequences belonging to Archaea were mainly affiliated to Lokiarchaeota (MBG-B) and Bathyarchaeota (MCG). Bacterial SSU rRNA genes were detected in samples from the full length of the CB core. Sequences obtained from each depth layer decreased in diversity with depth, with ~100 OTUs recovered in samples analyzed from between 6 and 31 mbsf, ~40 OTUs in samples from 346 and 524 mbsf, and fewer than 20 OTUs in samples from 634 to 1922 mbsf (Ciobanu et al. 2014). OTUs were mainly affiliated to *Chloroflexi* and *Proteobacteria* and exhibited a shift from *Chloroflexi* dominance above 600 mbsf to Proteobacterial dominance below 343 mbsf. Signatures of *Planctomycetes*, *Nitrospirae* and *Atribacteria* dominated the first layers, whereas *Acidobacteria* and *Firmicutes* were retrieved below 600 mbsf.

Other snapshots of DB diversity revealed close distributions in the patterns of particular taxa. For example, in deep coal-bed sediments up to 2457 mbsf (Inagaki et al. 2015) *Chloroflexi* sequences were also observed in quantity at shallow depths. *Chloroflexi* dominated bacterial communities above 600 mbsf for IODP Leg 317 (CB) samples and above 364 mbsf for IODP Leg 337 (Northwestern Pacific margin). In the deeper layers of the CB, a shift occured to *Actinobacteria*, *Proteobacteria* and *Firmicutes* as the dominant groups (Ciobanu et al. 2014; Inagaki et al. 2015).

Concerning Eukarya, few 18S sequences affiliated to bacterivorous protists were retrieved. Consistent with previous studies, Fungi appeared to be the most frequently detected micro-eukaryotes in the CB, particularly *Ascomycota* and *Basidiomycota*.

Eukaryotic richness dropped off gradually along the core, indicating a strong vertical structuring of the community.

A complementary approach specifically targeting fungal DNA confirmed low fungal diversity in CB sediments with poor overlap between fungal OTUs at the different depths suggesting a spatial differentiation of fungal communities (Rédou et al. 2014). Curiously, most fungal OTUs were phylogenetically related to ubiquitous terrestrial Fungi. Using the same samples, a culture-based approach generated 183 fungal isolates (Rédou et al. 2015). Consistent with molecular data, numerous fungal isolates were related to well-known Fungi in terrestrial environments, raising ecological questions regarding the ability of buried terrestrial Fungi to adapt to deep subsurface conditions. Indeed, this similarity points to the contribution of the surface biosphere to the subsurface biosphere and that terrestrial fungi "trail" into the sub-seafloor, undergoing a selection of the most adapted cells to burial and starvation (Starnawski et al. 2017).

We note that describing bacterial, archaeal or eukaryotic groups as abundant at specific depths does not mean they are dominant in the sediment core. During the dynamic burial process, species that are more resistant to starvation will survive (Starnawski et al. 2017) and, thus, communities will change their size and diversity (alpha diversity). The continuous selection of cells with burial makes the notion of microbial dominance relative and, therefore, extending the dominance of one group to other depths incongruent.

4.4 Activity *Versus* Viability

Given the biogeochemical heterogeneity of subsurface sediment horizons, crustal rocks and fluid flows, it stands to reason that there is great heterogeneity also in activity levels of subsurface microbiota, even at different depths along the same core length. From more active cells close to the surface to deeply buried cells, a community shift occurs with the selection of cells more adapted to subsurface starvation (Starnawski et al. 2017). Whether or not they are adapted to oligotrophy, subsurface cells can be more or less active in sediments. This question of activity *versus* dormancy remains a critical issue for understanding the microbial ecology of the subseafloor.

DNA and RNA preservation is possible in deep subsurface habitats, and hence signatures of active cells can be confused with signatures of inactive or even dead cells. To tease apart the signal of live cells from this pool of molecules, an integrated approach is required that blends information from "meta-omics" and culturing with physiological screening of microbial isolates and microscopy to visualize cells.

Cell activity in the deep subsurface biosphere has been detected using different approaches, including microscopy (dividing cells) (Pachiadaki et al. 2016), direct measurement of intermediates in cell processes (D'Hondt et al. 2002), analysis of intact polar lipids (Lipp et al. 2008) and DNA or RNA (Orsi et al. 2013a, b; Rédou et al. 2014; Pachiadaki et al. 2016). The flux of electron donors and acceptors, deriving notably from photosynthesis in overlying seawater or terrestrial environments, mainly controls the level of cell activity of subseafloor microorganisms. Organic carbon content in sub-seafloor provinces varies from 0.09 to >12 wt.% (Schrenk et al. 2010), is higher in continental margin and lower in the open ocean, leading to, respectively, higher and lower cell activity.

Successful cultivation of subsurface isolates (e.g., Batzke et al. 2007; Biddle et al. 2005) does not provide information about the in situ activity of cells but only on their viability once they have been brought to labs. Microbial groups in extremely low abundance in the subseafloor can be the dominant groups among isolates if they have simpler cultivation requirements and more easily reproduced. The description of cells on "physiological standby" maintaining the capacity to metabolize (Morono et al. 2011), supports the idea of a gap between in situ and cultivation-based diversities.

Metatranscriptomics helps to differentiate the metabolic signatures coming from living cells from those of living but inactive or dead cells. Transcripts involved in prokaryotic cell division were found in the Peru Margin subseafloor (Orsi et al. 2013b). Gene transcripts associated with motility may also be a signal of active cells, since motility in the subsurface may be limited by extremely low available energy (Jørgensen and Marshall 2016). In subsurface sediments from the Peru Margin, flagellar protein signatures were detected but, logically, appeared to decrease with depth in most, but not all, samples analysed (Orsi et al. 2013b). Similarly, transcripts of genes associated with cell-cell interactions, particularly those involved in competition for precious resources or pili formation, also indicated that not all cells are dormant in the subsurface. Pili are hair-like appendages on the surfaces of cells that involved in bacterial adherence, conjugation or movement (see Piepenbrink and Sundberg 2016 for a review) and are therefore another sign of interactions between cells and their environment.

4.4.1 The Canterbury Basin Subseafloor Case Study

Successful cultivation of prokaryotes and Fungi inhabiting CB sediments showed that at least a fraction of in situ microbes were viable.

Moreover, an integrated approach designed for studies of the subseafloor of the CB strongly supported the idea that part of the microbial community was not only present but also active. For example, a poly-A targeted metatranscriptome from a 345 mbsf sample revealed fungal transcripts involved in growth, cell division, sporulation, and catalytic activities within different classes of enzymes such as hydrolases, suggesting that Fungi play important roles in biogeochemical cycles of the DB. Bacterial and archaeal transcripts also showed that these buried cells are active in situ.

Finally, microscopic observations of reproductive fungal structures are direct evidence for in situ cell activity and growth (Pachiadaki et al. 2016).

4.5 Possible Metabolisms

Since microbiota exist in the majority of subseafloor locations investigated to date, they must have adaptations for survival in these energy-limited habitats. This is demonstrated by culture-based studies that show the ability of subsurface microbes to survive on extremely low energy fluxes (Jørgensen and Marshall 2016).

The activities of subsurface microorganisms were studied extensively in samples from the Peru Margin (PM) during ODP Leg 201. Since redox couples with highest free energy are first consumed, chemical profiles exhibit predictable zonation at the PM, with depletion of the highest energy yielding species in shallow depths (Fig. 4.3). Indeed, dissolved electron acceptors such as sulphate (SO_4^{2-}) and nitrate (NO_3^-) exhibit subsurface depletion, whereas metabolic products such as dissolved inorganic carbon ($CO_2 + HCO_3^- + CO_3^{2-}$), ammonia ($NH_3 + NH_4^+$), sulphide ($H_2S + HS^-$), methane ($CH_4$), manganese, and iron consistently exhibited concentration maxima deep in the drilled sediments (D'Hondt et al. 2004). As a consequence, processes thought to be active at the PM mainly include organic carbon oxidation, ammonification, methanogenesis, methanotrophy, sulphate reduction, manganese reduction and, to a lesser extent, iron reduction, production and consumption of formate, acetate, lactate, hydrogen, ethane, and propane. Dissimilatory

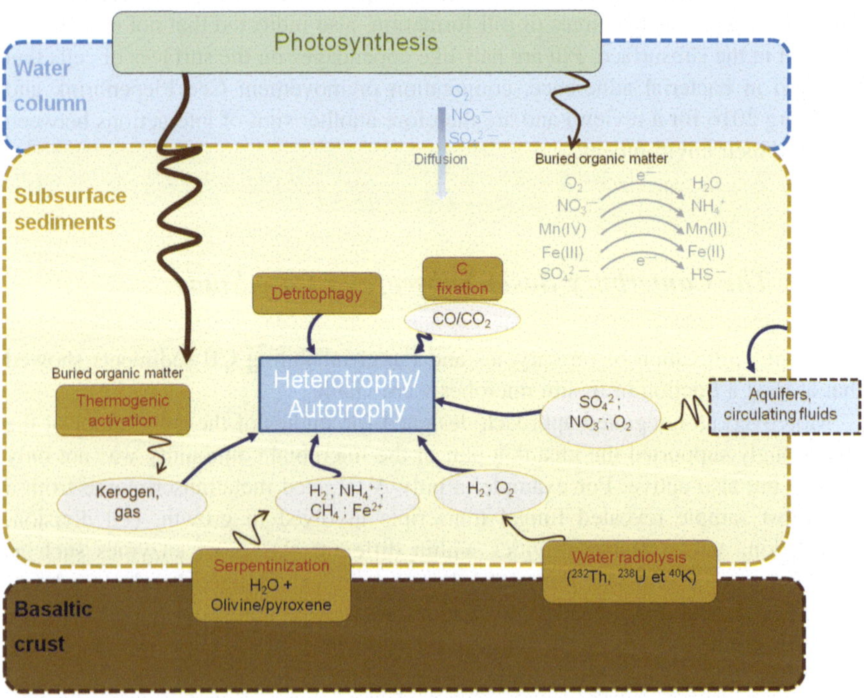

Fig. 4.3 Schematic representation of microbial metabolisms occurring in the subseafloor. Source: F Gaboyer

sulphate reduction may represent the dominant form of energy production in subseafloor sediments, as suggested by pore-water sulphate concentrations (D'Hondt et al. 2004) and by the occurrence of dissimilatory reductase transcripts that reflect biogenic sulphate reduction.

Concerning heterotrophy, with the exception of particularly organic-rich zones, microbial populations must survive on recalcitrant (low reactivity) and diminished concentrations of organic material that has escaped remineralisation by other cells during seafloor deposition and subsequent burial (Burdige 2007). The most labile molecules are rapidly removed after burial. Buried gas hydrates, organic rich layers, and sulphate/methane interfaces may provide new pools of labile organic molecules for cells. At depth, CH_4, hydrocarbons, acetate, H_2 and CO_2 may be released from buried organic material at thermogenic temperatures, providing additional sources of carbon and energy (Parkes et al. 2014). Serpentinisation, the aqueous alteration of ultramafic rocks, produces high-energy microbial substrates, such as organic compounds, H_2 and CH_4, and thus contributes to the maintenance of life in the DB (Fig. 4.3) (Schulte et al. 2006).

Metatranscriptomic analyses used to examine the microbial activities in samples from 6 to 95 mbsf at the PM (Pachiadaki et al. 2016) showed a majority of transcripts associated with enzymes involved in carbohydrate, amino acid and lipid metabolism. Most of these enzymes can participate both in the catabolism and anabolism of organic materials. The successful detection of transporters in all samples, considered as a proxy for active catabolism, indicated heterotrophy in the PM subseafloor. This strengthens previous suggestions that amino acids may be important sources for carbon and nitrogen metabolisms in the subsurface (Lloyd et al. 2013). The recovery of a wide range of transcripts for various carbohydrate transporters also suggests that carbohydrates serve as additional energy sources for microbes in the shallow samples at PM (Pachiadaki et al. 2016).

4.5.1 The Canterbury Basin Subseafloor Case Study

Metabolisms within the CB subsurface, as detected by metagenomics and metatranscriptomics, were mostly represented by heterotrophy, notably through fermentation and respiration of nitrite or sulphate, and by possible autotrophy based on CO fixation. Indeed, cultivation of Bacteria and Archaea from the CB highlighted capabilities for fermentation, while attempts to cultivate true methanogens and sulphate reducers were unsuccessful (Ciobanu et al. 2014). Cultivating subsurface microorganisms is very challenging and, hence, molecular approaches provide valuable information about metabolisms occurring in CB sediments.

Heterotrophy in the CB was also supported by metatranscriptomic and metagenomic data that revealed transcripts and genes for various sugar, amino acid and lipid transporters, as well as transcripts and genes for their degradation (Gaboyer et al. 2015; Pachiadaki et al. 2016). A possibly significant archaeal contribution to amino acid fermentation was indicated by metatransciptomic analyses. Metagenomic analyses showed the importance of extracellular substrate uptake by revealing the presence of more than 100 secreted extracellular peptidase genes. The expression of many genes involved in enzyme production, putative exoenzymes, indicates that these enzymes may be involved in organic carbon turnover in the DB and, more precisely, in degradation of refractory organic matter. This is consistent with recent microbial biomass quantification data on CB samples (Zhu et al. 2016), indicating that a significant proportion of microbial debris is preserved (e.g., prokaryotic necromass) in the CB and, thus, available for decomposers, including Fungi. Among other carbon sources, the presence of haloacid dehalogenase (*had*) genes suggests the potential for utilization of organohalide compounds.

Autotrophic pathways in the CB subseafloor have only been suggested by the detection of genes encoding the reductive acetyl-CoA pathway, based on monoxide carbon fixation. Determining the phylogeny of metabolic genes recovered within genomic fragments showed that Chloroflexi, Euryarcheota and Crenarcheota were all capable of fermentation, that Crenarcheota may use halogens as a carbon source, and that Euryarcheota may be capable of autotrophic CO fixation (Gaboyer et al. 2015).

Considering anoxygenic respiration, the description of *dsrA* (sulphate reduction) and *rdh* (reductive dehalogenase) genes as well, as transcripts for nitrite reductase (denitrification), indicated that sulphate, nitrite and organohalide compounds may, respectively, serve as electron acceptors within the CB subsurface (Pachiadaki et al. 2016).

4.6 Physiological Potential

The term "sociomicrobiology" reflects the idea that the Earth hosts subtle microbial lifestyles, leading to group behaviour and deep physiological changes in processes, such as quorum sensing, biofilm formation or motility. The possibility exists that such complex microbial lifestyles also occur in the DB. The few studies performed to date on subsurface samples combining metagenomics, metatranscriptomics and culture-based approaches are starting to provide answers to this question.

In the PM subseafloor, transcripts of genes associated with reproduction, cell-cell interactions, particularly those involved in competition for precious resources, for adhesin or pili all indicate complex cell-cell interactions. In all samples analysed from the PM, gene transcripts for toxin/antitoxin production, and with antibiotic production and resistance, were detected (Pachiadaki et al. 2016), such as transcripts for beta-lactamases used by bacteria to defend their peptidoglycan-synthesizing machinery against the toxic effects of penicillin derivatives (Fig. 4.4). Given the signaling and communicating roles of antibiotics in natural environments, this strengthens the idea of interacting cells in the subsurface.

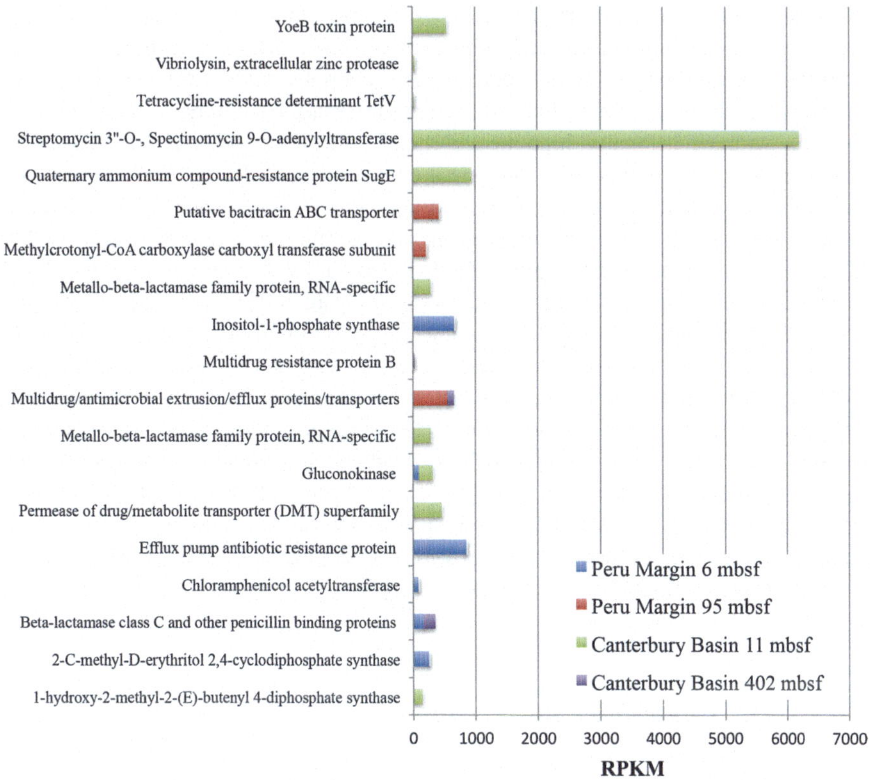

Fig. 4.4 Relative expression (presented as RPKM values) of genes associated with toxin and antimicrobial/antibiotic synthesis/resistance in the Canterbury Basin and Peru Margin subseafloors. Source: Reprinted with permission from Pachiadaki et al. (2016)

Beyond energy limitation and the subsequent selection of most adapted cells, there are many stressors for life in the deep subsurface, including pressure, temperatures, heavy metal toxicity, etc. Little is known about how microorganisms adapt to these co-occurring extremes, when the availabilities of electron acceptors/donors enables cells to deploy stress responses.

A study of two cultured hydrothermal vent Archaea reveals that growth depends on a delicate balance between stressors. As long as pH was mildly acidic or neutral and not more acidic than pH 5.5, *Thermococcus* and *Pyrococcus* could tolerate a wider range of pressures (up to 850 atm) and temperatures (up to 100 °C) than those likely to be found in the subsurface (Edgcomb et al. 2007). Pressure and elevated temperature changes are known to simultaneously induce a wide range of both heat shock and cold shock proteins, possibly as an attempt to repair the effects of changing pressure on membrane integrity/fluidity or macromolecule stability (Oger and Jebbar 2010). The extent to which these proteins play a role in the survival of DB communities is difficult to assess based on metatranscriptomics because depressurization and other physicochemical changes that occur in samples during recovery from the subsurface likely bias the expression of genes for these proteins.

Early studies of hydrothermal vents revealed the potential for high concentrations of heavy metals in vent fluids, hinting that the subsurface biosphere must also be able to cope with occasionally high heavy metal concentrations that are known to interfere with cell function by binding to vital macromolecules, and to control microbial community composition (Ravikumar et al. 2007). A study of three hydrothermal vent archaeal cultures revealed that the addition of sulphide improved the high toxicity of free metal cations of Zn, Co, and Cu by the formation of dissolved metal-sulphide complexes and precipitates (Lloyd et al. 2005). This suggests that the presence of sulphides in subsurface habitats may help microorganisms to cope with heavy metals concentrations. Metatranscriptome analyses of subsurface samples support the notion that heavy metals can stress subsurface microbial populations. The expression of general ion transport-related genes was detected in PM subsurface samples including genes affiliated with siderophore biosynthesis, magnesium and iron transporters, and chelatases that may be associated with detoxification activities (Pachiadaki et al. 2016). We also note that some microbes in anoxic sediments can respire through the coupled reduction of iron or sulphur and toxic metals, such as arsenic (Reyes et al. 2008).

4.6.1 The Canterbury Basin Subseafloor Case Study

Similar trends for complex cell-cell interactions and resistance to heavy metals and drugs were also highlighted in the CB subseafloor. For example, pilus assembly and flagellar motility are suggested by the occurrence of genes/mRNAs for pilus formation and flagellar assembly up to 345 mbsf, indicating that energy limitations do not completely prevent such processes from occurring in the CB. This is an important point considering that pore spaces offer support for cell adhesion (Gaboyer et al. 2015; Pachiadaki et al. 2016).

The ability of buried microorganisms to adapt their activity to changing environmental conditions is a crucial selective advantage. In the CB, numerous genes involved in environmental sensing and gene expression regulation were detected, including transcriptional regulators, sigma/anti-sigma factors, adenylate and diguanylate cyclases involved in the synthesis of signaling molecules, two-component systems and the GTP pyrophosphokinase involved in adaptation to starvation (stringent response) (Gaboyer et al. 2015). However, these DNA-data cannot be used as evidence for the in situ expression of adaptive genes but only for an in situ potential for such adaptations. Stress response strategies were more strongly suggested by mRNA-based studies (Pachiadaki et al. 2016). These included responses to pressure and low A_w (osmolytes accumulation and synthesis, regulation of membrane composition), and high temperatures (chaperonins, heat shock proteins) or oxidative stress (glutathione enzymes, thioredoxins). Pachiadaki et al. (2016) reported the apparent importance of drug/metal resistance genes (drug/metals membrane transporters, cobalt-zinc-cadmium proteins). Genes for sporulation were also detected in CB samples, suggesting that some microbes in the CB subseafloor can sporulate and wait for more favourable conditions. High-throughput physiological screening of fungal isolates in CB samples down to 1884 mbsf revealed a physiological shift from terrestrially-adapted to marine-adapted styles along the core for some specific species. Such an integrated approach clearly demonstrates that some Fungi may be able to adapt to subsurface conditions, while some may be dormant as spores (microbial zombies) and others may not survive. This corroborates the idea of a progressive selection of microbial groups adapted to long term starvation during the burial of surface communities in deeper sediments (Starnawski et al. 2017).

Polyketide synthases, nonribosomal peptide synthetases and terpene synthases are known as enzymes producing secondary metabolites, some with bioactive properties, involved in complex cell interactions. Genes coding for these enzymes were found in 167 of the 176 fungal isolates obtained from CB samples up to 765 mbsf, as well as their mRNAs up to 345 mbsf (Rédou et al. 2015; Fig. 4.5).

Detection of fungal genes/mRNAs for antibiotic production and prokaryotic genes/mRNAs for resistance suggest complex cell interactions, notably between Fungi and bacteria in the subsurface of the CB. This situation is not unique to the CB since similar microbial lifestyles were described in the subsurface of the well-studied PM (Orsi et al. 2013b). To confirm this hypothesis, fungal isolates from the CB were screened for their ability to synthesize antibacterial compounds (Navarri et al. 2016). A trend was observed whereby the proportion of fungal isolates producing antimicrobial compounds decreased with increasing depth and significantly correlated with bacterial diversity richness. From an ecological perspective, this suggests that, since shallow sediment depth layers are associated to higher microbial diversity, complex interactions between microorganisms are more pronounced in these regions than in deeper zones. This paves the way for more integrated studies implementing metabolomics to better understand interactions between microbial communities in the DB. The above-described results show that the deep subseafloor sediments are not free of sociomicrobiology, as genes and mRNAs for processes of microbial competition, adhesion or communication were detected. However, such interactions are only possible if there is sufficient energy and available nutrients. Indeed the main environmental constraints and limiting factor for life in the subseafloor remains long-term starvation. Here again, we note that the surface community composition progressively changes during burial, with community composition along the sediment core controlled by the selection of microorganisms adapted to oligotrophy, as experimentally confirmed (Starnawski et al. 2017). The lack of gene expression studies in the subseafloor still prevents the comparison of physiologies between depth and sites and scientists still have to wait for more investigations to be done in the future to clearly answer that point.

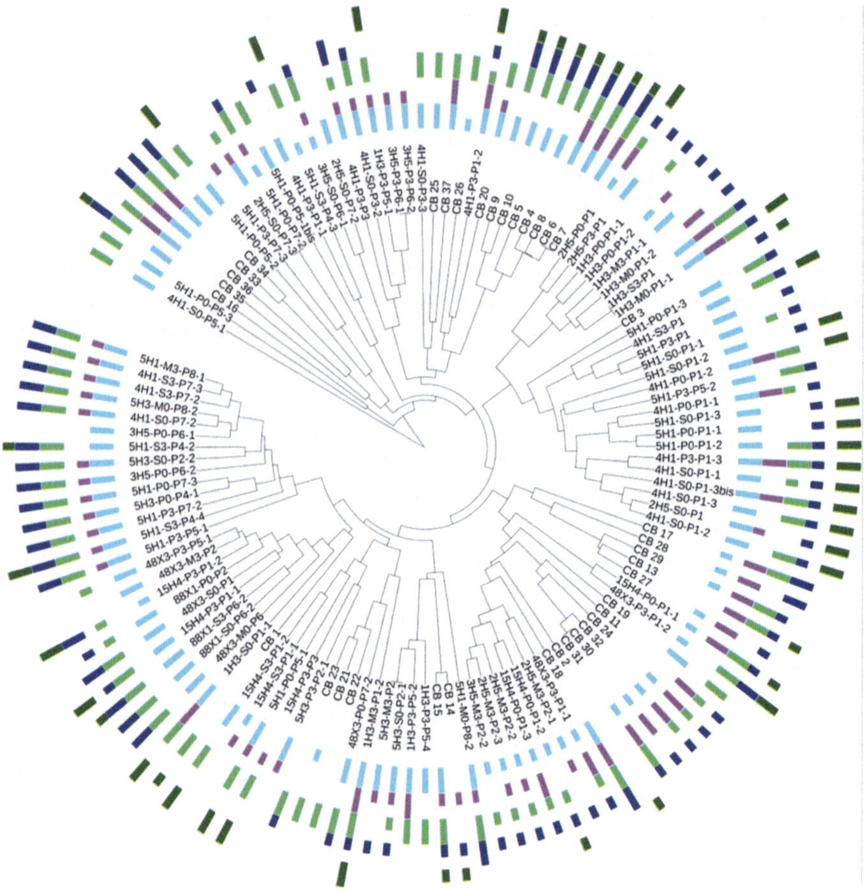

Fig. 4.5 Presence/absence of genes encoding type I and III Polyketide Synthase (PKSI, PKSIII), non ribosomal peptide synthetase (NRPSs), PKS-NRPS hybrids, and terpene synthase (TPS) in filamentous fungi of the CB subseafloor. PKSI, light blue; PKSIII, pink; NRPS, light green, PKS-NRPS hybrid, dark blue; TPS dark green. Occurrences are presented using an aligned multivalue bar chart (short bar, only one gene; long bar, several genes). Source: Adapted from Rédou et al. (2015)

References

Batzke A, Engelen B, Sass H et al (2007) Phylogenetic and physiological diversity of cultured deep-biosphere Bacteria from Equatorial Pacific Ocean and Peru Margin sediments. Geomicrobiol J 24:261–273

Biddle J, House CH, Brenchley JE (2005) Microbial stratification in deeply buried marine sediment reflects changes in sulfate/methane profiles. Geobiology 3(4):287–295

Biddle J, Lipp J, Lever M, Lloyd K, Sørensen K, Anderson R, Fredricks H, Elvert M, Kelly T, Schrag P (2006) Heterotrophic Archaea dominate sedimentary subsurface ecosystems off Peru. Proc Natl Acad Sci USA 103(10):3846–3851

Birrien J-L, Zeng X, Jebbar M et al (2011) *Pyrococcus yayanosii* sp. nov., an obligate piezophilic hyperthermophilic archaeon isolated from a deep-sea hydrothermal vent. Int J Syst Evol Microbiol 61:2827–2881

Blöchl E, Rachel R, Burggraf S et al (1997) *Pyrolobus fumarii*, gen. and sp. nov., represents a novel group of archaea, extending the upper *temperature* limit for life to 113 C. Extremophiles 1:14–21

Brochier-Armanet C, Boussau B, Gribaldo S, Forterre P (2008) Mesophilic Crenarchaeota: proposal for a third archaeal phylum, the Thaumarchaeota. Nature Reviews Microbiology 6 (3):245–252

Burdige DJ (2007) Preservation of organic matter in marine sediments: controls, mechanisms, and an imbalance in sediment organic carbon budgets? Chem Rev 107:467–485

Chivian D, Brodie EL, Alm EJ et al (2008) Environmental genomics reveals a single-species ecosystem deep within earth. Science 322:275–278

Ciobanu M-C, Burgaud G, Dufresne A et al (2014) Microorganisms persist at record depths in the subseafloor of the Canterbury Basin. ISME J 8:1370–1380

Cowen JP, Copson DA, Jolly J et al (2012) Advanced instrument system for real-time and time-series microbial geochemical sampling of the deep (basaltic) crustal biosphere. Deep-Sea Res I Oceanogr Res Pap 61:43–56

D'Hondt S, Rutherford S, Spivack A (2002) Metabolic activity of subsurface life in deep-sea sediments. Science 295:2067–2070

D'Hondt S et al (2004) Distributions of microbial activities in deep subseafloor sediments. Science 306:2216–2221

D'Hondt S, Spivack AJ, Pockalny R et al (2009) Subseafloor sedimentary life in the South Pacific Gyre. Proc Natl Acad Sci USA 106:11651–11656

DeLong E (2004) Microbial life breathes deep. Science 306:2198–2200

Deming J, Somers L, Straube W et al (1988) Isolation of an obligated barophilic bacterium and description of a new genus *Colwellia* Gen-nov. Syst Appl Microbiol 10:152–160

Durbin AM, Teske A (2011) Microbial diversity and stratification of South Pacific abyssal marine sediments. Environ Microbiol 13:3219–3234

Dziewonski AM, Anderson DL (1981) Preliminary reference Earth model. Phys Earth Planet Inter 25:297–356

Edgcomb VP, Molyneaux SJ, Böer S et al (2007) Survival and growth of two heterotrophic hydrothermal vent archaea, *Pyrococcus* strain GB-D and *Thermococcus fumicolans*, under low pH and high sulfide concentrations in combination with high temperature and pressure regimes. Extremophiles 11:329–342

Edgcomb VP, Beaudoin D, Gast R et al (2011) Marine subsurface eukaryotes: the fungal majority. Environ Microbiol 13:172–183

Engelhardt T, Sahlberg M, Cypionka H et al (2011) Induction of prophages from deep-subseafloor bacteria: phages in the deep-subseafloor. Environ Microbiol Rep 3:459–465

Engelhardt T, Sahlberg M, Cypionka H et al (2012) Biogeography of *Rhizobium radiobacter* and distribution of associated temperate phages in deep subseafloor sediments. ISME J 8:1503–1509

Engelhardt T, Kallmeyer J, Cypionka H et al (2014) High virus-to-cell ratios indicate ongoing production of viruses in deep subsurface sediments. ISME J 8(7):1503–1509

Foucher F, Westall F, Brandstätter F et al (2010) Testing the survival of microfossils in artificial martian sedimentary meteorites during entry into Earth's atmosphere: the STONE 6 experiment. Icarus 207:616–630

Foucher F, Ammar M-R, Westall F (2015) Revealing the biotic origin of silicified Precambrian carbonaceous microstructures using Raman spectroscopic mapping, a potential method for the detection of microfossils on Mars. J Raman Spectrosc 46:873–879

Friese A, Kallmeyer J, Kitte JA, et al the ICDP Lake Chalco Drilling Science Team and the ICDP Towuti Drilling Science Team (2017) A simple and inexpensive technique for assessing contamination during drilling operations: a simple and inexpensive technique. Limnol Oceanogr Methods 15:200–211

Fulthorpe C S, Hoyanagi K, Blum P et al (2011) Expedition 317 report. Proceedings of the IODP 317. Integrated Ocean Drilling Program, 2011. http://publications.iodp.org/proceedings/317/317title.htm
Gaboyer F, Burgaud G, Alain K (2015) Physiological and evolutionary potential of microorganisms from the Canterbury Basin subseafloor, a metagenomic approach. FEMS Microbiol Ecol 91:1–13
Harrison JP, Gheeraert N, Tsigelnitskiy D et al (2013) The limits for life under multiple extremes. Trends Microbiol 21:204–212
Hoehler TM, Jørgensen BB (2013) Microbial life under extreme energy limitation. Nat Rev Microbiol 11:83–94
Inagaki F, Takai K, Hirayama H et al (2003) Distribution and phylogenetic diversity of the subsurface microbial community in a Japanese epithermal gold mine. Extremophiles 7:307–317
Inagaki F, Hinrichs K-U, Kubo Y et al (2015) Exploring deep microbial life in coal-bearing sediment down to 2.5 km below the ocean floor. Science 349:420–424
Jones W, Leigh J, Mayer F et al (1983) *Methanococcus jannaschii* sp-nov, an extremely thermophilic methanogen from a submarine hydrothermal vent. Arch Microbiol 136:254–261
Jørgensen BB, Boetius A (2007) Feast and famine—microbial life in the deep-sea bed. Nat Rev Microbiol 5:770–781
Jørgensen BB, Marshall PG (2016) Slow microbial life in the seabed. Annu Rev Mar Sci 8:311–332
Kallmeyer J, Smith DC, Spivack AJ et al (2008) New cell extraction procedure applied to deep subsurface sediments. Limnol Oceanogr Methods 6:236–245
Kallmeyer J, Pockalny R, Adhikari RR et al (2012) Global distribution of microbial abundance and biomass in subseafloor sediment. Proc Natl Acad Sci USA 109:16213–16216
Lever MA, Alperin MJ, Engelen B et al (2006) Trends in basalt and sediment core contamination during IODP Expedition 301. Geomicrobiol J 23(7):517–530
Lipp JS, Morono Y, Inagaki F et al (2008) Significant contribution of Archaea to extant biomass in marine subsurface sediments. Nature 454:991–994
Liu C-H, Huang X, Xie T-N et al (2016) Exploration of cultivable fungal communities in deep coal-bearing sediments from ~1.3 to 2.5 km below the ocean floor. Environ Microbiol 2:803–818
Lloyd KG, Edgcomb VP, Molyneaux SJ et al (2005) Effects of dissolved sulfide, pH, and temperature on growth and survival of marine hyperthermophilic archaea. Appl Environ Microbiol 71:6383–6387
Lloyd KG, Schreiber L, Petersen DG et al (2013) Predominant archaea in marine sediments degrade detrital proteins. Nature 496:215–218
Lomstein BA, Langerhuus AT, D'Hondt S et al (2012) Endospore abundance, microbial growth and necromass turnover in deep subseafloor sediment. Nature 484:101–104
Morono Y, Terada T, Masui N et al (2009) Discriminative detection and enumeration of microbial life in marine subsurface sediments. ISME J 3:503–511
Morono Y, Terada T, Nishizawa M et al (2011) Carbon and nitrogen assimilation in deep subseafloor microbial cells. Proc Natl Acad Sci USA 108:18295–11830
Navarri M, Jégou C, Meslet-Cladière L et al (2016) Deep subseafloor fungi as an untapped reservoir of amphipathic antimicrobial compounds. Mar Drugs 14(3):50
Nunoura T, Soffientino B, Blazejak A et al (2009) Subseafloor microbial communities associated with rapid turbidite deposition in the Gulf of Mexico continental slope (IODP Expedition 308). FEMS Microbiol Ecol 69:410–424
Oger PM, Jebbar M (2010) The many ways of coping with pressure. Res Microbiol 161:799–809
Orcutt BN, Bach W, Becker K et al (2010) Colonization of subsurface microbial observatories deployed in young ocean crust. ISME J 5:692–703
Orcutt BN, Sylvan JB, Knab NJ et al (2011) Microbial ecology of the Dark Ocean above, at, and below the Seafloor. Microbiol Mol Biol Rev 75:361–422
Orsi WD, Biddle JF, Edgcomb V (2013a) Deep sequencing of subseafloor eukaryotic rRNA reveals active fungi across marine subsurface provinces. PLoS One 8:e56335

Orsi WD, Edgcomb VP, Christman GD et al (2013b) Gene expression in the deep biosphere. Nature 499:205–208

Pachiadaki MG, Rédou V, Beaudoin DJ et al (2016) Fungal and prokaryotic activities in the marine subsurface biosphere at Peru Margin and Canterbury Basin inferred from RNA-based analyses and microscopy. Front Microbiol 7:846

Parkes R, Cragg B, Wellsbury P (2000) Recent studies on bacterial populations and processes in subseafloor sediments: a review. Hydrogeol J 8:11–28

Parkes RJ, Cragg B, Roussel E et al (2014) A review of prokaryotic populations and processes in subseafloor sediments, including biosphere: geosphere interactions. Mar Geol 352:409–425

Piepenbrink KH, Sundberg EJ (2016) Motility and adhesion through type IV pili in Gram-positive bacteria. Biochem Soc Trans 44(6):1659–1666

Ravikumar S, Williams GP, Shanthy S et al (2007) Effect of heavy metals (Hg and Zn) on the growth and phosphate solubilising activity in halophilic phosphobacteria isolated from Manakudi mangrove. J Environ Biol 28:109–114

Rédou V, Ciobanu MC, Pachiadaki MG et al (2014) In-depth analyses of deep subsurface sediments using 454-pyrosequencing reveals a reservoir of buried fungal communities at record-breaking depths. FEMS Microbiol Ecol 90:908–921

Rédou V, Navarri M, Meslet-Cladière L et al (2015) Species richness and adaptation of marine fungi from deep-subseafloor sediments. Appl Environ Microbiol 81:3571–3583

Reyes C, Lloyd JR, Saltikov CW (2008) Geomicrobiology of iron and arsenic in anoxic sediments. In: Ahuja S (ed) Arsenic contamination of groundwater. Wiley, Hoboken, pp 123–146

Schrenk M, Huber JA, Edwards KJ (2010) Microbial provinces in the subseafloor. Ann Rev Mar Sci 2:279–304

Schulte M, Blake D, Hoehler T et al (2006) Serpentinization and its implications for life on the early Earth and Mars. Astrobiology 6:364–376

Schouten S, Hopmans EC, Damsté JSS (2013) The organic geochemistry of glycerol dialkyl glycerol tetraether lipids: a review. Organic geochemistry 54:19–61

Smith DC, Spivack A, Fisk MR et al (2000) Methods for quantifying potential microbial contamination during deep ocean coring. ODP Technical Note 28

Starnawski P, Bataillon T, Ettema TJG et al (2017) Microbial community assembly and evolution in subseafloor sediment. Proc Natl Acad Sci USA 114(11):2940–2945

Stevenson A, Cray J, Williams J et al (2015) Is there a common water-activity limit for the three domains of life? ISME J 9:1333–1351

Sturt HF, Summons RE, Smith K et al (2004) Intact polar membrane lipids in prokaryotes and sediments deciphered by high-performance liquid chromatography/electrospray ionization multistage mass spectrometry—new biomarkers for biogeochemistry and microbial ecology. Rapid Commun Mass Spectrom 18:617–628

Takai K, Nakamura K, Toki T et al (2008) Cell proliferation at 122°C and isotopically heavy CH_4 production by a hyperthermophilic methanogen under high-pressure cultivation. Proc Natl Acad Sci USA 105:10949–10954

Teske A, Sørensen KB (2008) Uncultured archaea in deep marine subsurface sediments: have we caught them all? ISME J 2:3–18

Turcotte DL, Schubert G (2002) Geodynamics, 2nd edn. Cambridge University Press, Cambridge

Vreeland RH, Rosenzweig WD, Powers DW (2000) Isolation of a 250 million-year-old halotolerant bacterium from a primary salt crystal. Nature 407(6806):897–900

Whitman WB, Coleman DC, Wiebe WJ (1998) Prokaryotes: the unseen majority. Proc Natl Acad Sci USA 95:6578–6583

Zhu R, Versteegh GJM, Hinrichs K-U (2016) Detection of microbial biomass in subseafloor sediment by pyrolysis–GC/MS. J Anal Appl Pyrolysis 118:175–180

Chapter 5
A Systematic Way to Life Detection: Combining Field, Lab and Space Research in Low Earth Orbit

Jean-Pierre de Vera and The Life Detection Group of BIOMEX/BIOSIGN

Abstract The characterization and detection of biosignatures is a challenging task, but one that needs to be solved before instruments are used for life detection missions on other planets and moons. A complex logistical effort is needed to support such exploration missions and a significant amount of preparation and investigation is required to prevent and eliminate pitfalls and errors, which may occur during the technical and scientific operations. Herein is suggested a systematic approach to prepare for "life-detection" missions, and an overview is given on the necessary steps in order to search for life *in-situ* on another planet or moon. Results obtained from research performed in the field, in the lab and in space will help to enhance our knowledge regarding the traces and signatures of life, and how to recognize life itself.

5.1 Introduction

Future space exploration missions with a primary focus to search for life elsewhere in the Solar System will have a number of challenging tasks. In addition to investigating the habitability of planetary objects for past and even present life, i.e. the ability of an environment to support the activity of at least one known organism (Cockell et al. 2016), it is important to find out which kind of life forms, or remnants thereof, should be looked for on potentially habitable worlds, subsequently identifying which of them could be detected by specific technologies in extraterrestrial environments. In this chapter, we aim to determine the necessary systematic preparations, investigations and logistic operations needed before starting a life detection

Life Detection Group of BIOMEX/BIOSIGN: Mickaël Baqué, Daniela Billi, Ute Böttger, Charles S. Cockell, Rosa de la Torre, Bernard H Foing, Franziska Hanke, Stefan Leuko, Jesús Martinez-Frías, Ralf Moeller, Karen Olsson-Francis, Silvano Onofri, Petra Rettberg, Susanne Schröder, Dirk Schulze-Makuch, Laura Selbmann, Dirk Wagner, Laura Zucconi

J.-P. de Vera
German Aerospace Center (DLR), Institute of Planetary Research, Management and Infrastructure, Astrobiological Laboratories, Berlin, Germany
e-mail: jean-pierre.devera@dlr.de

B. Cavalazzi, F. Westall (eds.), *Biosignatures for Astrobiology*, Advances in Astrobiology and Biogeophysics, https://doi.org/10.1007/978-3-319-96175-0_5

mission. In addition to a theoretical approach, accompanying experimental procedures are essential in this kind of research discipline. For example, ionizing and UV radiation can modify biosignatures in the first meters beneath the surface, and may modify their detectability by Raman or fluorescence methods (Dartnell et al. 2012; Dartnell and Patel 2014). Therefore it is necessary to perform preparatory ground-based studies in planetary simulation facilities and at spectroscopy laboratories (such as facilities which can offer Raman, Infrared (IR), Ultraviolet/Visible (UV/VIS), and fluorescence spectroscopy). Field astrobiology research can be conducted in terrestrial analogues to validate instruments and measurements of biomarkers in their *in situ* context (Foing et al. 2011), as well as laboratory analytical methods on retrieved samples (see Ehrenfreund et al. 2011; Martins et al. 2011; Orzechowska et al. 2011; Direito et al. 2011). Furthermore, some part of the work has to be done in Low Earth Orbit (LEO) or on the Moon (de Vera et al. 2012) where samples are exposed to space conditions, which can mimic Mars-like conditions or approach the conditions observed on the icy moons of Jupiter and Saturn. The reason for this is that no laboratory on Earth is able to simulate simultaneously all of the conditions that biomolecules might face in real space conditions. Knowledge about the stability of biomolecules and the response of microorganisms exposed to extreme space environments, or their degradation products, might significantly facilitate the detection of biosignatures elsewhere in our Solar System. These data have to be systematically analysed and arranged in a planetary biosignature database taking into account all protocols of environmental conditions, measurement regimes, obtained spectra, diagrams, tables and images. This database would provide, for the first time, an easily accessible repository of planetary-relevant biosignatures from all investigations performed by instruments on planetary analogue field sites, together with data obtained by simulated planetary condition experiments, including research performed directly on different space platforms. Combining these approaches will provide extensive insight on the expected effects when biomolecules are exposed to the environmental conditions present on potential habitable planets in our Solar System. The collected and arranged data could be made available to the international community and may serve as a biosignature backup database and guide for future life detection missions on other planets and moons.

5.2 Promising Locations in the Solar System for Life Detection: Mars, Europa and Enceladus

The definition and identification of unambiguous biosignatures is a necessary endeavour in order to develop a database for future space exploration programmes searching for life in potentially habitable extraterrestrial environments. The ever-growing databases of "fingerprints" of life contextualized to the search for life on Mars have been the focus of the astrobiological community for several decades (Banfield et al. 2001; Cady et al. 2003; Serrano et al. 2015; Westall et al. 2015), but more effort is still needed. For example, we still do not know much about the types of biosignatures, which could be expected on the icy moons of the gas giants within our Solar System.

5.2.1 *The Search for Habitable Worlds*

The search for life usually begins with the search for habitable extraterrestrial locations. Liquid water is essential for life as we know it and can only exist within the subsurface or, protected by a suitable atmosphere, on the surface of planetary bodies. The study of our Solar System suggests that subsurface environments containing liquid water are more widespread than surface hydrospheres. For example, there is evidence that below Mars' cryosphere resides a global aquifer (Head et al. 2003), and that small icy satellites such as Europa (Jupiter) or Enceladus (Saturn) harbour liquid oceans or lakes beneath their water-ice crust. These bodies, however, do not lie within the habitable zone with respect to its traditional definition, lack a dense and global magnetic field, and have no atmosphere. Nevertheless, such subsurface aqueous environments are protected against extreme radiation fluxes and would have access to nutrients and energy, for example, through hydrothermalism (e.g. hydrothermal vents) or geochemical reactions (e.g. serpentinisation) where there is direct contact between the lithosphere and the subsurface ocean (Hsu et al. 2015; McMahon et al. 2013).

5.2.2 *Promising Potentially Habitable Targets of Relevance to Life Detection Missions*

Excluding the possibility of probing the subsurface of the icy moons in the near future, the ability to detect biosignatures of putative biospheres is quite limited and unrealistic for exoplanets in the near future. However, in the case of Enceladus, the analysis of the South Polar Plume by the *Cassini* mission has led to the detection of organic compounds (Waite et al. 2009). For Europa, it has been suggested that the remote detection of biosignatures might be possible through spectroscopy of ocean material that reaches the surface through fractures or *via* briny diapirs. If biogenic volatiles can reach the surface and accumulate in the atmosphere, as has been speculated for methane on Mars, it is conceivable that such biosignatures could be detected on exoplanets in the long-term future (Rauer et al. 2011). In the presence of an outer ice shell, it will be difficult to detect biosignatures of putative biospheres below. However, *in situ* analysis of hydrocarbons in eruption plumes emanating from the surface of active satellites (e.g., Enceladus' south pole terrain; Waite et al. 2009) could be performed during close spacecraft encounters to determine the ratio of non-methane hydrocarbons to methane (which is extremely low for biological sources and higher for non-biological sources; McKay et al. 2014). These authors further suggested that several cold, oceanic, methanogenic subsurface ecosystems on Earth may serve as analogues for possible habitable environments on Enceladus and Europa. The remote detection of biosignatures in the harsh electron irradiation environment of Europa might be possible through spectroscopy of radio-chemically modified surface areas and liquids deriving from the subsurface ocean, which could be exchanged between the surface and the ocean through fractures or *via* briny diapirs (Hand et al. 2007). Brines could even form in the Martian Polar Regions and they may temporarily make liquid water available at very low temperatures, an essential ingredient to be a habitable environment for microorganisms (Fischer et al. 2016).

5.3 Finding Stable and Space-Resistant Biomolecules as Potential Biosignatures

Extraterrestrial niche habitats, which might harbour extant life forms or retain traces of ancient life (as suggested for Mars), should be detectable in future space exploration missions. Some of these niches can be characterized by the porosity of their lithic substrates or ice, or by the presence of liquid-water veins in ices, cracks and fissures in the outermost crust layers of the icy moons. These niche habitats could store possible ingredients or deposits of the geysers and plumes of the icy moons.

5.3.1 The Content of a Life Detection Database

A useful life detection database requires the collection of all relevant information concerning robust organic compounds that are unambiguously produced by (micro-) organisms. For this database, a collection of various species from a broad range of Earth environments ranging from deep sea to permafrost areas in the Polar Regions, together with an extraction of their related isolated cell components, such as secondary metabolites, is of principal interest. The results and protocols of their responses to different experiments, such as exposure to space conditions, is essential because space conditions with different radiation sources, e.g. ionizing radiation (X-rays, gamma-ray), UV-radiation, heavy ions from galactic radiation, vacuum and space weathering by micro-dust, etc., could significantly influence their identification by different detection methods. Since these conditions cannot be simultaneously simulated in even the best planetary simulation chambers on Earth, we need results from space experiments to know more about the stability of organisms and their constituent organic parts. This is especially important if we take into account evolutionary processes on planets. For example, on Mars, the availability of liquid water and the thickness of the atmosphere have decreased during its history. In the case of liquid-water oceans beneath the ice crusts of the outer planets satellites, water charged with unknown organic material is emitted by plumes into the space environment (Waite et al. 2009; Dong et al. 2011; Postberg et al. 2011; Iess et al. 2014). Traces of extinct or extant life forms on the surface of Mars might have been altered due to a changing radiation environment (decaying atmosphere and collapse of the early geomagnetic field). Irradiation and vacuum conditions are expected to affect the stability of organics at the surface and certainly any putative life forms ejected into space, for instance through the venting plumes discovered on Enceladus (Waite et al. 2009) and possibly Europa (Roth et al. 2014) that are fed by shallow liquid reservoirs. Ideal targets for the classification of stable biosignatures that could serve as usable "fingerprints" are both organisms and organics, which might be stable even in a fossilized state or under intense radiation environments, partly due to physical and chemical interactions in the rock/soil/salt/ice environment. Currently, the ideal method to study this and to develop a sophisticated biosignature database including data with relevance to Mars and the icy moons in the outer Solar System is to use planetary simulation facilities and exposure platforms on the International Space Station (ISS) such as EXPOSE (Europe, ESA) and Tanpopo (Japan, JAXA) and the up-coming micro-, nano-, and cube satellite generation.

5.3.2 *Low Earth Orbit Missions Supporting Exploration Missions to Search for Life*

The on-going BIOMEX project (BIOlogy and Mars EXperiment; de Vera et al. 2012) is an example of combining planetary analogue fieldwork, the use of simulation facilities and the EXPOSE platform. Its follow up is foreseen in the BIOSIGN project (BIOSIGnatures and habitable Niches—ILSRA-2014-0019) as a new experiment in preparation with ESA. Like BIOMEX, the BIOSIGN project will be an international and interdisciplinary experiment in LEO where the four main scientific disciplines of Polar Research, Deep Sea/Ocean Research, Deep Biosphere Research and Space Research, as well as their exploration programs, will cooperate to work on the establishment of a database. This would bring together all data relevant to spectroscopy and life detection, particularly with regard to minerals, salts, ices and organic biosignatures relevant to Mars and the icy satellites of the outer Solar System.

5.4 Relevance to Biosignature, Bio-Trace and Bio-Fingerprint Research

A number of experiments were carried out on selected pigments and membrane components in preparation for the BIOMEX experiment. Due to the fact that biosignatures have to be detectable in the context of background material (environmental parameters, mineral matrix etc.), the first steps were to characterize Mars-analogue mineral mixtures and isolated biological samples (e.g. cyanobacteria, archaea, fungi and lichens; Baqué et al. 2016; Böttger et al. 2012, 2013a, b; Serrano et al. 2014, 2015) and minerals intermixed with bio-relevant organic material. At the time of the writing, new papers are in preparation to show the differences between irradiated and non-irradiated samples which were done during the Environmental Verification Tests (EVTs), the Scientific Verification Tests (SVTs) of BIOMEX and the real space experiments on EXPOSE-R2 aboard the ISS. Moreover, the extracted pigments, membrane and cell wall components have been characterized and are available in the DLR internal preliminary Raman database. Other work is still on-going using UV/VIS-spectrometry and IR-spectroscopy. However, much more work is needed in this context, in particular in reference to potential microfossils of extinct Martian life and in regard to biosignatures of potential extant life present in the subsurface oceans of the icy moons and detectable through plume activity. Although, in this chapter, we mainly consider physical and biochemical biosignatures, it is also important to emphasize that, for detecting life, bio-morphological structures have to be taken into account as well and should not be excluded from the biosignature data base. This is especially relevant if different microscope techniques will be used in future space exploration missions.

5.5 Experimental Approach

Previous experiments paid some attention to the high level of structural liability and stability of most of the biological components, especially on the protection of surface coats, membranes, proteins or secondary metabolite deposits, and even on biomolecules of the cell interior, such as DNA and RNA. These organic biosignatures were studied with reference to Mars and space conditions (Baqué et al. 2016; Onofri et al. 2012, 2015; Pacelli et al. 2016, 2017). However, little is known for samples that are relevant to the icy moons in our Solar System (Gleeson et al. 2012). Possible Mars biosignatures are better investigated, but even here knowledge is lacking. The intention of future experiments, e.g. BIOSIGN, is therefore to develop a further understanding and to analyse the effects of the space environment. A number of samples embedded in salty material or sediments of the deep sea will be used, in addition to the well-known Mars analogue mixture with characteristics from the early and late Martian evolutionary stages. The Museum für Naturkunde in Berlin has developed Mars-analogue mixtures based on insights gained from remote spectroscopy (Bibring et al. 2005, 2006; Chevrier and Mathé 2007; Poulet et al. 2005; Schirmack et al. 2015) within the framework of the founded German Helmholtz-Alliance "Planetary Evolution and Life" (Böttger et al. 2012). For example, the detection of silica in the plumes of Enceladus during the Cassini mission can only originate from the rocky core of the icy moon (Hsu et al. 2015). Thus, it is possible that the detected organic components could be a hint of prebiotic molecules or life forms from Enceladus' deep-sea regions, which might reach the surface of the ice crust of the moon through convective transport within its ocean. Therefore, it would be very interesting to analyse what happens to terrestrial analogue material from the deep sea, including life forms from these deep ocean areas. This culminates with the question: what could happen if life-related sample material is suddenly exposed to vacuum conditions (as might happen in the plumes of the icy moons), solar/space radiation and deep freezing conditions of simulated open space? It would be desirable to test this under space conditions on the ISS, on a new generation of satellites, or on the Moon. The effects of space, Enceladus- and Europa analogues, and Mars-like environments in their present and past epochs, particularly on minerals, must be tested in parallel with the biological investigations. This kind of programme of investigation, wherein the evolutionary aspects of planets and moons of interest are integrated, could clarify potential changes and alterations of biomolecules identified as putative biosignatures over short or long-term time scales. A welcome side-effect of these space and time experiments is that the investigated samples would also serve for viability and space resistance analysis and would result in a replicate space experiment producing valuable data for the question of the probability of habitable niches in the Solar System. The experiment would also determine the likelihood of lithopanspermia occurring today and/or in the past. Initial experiments had previously been carried out on FOTON/BIOPAN some years ago (Demets et al. 2005; de la Torre et al. 2010; Raggio et al. 2011), the STONE experiment (Cockell et al. 2007; Foucher et al. 2010), on the EXPOSE-E (Rabbow et al. 2012; Onofri et al. 2012, 2015), the EXPOSE-R mission (Rabbow et al. 2015), the recent EXPOSE-R2 mission (de Vera et al. 2012), and the Exposure Facility (EF) of the Japanese Experiment Module (JEM; Kawaguchi et al. 2016) on the ISS.

5.6 Analysis Strategy and Biosignature Database Concept

A necessary first investigative step is the thorough study of planetary analogue field sites. Here, information regarding potential biomolecules, which could serve as characterised biosignatures, can be obtained. Due to the fact that Mars and the icy satellites are classified as desert icy worlds, a selection of specifically cryophilic, piezophilic, halophilic, radiation-, desiccation and starvation-resistant microorganisms is needed. Through fieldwork undertaken in caves, in deserts, in the deep sea or in polar planetary analogue field sites, the collection of samples with planetary relevance, as well as testing of the instrumentation for their detection, can be performed (Fig. 5.1). This planetary analogue fieldwork also includes searching for remnants of life forms (fossils) and for biogenic molecules (biomarkers). The second step of the investigation aims to test the robustness of the collected organisms, fossils and biomolecules and to study life detection methods under simulated planetary conditions in specific planetary simulation laboratories. This experimental work will give insights on life processes, survival, resistance and stability of the samples, as well as on result quality, possible pitfalls and artefacts, which might

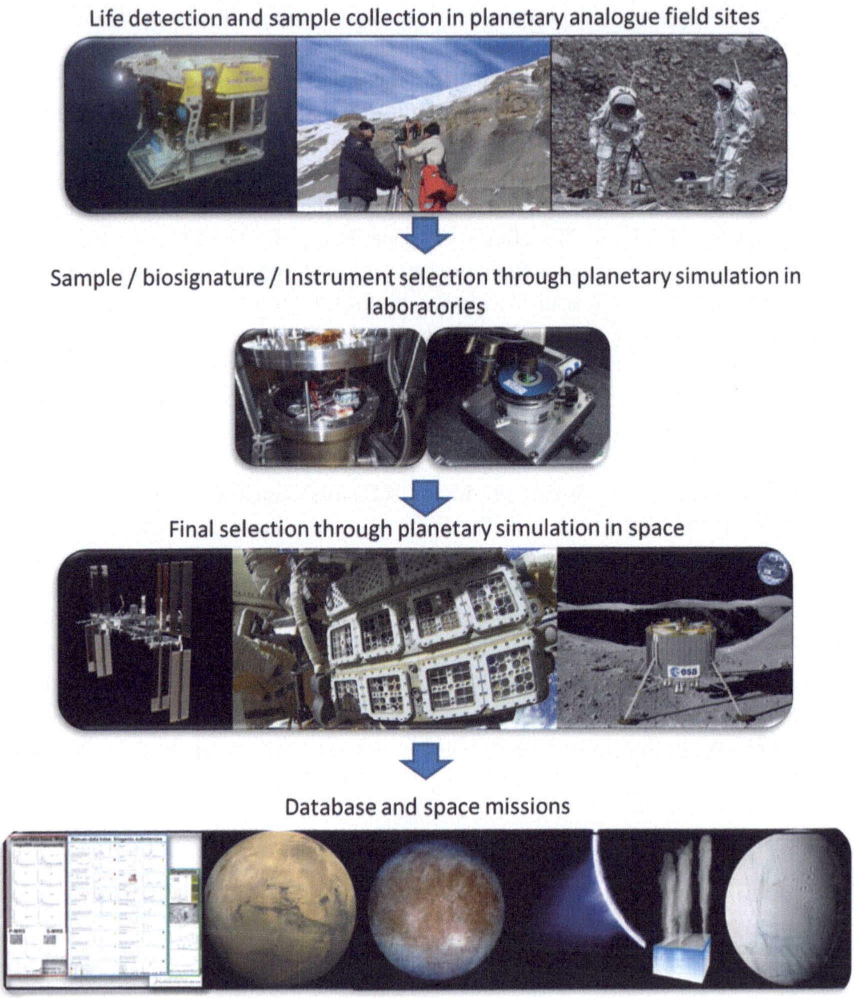

Fig. 5.1 Workflow of the biosignature analysis strategy, showing the sequence of analyses from planetary analogue field sites to space missions with a principal focus to detect life on Mars, Europa or Enceladus. Source: Adapted from DLR, GEOMAR (ROBEX), NASA, ESA, Roscosmos

occur during the measurements and detection operations. After successful selection of promising detectable and stable samples, as well as selection of instruments capable of reliable life detection, the third step is to conduct further work directly in space. This may use existing or future space exposure platforms in LEO, on space stations orbiting the Moon, or directly on Lunar landers at the surface of the Moon (Fig. 5.1). Although these space experiments are costly and complex, they are indispensable since their outcome cannot be identically imitated in any planetary simulation laboratory on Earth. Measurements performed in space can provide insights on the effects of combined space conditions, e.g. of Mars and the icy moons, on life detection instruments and potential life forms, remnants of life or biomolecules. Further valuable results can be obtained if the bio-relevant samples are embedded in planetary analogue minerals, salts or ice and exposed to atmospheric conditions of the planet or moon around which a future exploration project is focused.

Finally, the results which have relevance to Mars and the icy ocean worlds need to be systematically implemented into a mission taking into account all protocols of environmental conditions, measurement regimes, obtained spectra, diagrams, tables and pictures (Fig. 5.2). This will ensure a systematic investigation and support for future space missions whose main goal is to search for life in the Solar System.

The systematic organization of data, which is needed to create a well-developed and comprehensive biosignature database relevant for different life detection missions, can broadly be described in the following (see also Fig. 5.2):

(i) the type of mission has to be clear. While in principle a true life detection mission has to include close-up observations and analyses, and not only remote sensing, at least for the immediate future, there are three different categories of targets suitable for life detection missions, which include:

   1. exoplanets (remote sensing, e.g. planets orbiting dwarf stars),
   2. terrestrial planets (like Mercury, Venus, Earth, Mars; remote sensing and *in situ),*
   3. icy moons and dwarf planets (like Europa, Enceladus, Titan, Ceres, Pluto, remote sensing and *in situ*);

(ii) the detection method has to be chosen for the three categories (e.g. spectroscopy: fluorescence, IR-, UV/VIS, Raman, gas chromatograph-mass spectrometer (e.g., MOMA on ExoMars2020) or LD-MS etc.; microscopy: raster electron microscopy (REM), confocal laser scanning microscopy (CLSM); antibody detection etc.) including the maximum possible detail about instrumental operation and the lower limits of detection, controls, and calibration routines;

(iii) the planetary environmental conditions, which could have affected the tested samples and detection instruments during measurements, and therefore their influence on the resulting signatures of life, must be listed. Then, the obtained data have to be arranged according to the tested environmental parameters (environment can be specified by planetary simulation experiments from ground facilities and in space approaching the conditions of the planets/moons of interest).

This well-developed database could be used for the interpretation of observations in future space missions to the planets and moons in the Solar System and beyond.

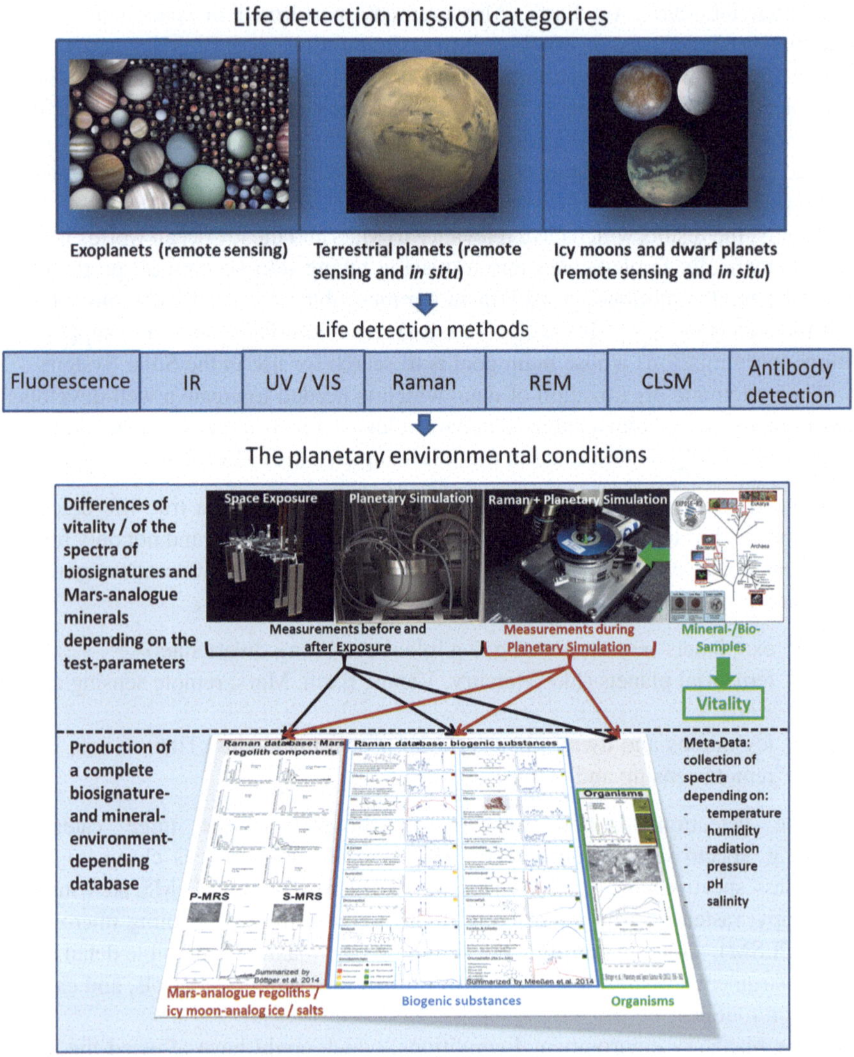

Fig. 5.2 Concept for the organisation of a biosignature database considering the methods planned to be used in future space exploration missions with the goal of finding life in space. The data will take into account all metadata of environmental conditions before, during and after the planetary simulation experiments. Source: Adapted from DLR, NASA, and ESA

References

Banfield JF, Moreau JW, Chan CS et al (2001) Mineralogical biosignatures and the search for life on Mars. Astrobiology 1:447–465

Baqué M, Verseux C, Böttger U et al (2016) Preservation of biomarkers from cyanobacteria mixed with Mars-like regolith under simulated Martian atmosphere and UV flux. Orig Life Evol Biosph 46:289–310

Bibring J-P, Langevin Y, Gendrin A et al (2005) Mars surface diversity as revealed by the OMEGA/Mars express observations. Science 307:1576–1581
Bibring J-P, Squyres SW, Arvidson RE (2006) Merging views on Mars. Science 313:1899–1901
Böttger U, de Vera J-P, Fritz J et al (2012) Optimizing the detection of carotene in cyanobacteria in a Martian regolith analogue with a Raman spectrometer for the ExoMars mission. Planet Space Sci 60:356–362
Böttger U, de la Torre R, Frias J-M et al (2013a) Raman spectroscopic analysis of the oxalate producing extremophile Circinaria Gyrosa. Int J Astrobiol 13:19–27
Böttger U, de Vera J-P, Hermelink A et al (2013b) Application of Raman spectroscopy, as in situ technology for the search for life. In: de Vera JP, Seckbach J (eds) Cellular origins, life in extreme habitats and astrobiology 28: habitability of other planets and satellites. Springer, Berlin, pp 333–345
Cady SL, Farmer JD, Grotzinger JP et al (2003) Morphological biosignatures and the search for life on Mars. Astrobiology 3:351–368
Chevrier V, Mathé PE (2007) Mineralogy and evolution of the surface of Mars: a review. Planet Space Sci 55:289–314
Cockell CS, Brack A, Wynn-Williams DD et al (2007) Interplanetary transfer of photosynthesis: an experimental demonstration of a selective dispersal filter in planetary island biogeography. Astrobiology 7:1–9
Cockell CS, Bush T, Bryce C et al (2016) Habitability: a review. Astrobiology 16:89–117
Dartnell LR, Patel MR (2014) Degradation of microbial fluorescence biosignatures by solar ultraviolet radiation on Mars. Int J Astrobiol 13:112–123
Dartnell LR, Page K, Jorge-Villar SE et al (2012) Destruction of Raman biosignatures by ionising radiation and the implications for life detection on Mars. Anal Bioanal Chem 403:131–144
de la Torre R, Sancho L, Horneck G et al (2010) Survival of lichens and bacteria exposed to outer space conditions-results of the Lithopanspermia experiments. Icarus 208:735–748
de Vera J-P, Böttger U, de la Torre R et al (2012) Supporting Mars exploration: BIOMEX in Low Earth Orbit and further astrobiological studies on the Moon using Raman and PanCam technology. Planet Space Sci 74:103–110
Demets R, Schulte W, Baglioni P (2005) The past, present and future Biopan. Adv Space Res 36:311–316
Direito SOL, Ehrenfreund P, Marees A et al (2011) A wide variety of putative extremophiles and large beta-diversity at the Mars Desert Research Station (Utah). Int J Astrobiol 10:191–207
Dong Y, Hill TW, Teolis BD et al (2011) The water vapor plumes of Enceladus. J Geophys Res 116:A10204
Ehrenfreund P, Röling WFM, Thiel C et al (2011) Astrobiolgy and habitability studies in preparation for future Mars missions: trends from investigating minerals, organics and biota. Int J Astrobiol 10:239–253
Fischer E, Martínez GM, Renno NO (2016) Formation and persistence of brine on Mars: experimental simulations throughout the diurnal cycle at the phoenix landing site. Astrobiology 16:937–948
Foing BH, Stoker C, Zavaleta J et al (2011) Field astrobiology research in Moon-Mars analogue environments: instruments and methods. Int J Astrobiol 10:141–160
Foucher F, Westall F, Brandstätter F et al (2010) Testing the survival of microfossils in artificial martian sedimentary meteorites during entry into Earth's atmosphere: the STONE 6 experiment. Icarus 207:616–630
Gleeson DF, Pappalardo RT, Anderson MS et al (2012) Biosignature detection at an Arctic analog to Europa. Astrobiology 12:135–150
Hand KP, Carlson RW, Chyba CF (2007) Energy, chemical disequilibrium, and geological constraints on Europa. Astrobiology 7:1006–1022
Head JW, Wilson L, Mitchell KL (2003) Generation of recent massive water floods at Cerberus Fossae, Mars by dike emplacement, cryospheric cracking, and confined aquifer groundwater release. Geophys Res Lett 30:1577
Hsu H-W, Postberg F, Sekine Y et al (2015) Ongoing hydrothermal activities within Enceladus. Nature 519:207–210

Iess L, Stevenson DJ, Parisi M et al (2014) The gravity field and interior structure of enceladus. Science 344:78–80

Kawaguchi Y, Yokobori S, Hashimoto H et al (2016) Investigation of the interplanetary transfer of microbes in the tanpopo mission at the exposed facility of the international space station. Astrobiology 16:363–376

Martins Z, Sephton MA, Foing BH et al (2011) Extraction of amino acids from soils close to the Mars Desert Research Station (MDRS), Utah. Int J Astrobiol 10:231–238

McKay CP, Anbar AD, Porco C et al (2014) Follow the plume: the habitability of enceladus. Astrobiology 14:352–355

McMahon S, O'Malley-James J, Parnell J (2013) Circumstellar habitable zones for deep terrestrial biospheres. Planet Space Sci 85:312–318

Onofri S, de la Torre R, de Vera J-P et al (2012) Survival of rock-colonizing organisms after 1.5 years in outer space. Astrobiology 12:508–516

Onofri S, de Vera J-P, Zucconi L et al (2015) Survival of Antarctic cryptoendolithic fungi in simulated martian conditions on board the International Space Station. Astrobiology 15:1052–1059

Orzechowska GE, Kidd RD, Foing BH et al (2011) Analysis of Mars analogue soil samples using solid-phase microextraction, organic solvent extraction and gas chromatography/mass spectrometry. Int J Astrobiol 10:209–219

Pacelli C, Selbmann L, Zucconi L et al (2016) BIOMEX experiment: ultrastructural alterations, molecular damage and survival of the fungus *Cryomyces antarcticus* after the experiment verification tests. Orig Life Evol Biosph 47:187–202

Pacelli C, Selbmann L, Zucconi L et al (2017) Survival, DNA integrity, and ultrastructural damage in Antarctic cryptoendolithic eukaryotic microorganisms exposed to ionizing radiation. Astrobiology 17:126–135

Postberg F, Schmidt J, Hillier J et al (2011) A salt-water reservoir as the source of a compositionally stratified plume on Enceladus. Nature 474:620–622

Poulet F, Bibring JP, Mustard JF et al (2005) Phyllosilicates on Mars and implications for early Martian climate. Nature 438:623–627

Rabbow E, Rettberg P, Barczyk S et al (2012) EXPOSE-E: an ESA astrobiology mission 1.5 years in space. Astrobiology 12:374–386

Rabbow E, Rettberg P, Barczyk S et al (2015) The astrobiological mission EXPOSE-R on board of the International Space Station. Int J Astrobiol 14:3–16

Raggio J, Pintado A, Ascaso C et al (2011) Whole lichen Thalli survive exposure to space conditions: results of lithopanspermia experiment with *Aspicilia fruticulosa*. Astrobiology 11:281–292

Rauer H, Gebauer S, Pv P et al (2011) Potential biosignatures in super-Earth atmospheres – I Spectral appearance of super-Earths around M dwarfs. Astron Astrophys 529:A8

Roth L, Saur J, Retherford KD et al (2014) Transient water vapor at Europa's south pole. Science 343:171–174

Schirmack J, Alawi M, Wagner D (2015) Influence of Martian regolith analogs on the activity and growth of methanogenic archaea, with special regard to long-term desiccation. Front Microbiol 6:210

Serrano P, Hermelink A, Böttger U et al (2014) Biosignature detection of methanogenic archaea from Siberian permafrost using confocal Raman spectroscopy. Planet Space Sci 98:191–197

Serrano P, Hermelink A, Lasch P et al (2015) Confocal Raman microspectroscopy reveals a convergence of the chemical composition in methanogenic archaea from a Siberian permafrost-affected soil. FEMS Microbiol Ecol 91:fiv126

Waite JH Jr, Lewis WS, Magee BA et al (2009) Liquid water on Enceladus from observations of ammonia and 40Ar in the plume. Nature 460:487–490

Westall F, Foucher F, Bost N et al (2015) Biosignatures on Mars: what, where, and how? Implications for the search for Martian life. Astrobiology 15:998–1029

Chapter 6
Mineralogical Identification of Traces of Life

Karim Benzerara, Sylvain Bernard, and Jennyfer Miot

Abstract Many organisms impact mineral nucleation and growth. This results in the formation of biominerals with chemical, structural and textural properties providing clues to their biogenicity. However, ageing modifies these properties to some extent. Moreover, some abiotic processes form minerals with similar properties. Therefore, decoding traces of life in minerals requires caution, and one prerequisite is a reliable estimation of the geochemical conditions under which a biomineral formed. Here we discuss several examples of biominerals which illustrate these different ideas.

6.1 Introduction

Minerals are defined as naturally occurring solids, usually crystalline, with unique chemical and crystallographic properties. Presently, more than 5000 species of minerals have been approved by the International Mineralogical Association (Nickel and Grice 1998). A broad diversity of organisms can form minerals (e.g., Weiner and Dove 2003). Hereafter, we use the term "biominerals" for all minerals that are formed by life. Since their morphology, atomic structure, texture and/or chemical composition are sensitive to their formational conditions, biominerals may retain some information about their biogenicity (Miot et al. 2016). This is why minerals may serve as valuable targets in the search for ancient traces of life in an astrobiological context (Banfield et al. 2001; Benzerara and Menguy 2009; Benzerara and Miot 2011).

The recent, present and near-future missions to Mars have provided and will continue to provide a large amount of mineralogical information about the surface of the Red Planet (e.g., Maurice et al. 2016). It is therefore crucial to assess what such information may tell us about potential past Martian life. Some studies have

K. Benzerara (✉) · S. Bernard · J. Miot
Muséum National d'Histoire Naturelle, Sorbonne Université, UMR CNRS 7590, IRD, Institut de Minéralogie, de Physique des Matériaux et de Cosmochimie, Paris, France
e-mail: karim.benzerara@upmc.fr; sylvain.bernard@mnhn.fr; jennyfer.miot@mnhn.fr

B. Cavalazzi, F. Westall (eds.), *Biosignatures for Astrobiology*, Advances in Astrobiology and Biogeophysics, https://doi.org/10.1007/978-3-319-96175-0_6

suggested that traces of life may be left in minerals by microbially induced dissolution (e.g., McLoughlin and Grosch 2015), however, we will instead focus here on traces of life left by biomineralization.

Hazen et al. (2008) suggested that many known mineral species on Earth appeared following major environmental changes due to life, such as the Great Oxidation Event. Moreover, Hazen and Ausubel (2016) reported that more than half of all mineral species are known from five or fewer localities and proposed that the rarity of some species may be attributed to the diversification of chemical conditions associated, in part, with biology. Hence, this rarity may serve as a biosignature. Yet, the relationship between the involvement of life and the formation of minerals with given properties is not necessarily a bijective function, i.e., the fact that organisms form minerals with particular features does not mean that such minerals can only be formed by life.

It is clear that life can catalyse many reactions and therefore generate conditions prone to the formation of minerals that would not form otherwise in a given environment. But the same physical and chemical conditions may also be encountered in some other environments devoid of life. Therefore, there is likely no mineral phase that is formed only by life. In an apparent contradiction to this statement, hazenite, a struvite-like K-, Na- and Mg-containing phosphate phase, has been described as a purely biogenic mineral phase (Yang et al. 2011). However, although the single reported occurrence of this phase is indeed in a biogenic context, there is little evidence to suggest that it could not also form under abiotic conditions.

A biosignature is classically defined as a morphological, chemical or crystallographical feature that is unequivocally related to life, irrespective of the chemical and physical conditions under which it formed. Here, we assume that any property of a mineral may be produced under some appropriate abiotic conditions. Nevertheless, in a given environment, life may change, at least locally, the chemical and/or physical conditions and induce the formation of minerals with properties that are different from minerals formed abiotically in the same environment. Life may therefore leave some traces in minerals but it is necessary to understand the environmental context to identify them.

In this chapter, we will review the processes involved in mineral formation and mention how they are impacted by life. Then, we will detail several examples of mineralogical features affected by life and describe how they degrade upon ageing, i.e. all post-mortem structural and geochemical alteration processes, including those occurring during diagenesis, metamorphism or irradiation at a planetary surface. Finally, we will explain how some mineralogical features found in the geological record, and usually attributed to biological activity, can be produced by abiotic processes. Overall, this will highlight the potential pitfalls encountered when searching for past mineralogical traces of life in rocks.

We will refer to many examples of microbial mineralization, which is likely the most relevant for astrobiology. Moreover, we advocate that, regarding ageing processes, there is much to learn from the numerous studies focusing on biominerals produced by non-microbial organisms since the mechanisms affecting their structure and chemical compositions are most likely very similar to the processes degrading microbial biominerals.

6.2 How Does Life Influence the Formation of Minerals?

In order to understand how life leaves identifiable traces in a mineral, we can consider the different stages of mineral formation.

Mineral formation starts by nucleation i.e., a stochastic process forming a nucleus with a critical size large enough to overcome spontaneous destabilization. This stage is followed by mineral growth. Both stages affect the morphology, structure and texture of the resulting minerals. Life can impact nucleation and/or growth by two major mechanisms: (1) by modifying, at least locally, the activities of the dissolved chemical species which compose mineral phases, and (2) by forming organic molecules which interact with growing nuclei/crystals.

Many mineralogical features have been successfully interpreted by classical models of nucleation and growth (e.g., De Yoreo and Vekilov 2003). In these models, chemical species or molecules composing the mineral phase are added one by one to the nucleus and contribute to the growth of crystals layer by layer. However, it has been increasingly reported that crystals can also form and grow by the aggregation of nanoparticles, sometimes involving amorphous precursor phases or sometimes attaching to each other in a crystallographically oriented manner (e.g., Cölfen and Antonietti 2008). These non-classical pathways to crystallization satisfactorily explain the formation of minerals with unusual morphologies, structures and textures, previously considered as enigmatic (De Yoreo et al. 2015).

The mechanisms governing crystallization by particle attachment are not yet completely understood (De Yoreo et al. 2015) but these non-classical pathways are clearly involved in the formation of some biominerals (e.g., Oaki and Imai 2005; Benzerara et al. 2011). However, they also occur in abiotic systems and, thus, "strange" mineral morphologies produced through these processes are not specific to biology (e.g., De Yoreo et al. 2015).

6.2.1 *Impact of Life on Solution Saturation*

The precipitation of a mineral phase in a solution occurs only if this solution is supersaturated with this phase. This is true in both biotic and abiotic situations. However, supersaturation of a solution is not sufficient for precipitation to occur as mineral formation may nonetheless be very slow (e.g., Shiraki and Brantley 1995). By modifying the chemical composition of a solution, at least locally, a metabolizing organism can drive the saturation of a solution from below to above the equilibrium between the mineral phase and the solution, making precipitation possible. In many cases, organisms thrive in solutions, which are already globally saturated/

supersaturated with some mineral phases. In these cases, the increase of the solution saturation by biological activity speeds up the kinetics of the reaction and makes precipitation occur more effectively (e.g., Burton and Walter 1987). Moreover, the increase of saturation by biological activity can also modify nucleation/growth mechanisms and impact the morphology and texture of crystals (e.g., De Yoreo et al. 2015).

How microorganisms impact the precipitation of carbonate microbialites is one particularly interesting illustration of these fundamental considerations (e.g., Burne and Moore 1987). These rocks form partly by in situ precipitation of carbonate minerals in association with benthic microbial communities in marine and freshwater environments (Fig. 6.1).

Large chemical gradients can occur within the biofilms forming modern microbialites with, for example, transitions from oxic to anoxic conditions, allowing the development of diverse microbial communities. Studies have classically stressed on the role of oxygenic photosynthesis and cyanobacteria as drivers of chemical changes, in particular an increase of pH (e.g., Ludwig et al. 2005). Indeed, oxygenic photosynthesis raises the local saturation of the solution with respect to carbonate phases and induces their precipitation (Aloisi 2008), however, the microbial diversity of microbialite-forming biofilms generally includes many other microorganisms apart from cyanobacteria, which can also contribute to carbonate precipitation by metabolically modifying the chemical composition of their extracellular environment (e.g., Saghaï et al. 2015, 2016). For example, sulphate-reducing bacteria have often been mentioned as important contributors to carbonatogenesis in modern marine microbialites (Braissant et al. 2007). Dupraz et al. (2009) pointed out that, in a microbial mat, several metabolisms tend to increase carbonate alkalinity, hence promoting the precipitation of carbonate phases, whereas other metabolisms decrease carbonate alkalinity, hence favouring carbonate dissolution. They concluded that the precipitation of carbonates depends, overall, on the balance between these two groups of metabolisms. They also mentioned that extrinsic, abiotic processes such as degassing or evaporation contribute to the promotion of carbonate precipitation.

This has several implications when interpreting the origin of these so-called "microbial carbonates". Firstly, microbialites as a whole cannot be considered as entirely biotic. Some variations in texture and morphology can be observed in carbonate crystals forming microbialites. For example, Benzerara et al. (2010) described the co-existence of aragonite nanoglobules and fibers as well as clusters of crystallographically oriented nanoglobules, i.e. mesocrystals. Some properties, e.g., the micritic texture of carbonates, may be related to a higher biological influence but there are still uncertainties regarding this. Secondly, in this framework, one cannot infer that one single metabolism induced the formation of carbonates; in contrast, the formation of microbialites likely results from the activity of a metabolically complex community. To further complicate this view, Dupraz et al. (2009)'s model is most likely a little simplistic. It only considers energy metabolisms but secondary metabolisms, i.e. metabolisms that are not absolutely required for the survival of organisms, also impact the chemical composition of the extracellular

Fig. 6.1 Modern carbonate microbialites from Lake Alchichica (Mexico). Hand-sized samples (top) show the various colours of microbial populations associated with these rocks. Microbialites are covered by biofilms when located beneath the water surface (bottom) but surface biofilms disappear once microbialites emerge. Source: K Benzerara

environment and, thus, influence mineral precipitation. For example, *Bacillus subtilis*, a known aerobic respiration, should favour carbonate dissolution considering its primary metabolism, yet it can induce calcium carbonate precipitation by increasing pH due to a fatty acid secondary metabolism (Barabesi et al. 2007). Moreover, the model assumes that all cyanobacteria should favour carbonate precipitation. However, it has been shown that some species are considerably more active in this process (e.g., Couradeau et al. 2013), while other species even favour carbonate dissolution (e.g., Guida and Garcia-Pichel 2016). Overall, generalization should be avoided when inferring metabolisms based on the identity of a mineral.

Fig. 6.2 Scanning electron microscopy image of a biofilm populating the surface of a microbialite. The sample was prepared by critical point drying. The image shows an abundant fibrous network of EPS extending between organisms. Minerals in the microbialites, here hydromagnesite plates (arrows), form in this network (see Couradeau et al. 2013). Scale bar is 1 µm. Source: K Benzerara

6.2.2 Impact of Organic Molecules on Crystal Nucleation and Growth

By producing diverse organic polymers interacting with mineral phases, organisms can impact both nucleation and crystal growth (Fig. 6.2). Classical models for nucleation and crystal growth have successfully explained how this happens. For example, Giuffre et al. (2013) quantified the extent to which some polysaccharides decrease the energy barrier of calcite nucleation.

Depending on the functional groups composing them, diverse polysaccharides more or less favour heterogeneous nucleation by affecting the net surface free energy of the crystal-polysaccharide-liquid system. In other words, this heterogeneous nucleation has a barrier energy lower than that of homogeneous nucleation, and can therefore occur in slightly saturated solutions, in which homogeneous nucleation would be kinetically limited.

As a result, organic molecules may impact the texture of carbonate crystals. For example, extracellular polymeric substances (EPS), mostly polysaccharides, have been repeatedly associated with micritic textures in microbialites (e.g., Dupraz and Visscher 2005; Benzerara et al. 2006).

Experiments have also shown that the precipitation of carbonates in some organic matrices results in the formation of numerous and small crystals due to high nucleation rates (Sethmann et al. 2005). Organic molecules may also affect crystal growth and therefore modify crystal morphology. This can be due to the preferential adsorption of these molecules to some crystal faces with a subsequent poisoning of the growth of these faces. Eventually, this results in a crystal having morphology different from the equilibrium shape with well-developed faces (e.g., Meldrum and Cölfen 2008). Alternatively, some organic polymers can template crystal nucleation and growth, resulting in the formation of minerals with very unique width/length ratios (e.g., Chan 2004). Cell surfaces have been particularly scrutinized as promoters of mineral nucleation. Indeed, at neutral pH, they are usually negatively charged and bind cations, which participate in mineral formation (Schultze-Lam et al. 1996).

6.2.3 *Extracellular Versus Intracellular Biomineralization*

Biomineralization processes have been categorized in several ways. Biologically induced mineralization refers to cases in which organisms indirectly modify the chemical composition of a solution due to their metabolic activity. In contrast, biologically controlled mineralization encompasses all biomineralization processes involving specific genes (Bazylinski and Frankel 2004). Biomineralization is active when there is an energy cost and, therefore, involves some cellular activity; otherwise, it is passive. Another important distinction can be made between intracellular *versus* extracellular biomineralization. In the case of extracellular biomineralization, microbes increase supersaturation locally and/or produce extracellular polymers, impacting mineral nucleation and growth. The resulting biominerals may have morphologies or textures different from crystals forming at distance from the cells and under no biological influence. However, their chemical composition will usually be relatively similar. In contrast, biominerals forming intracellularly may be very different from minerals forming outside the cells because the chemical conditions within cells are different.

Cyanobacterial carbonatogenesis provides a useful illustration of this. Cyanobacteria have long been known to form diverse carbonate phases by extracellular and biologically induced mineralization in a broad diversity of environments, including marine water (e.g., Arp et al. 2001). The chemical composition of these carbonate phases depends mostly on the extracellular chemical composition: e.g., Mg-carbonates (e.g., hydromagnesite, Fig. 6.2) in Mg-rich solutions, or Sr-carbonates in Sr-rich solutions. There are a few exceptions. For example, in Lake Alchichica, Mexico, which is Mg-rich and where hydromagnesite is the major precipitating carbonate phase, some cyanobacteria of the Pleurocapsales order orient precipitation toward aragonite (Saghaï et al. 2015). This might result from a kinetic control due to some local shift in the chemical composition of the solution, possibly pH (Couradeau et al. 2013). Alternatively, there are some cyanobacterial species forming intracellular carbonates (Couradeau et al. 2012; Benzerara et al. 2014). These carbonate phases are very different from those forming outside the cells. Firstly, they are relatively small, with a narrow size distribution, and they are amorphous (Li et al. 2016). Secondly, these cyanobacteria can form Ca-carbonates even in globally undersaturated growth solutions (Cam et al. 2018). Finally, it has been shown that, in solutions with low Sr/Ca and Ba/Ca, one cyanobacterial species was able to form Sr-rich and Ba-rich amorphous carbonates after the selective uptake of Sr and Ba (Fig. 6.3; Cam et al. 2016).

In the same solutions, carbonate phases forming outside the cells have low Sr/Ca and Ba/Ca ratios. As a result, there are two populations of carbonate phases in such an environment with distinct morphologies, structures and chemical compositions:

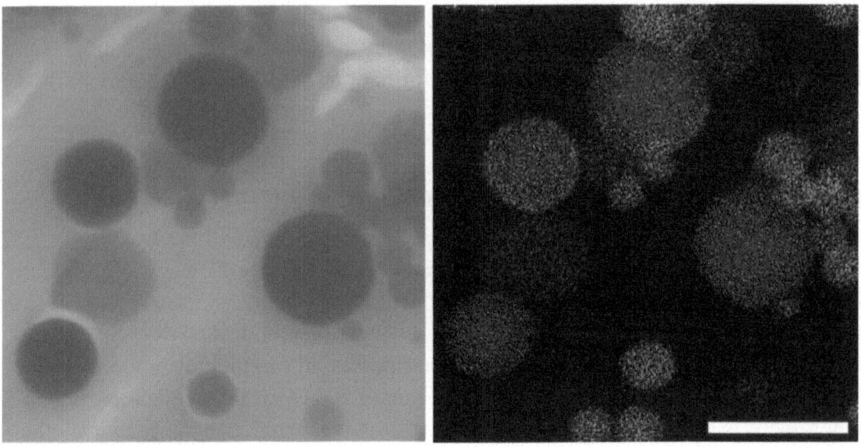

Fig. 6.3 Scanning transmission electron microscopy image in bright field mode (left) and chemical mapping (right) of barium (blue), strontium (red) and calcium (green) in intracellular amorphous carbonates formed by the cyanobacterium *Gloeomargarita lithophora*. Cells incorporate Ba and Sr preferentially to Ca and form carbonates with chemical compositions very different from those forming outside the cells as shown by Cam et al. (2016). Scale bar is 1 μm. Source: K Benzerara

one population formed intracellularly and another one formed extracellularly either by biologically induced mineralization or completely abiotic processes.

6.3 Examples of Biominerals with Particular Physico-Chemical and Structural Properties

As mentioned above, due to conditions prevailing upon their formation—sometimes influenced by microbial activity or controlled by microorganisms down to the genetic level—biominerals can exhibit particular chemical, physical and structural properties. In this section, we present emblematic biominerals and highlight the impact of life on their properties, i.e. the structural and chemical traces left by biological activity.

6.3.1 Genetic Control Over Biomineral Properties: Example of Magnetite Formed by Magnetotactic Bacteria (MTB)

Magnetite (Fe_3O_4) is a mixed valence iron oxide widespread in the geological record (Kopp and Kirschvink 2008). Apart from abiotic routes and extracellular biomineralization, magnetite, can be synthesized at ambient temperature by magnetotactic bacteria (MTB). Hence, specific properties of magnetites formed by MTB can be used to trace them in the geological record. MTB synthesize magnetosomes, i.e. intracellular magnetite crystals individually surrounded by a lipid bilayer membrane and organized in one or multiple chains (Bazylinski and Frankel 2004). These magnetite crystals display a narrow size distribution (35–120 nm), in the single-domain size range, hence maximizing the magnetic dipole moment of the cell, which can align along magnetic field lines (Faivre and Schüler 2008). Since these magnetic properties are rare for natural abiotic magnetites as we know them, they have been often used to identify bacterial magnetofossils in sediments (Li et al. 2013a). In addition, magnetites from MTB are usually very poor in chemical impurities and crystal defects (Bazylinski and Frankel 2004). Such unusual properties can be attributed to the way MTB control their formation, involving diversity genes that code the formation of the magnetosome membrane and proteins controlling crystal size and morphology (e.g., Murat et al. 2010; Lefevre and Bazylinski 2013).

Two additional properties may help in discriminating biogenic MTB magnetites from their abiotic counterparts. Firstly, Fe in magnetites produced by *Magnetospirillum magneticum* strain AMB-1 exhibits a mass independent fractionation (MIF) (Amor et al. 2016). The origin of this MIF signature is not yet well-understood but it might result from magnetic isotope effects potentially due to spin transitions upon Fe(III) reduction catalyzed by cytochromes and other proteins. An elucidation of the mechanisms and environmental conditions responsible for these isotopic compositions and an assessment of whether MIF can also be detected in other MTB will be required in the future. Secondly, it has been observed that the partition coefficients of most trace elements in magnetites produced by AMB-1 are 100 times lower than partition coefficients of the same elements in abiotic magnetites (Amor et al. 2015). In particular, the Sr/Ca ratio strongly differs between extracellular and MTB intracellular magnetites. One issue is the assessment of the chemical composition of the fluids in which MTB were living. One approach might be to measure the Sr/Ca ratio of carbonates that are frequently associated with magnetites: this would allow calculation of the Sr/Ca ratio of the past extracellular fluids and its comparison with the Sr/Ca of the magnetites. Based on the calculated partition coefficients between the observed magnetites and the fluids, assumptions can be made regarding the possible MTB origin of the magnetites. Again, it will be important to prove that these partitioning properties are widespread among biomagnetites from different MTB species.

6.3.2 Properties of Extracellular and Cell Wall-Associated Iron Biominerals

While only MTB precipitate magnetite intracellularly with regulated properties, several other bacteria precipitate extracellular iron oxides, including magnetite, by biologically induced mineralization. For example, dissimilatory iron(III) reducing bacteria (DIRB) induce the formation of extracellular magnetite by reducing ferrihydrite or lepidocrocite (e.g., Lovley et al. 1987; Miot and Etique 2016). Although most DIRB magnetites are superparamagnetic, the formation of single-domain magnetite has been reported under several conditions (e.g., Vali et al. 2004). Similarly, microbial anaerobic nitrate-dependent Fe(II) oxidation at a slightly basic pH can form extracellular single-domain magnetite (Miot et al. 2014a). Iron oxides produced by DIRB and Fe(II)-oxidizing bacteria (IOB) exhibit additional unique properties: first, some iron biominerals exhibit specific morphologies and textures. For example, Fe biomineralization by Fe(II)-oxidizing Gram-negative bacteria sometimes encrusts cells, with mineral precipitation filling the periplasm and coating the outer cell surface. As a consequence, the biominerals assemble into empty shells with the same dimensions as the initial microbes. This has been observed in laboratory cultures (Miot et al. 2009a) as well as in nature (Cosmidis et al. 2014). Additionally, crystals resulting from this precipitation have been observed to be crystallographically oriented within the periplasm of the cells (Miot et al. 2011, 2014b). Similarly, some iron minerals precipitated by IOB are anisotropic, resulting from a crystal aggregation-based growth (Banfield 2000; Chan 2004; Miot et al. 2014b). Periplasmic encrustation has also been observed for calcium phosphate biomineralization in cultures grown in the laboratory, as well as in 60 Myr-old samples (Benzerara et al. 2004; Cosmidis et al. 2013). Also, Fe-biominerals are usually associated with organic matter, either polysaccharidic polymers (Chan et al. 2009, 2011; Miot et al. 2009a), protein globules, or peptidoglycan (Miot et al. 2011). Finally, Fe biominerals sometimes display strong redox heterogeneities down to the nanometer scale. For instance, the photoferrotroph *Rhodobacter sp.* strain SW2 precipitates goethite nanoparticles at the surface of few μm-long lipopolysaccharidic fibers. These nanominerals display a FeIII/(FeII+FeIII) ratio decreasing along the fiber with distance from the cell (Miot et al. 2009b). In contrast, iron minerals precipitated by the DIRB *Shewanella oneidensis* shift from magnetite (i.e. mixed Fe valence: Fe_3O_4) at the cell contact to maghemite (i.e. pure Fe(III): Fe_2O_3) at distance (Coker et al. 2012).

6.4 The Loss of the Biological Information Contained by Biominerals Upon Ageing

As mentioned above, biominerals have chemical, physical and/or structural properties that result from metabolic activity and/or the influence of organic polymers. However, when dealing with fossil traces, one needs to consider ageing processes carefully since they may degrade "biogeochemical" and "paleoenvironmental" signals. Biominerals are generally reactive and their chemical, structural and isotopic properties inevitably undergo modification after formation (e.g., Reynard and Balter 2014; Keenan 2016). This obscures the original chemistry and structure of biominerals, and increases the difficulties inherent in assessing their potential biogenic origin in ancient rocks (Shapiro and Konhauser 2015). Even though astrobiological studies are usually more concerned with microbial systems, there is much to learn from the ageing of metazoan biominerals, which likely occurs in a similar manner and for which we have a much larger record of investigations. We therefore use some examples from this record in the following.

6.4.1 Structural Alteration of Biominerals in Natural Settings

The degree of structural order in biominerals is usually lower than that in their abiotic equivalents. Upon ageing, there is (1) a preferential dissolution of small disordered phases and/or (2) recrystallization and growth of new authigenic mineral phases. As a result, the overall "crystallinity" of "fossil" biominerals can be used to gauge their alteration degree (Pucéat et al. 2004; Keenan 2016).

Bone is an attractive material in which to study ageing processes as it comprises up to 70 wt.% nanometric highly disordered crystallites of nonstoichiometric carbonated apatite and 30 wt.% organic macromolecules (Grunenwald et al. 2014; Keenan 2016). The extensive post-mortem recrystallization of disordered bone crystallites was, for instance, evidenced by studying a suite of modern bones exposed for 40 years at the surface of soils (Trueman et al. 2004). Recent studies also identified increases in the mean crystal size of bone apatite with increasing ageing (Rogers et al. 2010).

Unlike bone, tooth enamel is quite resistant to ageing processes because it has a lower organic matter content and a more compact microstructure. Tooth enamel is made of almost stoichiometric hydroxyapatite microcrystals (~ 97 wt.%) and thus exhibits a relatively high thermodynamic stability compared to the nonstoichiometric apatite of bones (Grunenwald et al. 2014). Yet, Yi et al. (2014) showed that the primary bioapatites of tooth enamel of 4–6 Ma old fossils of large herbivorous mammals had been replaced by carbonate-fluorapatite phases, thereby increasing their bulk crystallinity.

6.4.2 (Geo)chemical Alteration of Biominerals in Natural Settings

In addition to structural changes, biominerals sometimes undergo significant chemical modification upon ageing, including trace element and isotopic exchanges with the surrounding environment. These processes occur through a combination of adsorption, diffusion and dissolution/recrystallization. Identifying and quantifying these modifications within fossil biominerals is essential for separating palaeobiological and palaeoenvironmental information from burial-induced transformations (Reynard and Balter 2014).

Trace elements have long been known to enter biominerals during post mortem ageing processes (Parker and Toots 1970). Partitioning occurs due to kinetic controls on dissolution, precipitation, and transport of the elements in the complex biomineral structure (Reynard and Balter 2014). Diffusion-limited uptake together with the adsorption of the trace elements onto the crystal surfaces, are the dominant processes (Millard and Hedges 1996), but interactions with chelating agents or organic complexes may also play a role (Pourret et al. 2007). For example, bioapatite can incorporate a wide range of rare earth elements (REE) into its lattice, largely as substitutions for Ca^{2+}. As a result, fossil bones, teeth and skeletal tissues sometimes contain REE concentrations between 2 and 5 orders of magnitude higher than modern specimens (MacFadden et al. 2010; Herwartz et al. 2013).

Ageing processes can also modify the original isotopic composition of biominerals, which is sometimes suggested as a record of biogenicity in a microbial context. For example, dissolution patterns and secondary calcite overgrowths impact the oxygen isotope signal carried by foraminifera (e.g., Edgar et al. 2015). Additionally, carbonates can undergo isotope re-equilibration even at quite low temperatures (e.g., Mavromatis et al. 2015).

In a more optimistic view, skeletal tissues with different physico-chemical properties are affected differently by ageing processes. For instance, carbonates in calcinized bone seem to be particularly resistant to isotope exchange (Zazzo and Saliège 2011). Moreover, even the total chemical transformation of a biomineral does not preclude the inference of a primary biogenic origin. For example, Zeyen et al. (2015) showed the replacement of silica composing the frustules of diatoms by talc-like phases with a perfect preservation of the frustule shape.

6.4.3 Constraining Biomineral Ageing Processes in the Laboratory

Since geological timescales cannot be replicated in the laboratory, extrapolating laboratory results to natural settings remains difficult (Li et al. 2014; Bernard et al. 2015; Alleon et al. 2016). However, experimental approaches have provided valuable insights into biomineral ageing processes (Li et al. 2013b). For instance, thermal treatments induce important compositional changes in bioapatites (e.g., Lebon et al. 2010). From 250 to 300 °C, the carbonate content decreases rapidly while the organic matrix is progressively carbonized. This carbonization coupled to the loss of carbonates is responsible for the increase of apatite crystallinity observed in natural settings (Lebon et al. 2010). Consistently, Alleon et al. (2016) recently showed that organic compounds retard quartz crystallization depending on their chemical nature. From another perspective, Li et al. (2014) showed that bacterial cells encrusted by hydroxyapatite suffered limited morphological and chemical alteration compared to non-encrusted cells, as attested by electron microscopy,

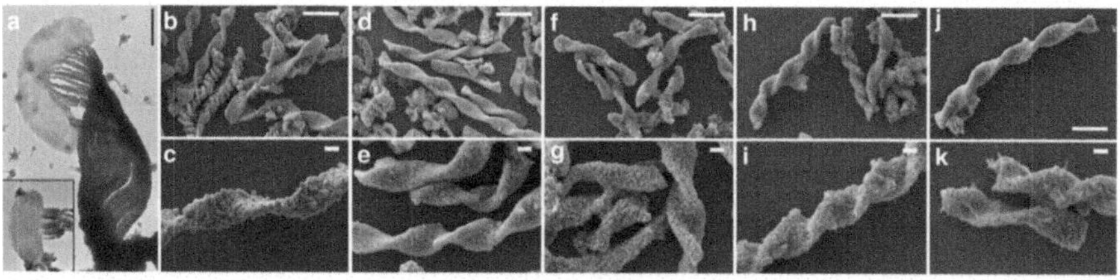

Fig. 6.4 Experimental diagenesis of twisted stalks. (**a**) TEM images of stalks of *M. ferrooxydans* composed of individual filaments. Inset: smaller cell and stalk, displayed at the same scale. Scale bar 500 nm. Source: Adapted from Chan et al. (2011). (**b–k**) SEM images of twisted stalks before (original sample: **b**, **c**) and after incubations of 1 week at 130 °C and 100 MPa (**d**, **e**), at 170 °C and 120 MPa (**f**, **g**), and at 250 °C and 140 MPa (**j**, **k**) and of 16 weeks at 170 °C and 120 MPa (**h**, **i**). Scale bars are 5 μm (**b**, **d**, **f**, **h**, **j**) and 1 μm (**c**, **e**, **g**, **i**, **k**). Source: Adapted from Picard et al. (2015)

FT-IR and EPR spectroscopies, thus, confirming that biominerals can favour the preservation of traces of life.

Picard et al. (2015) submitted organo-mineral structures produced by Fe(II)-oxidizing bacteria (i.e., twisted stalks covered with ferrihydrite) to pressure and temperature conditions typical of diagenetic settings. Although the general twisted morphology of the stalks was not altered during the experiments (Fig. 6.4), the authors documented some textural and mineralogical changes, such as an increase in crystallinity with increasing pressure and temperature conditions and the transformation of ferrihydrite into hematite and magnetite above 150 °C (Picard et al. 2015). The transformation of biogenic goethite into hematite and magnetite under pressure and temperature was limited in the presence of Si (Picard et al. 2016), thereby showing that external conditions can favour the preservation of biominerals.

Some experimental studies have dealt with the preservation/degradation of the isotopic signatures of biominerals during burial. For example, heating experiments demonstrated that bioapatite isotope compositions can change substantially even at a relatively low temperature and modest changes in crystallinity (Zazzo et al. 2004). Recently, Riechelmann et al. (2016) showed that magnesium isotope values of brachiopod shell carbonates change in hydrothermal alteration experiments as a function of fluid temperature, chemistry, and experimental duration, depending on the original geochemistry of these biominerals. This was likely due to the complexation of Mg by water-soluble and -insoluble organic matter.

6.5 The Thin Frontier Between Biominerals and Abiotic Minerals

Some biominerals exhibit properties that may help to identify their biogenic origin if these properties withstand ageing. However, this is usually not straightforward since some abiotic reactions can form minerals sharing similarities with biominerals. As further described in Chap. 8, magnetites observed in the Martian meteorite ALH84001 offer one example of this thin frontier between abiogenic and biogenic minerals. While Thomas-Keprta et al. (2001) interpreted them as biogenic based on their purity and their purportedly unique crystal habit, Barber and Scott (2002), Golden et al. (2004), and Bell (2007) proposed a purely abiotic formation model related to shock metamorphism. Here we detail additional cases where distinguishing between biogenic and abiogenic minerals has been a challenge.

6.5.1 Biogenic and Abiotic Framboidal Pyrite

Framboids of pyrite (FeS_2), so-called because of their raspberry-like morphology, are spherical to sub-spherical structures composed of multiple microcrystals. They usually have a very organized texture (e.g., Ohfuji and Akai 2002), but are sometimes polyhedral or disordered (e.g., Butler and Rickard 2000). Microcrystals within a single framboid display a narrow size distribution (0.2–2 μm) and framboids usually measure 1–10 μm in diameter, i.e. each framboid contains 10^2 to 10^5 microcrystals (Wilkin et al. 1996). Framboids are commonly observed in the fossil record and their structure and size distribution have been used as a proxy for palaeoredox conditions.

It has frequently been proposed that framboidal pyrites are purely abiotic (e.g. Rust 1935). They can be synthesized in a wide range of conditions, even at room temperature, in the absence of life (Ohfuji and Rickard 2005). Usually, the precipitation pathway involves a mackinawite FeS precursor, which reacts with elemental sulphur to form an intermediate, but not obligatory, greigite phase (Fe_3S_4), evolving *via* solid-state reaction towards pyrite (Wilkin and Barnes 1997). Formation of framboids instead of euhedral pyrite is mainly driven by high nucleation rates (Ohfuji and Rickard 2005). This condition is met in solutions highly supersaturated with respect to pyrite, i.e. under conditions of either high S(0) content, and/or high Eh (e.g., in the presence of O_2), or at elevated temperature (Ohfuji and Rickard 2005). Additional models of framboid formation have also been suggested, involving coacervation (Kalliokoski and Cathles 1969) or aggregation of intermediate greigite crystals driven by magnetic forces (Wilkin and Barnes 1997).

Since organic molecules have been frequently observed in association with pyrite framboids in natural samples, an alternative pathway of formation has been proposed consisting of the nucleation of pyrite microcrystals at the contact with organic molecules, such as Fe-rich humic substances. Some early models of framboid formation proposed the involvement of microorganisms (Love 1957; Folk 2005) since the framboid structure evokes microbial colony mineralization. Moreover, pyrite crystals in framboids display low $\delta^{34}S$, consistent with a microbial origin of their sulfur content (e.g., Kohn et al. 1998). More recently, extracellular polymeric substances (EPS) and possibly extracellular DNA were shown to be associated with microcrystals of pyrite in a natural biofilm (MacLean et al. 2008). It was proposed that organics migrated from the periphery of the microcrystals outward into the framboids or were trapped within the microcrystals. This scenario bolstered a biogenic interpretation of Precambrian pyrite framboids in which, CN enrichments were revealed by NanoSIMS (Wacey et al. 2015).

Yet, at the time of writing, there is no report of experimental biomineralization of pyrite under controlled conditions. Such an experimental approach could describe the different steps of framboid formation within a microbial biofilm and would provide valuable insights into the putative roles played by microorganisms in the formation of framboidal pyrite. Microbial activity may locally increase the ionic activity product of pyrite (e.g. through sulphate reduction) and provide nucleation sites. However, since abiotic routes of framboidal pyrite formation do exist and result in textures similar to those described for purportedly biogenic framboids, framboidal texture or size distribution alone should not be used as a definitive trace of life.

6.5.2 Twisted Stalks

Some biominerals exhibit very recognizable, purportedly "complex" morphologies, which are commonly used to identify the microbes that produced them. This is the case for twisted stalks and tubes of iron (oxyhydr)oxides, usually produced by microaerophilic bacteria (Emerson et al. 2010; Miot and Etique 2016), or elemental sulphur filaments produced by sulphur-oxidizing bacteria (Thar and Kuhl 2001). The formation of stalks by *Mariprofundus sp.* starts with the production of a polysaccharidic matrix at the surface of the cell, evolving towards a twisted stalk serving as a nucleation template for Fe oxide deposition and growth (Chan et al. 2011). The resulting material is a polysaccharide-iron oxide composite (Chan et al. 2009), with minerals growing through a non-classical pathway of oriented aggregation (Banfield 2000). In addition, the fine-scale morphology of the stalks is directly influenced by the microbial physiology (Krepski et al. 2013). Branching results from cell division, and directionality is inherited from an adaptation to the O_2 gradient since stalks grow perpendicular to the anoxic-oxic transition zone. Finally, the twisted morphology results from cell rotation during stalk elongation (Krepski et al. 2013). This scenario is in agreement with models of S-oxidizing bacteria motility in O_2/sulphur gradients (Thar and Kuhl 2003). However, structures with similar, sometimes twisted, morphologies (termed "biomorphs") can also be produced under purely abiotic conditions (Fig. 6.5). For example, Garcia-Ruiz et al. (2003) produced a range of biomorphs by precipitating barium carbonate in a silica-rich solution at room temperature and $pH > 10$. Crystal growth is explained by the propagation of a growth front in directions that depend upon the distribution of chemical impurities at this front. This distribution results from oscillations of the pH induced by successive carbonate and silica precipitation, such that oversaturation of the solution with respect to either mineral finally leads to their alternate deposit.

Different morphologies are explained by the modes and directions of propagation of one or multiple curling fronts during crystal growth (Garcia-Ruiz et al. 2009). In

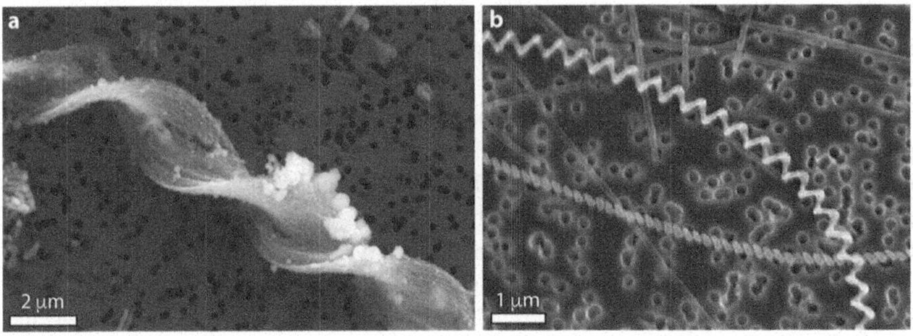

Fig. 6.5 Stalks can be produced biotically or abiotically. (**a**) Amorphous Fe phosphate *Galionella*-like stalk observed in Lake Pavin (Miot et al. 2016). (**b**) Abiotic helical filaments of S and C formed in a sulfide gradient tube in the presence of yeast extract. Source: Reprinted with permission from Cosmidis and Templeton (2016)

another challenge to the biogenic interpretation, Cosmidis and Templeton (2016) synthesized carbon/sulphur microstructures with a variety of morphologies, including twisted stalks and filaments, in an inverse O_2/sulphur gradient in the presence of simple organic molecules under abiotic conditions.

Not only the morphology, but also the size and chemical composition converged with those of biominerals, such as elemental sulphur filaments formed by sulphur-oxidizing bacteria (Sievert et al. 2007). Although some of these biomorphs (Garcia-Ruiz et al. 2009) have chemical compositions different from some similar biominerals, these abiotic syntheses shed light on the fact that morphology alone is not diagnostic of biogenicity. Moreover, it indicates that some biominerals acquire morphologies due to the same abiotic mechanisms involved in the formation of biomorphs, suggesting self-organization processes. Microorganisms may create the appropriate conditions to start self-organization, then abiotic nucleation and growth may proceed without further impact from the organisms.

6.6 Conclusions

We have reported that life can impact the formation of minerals and affect the morphological, textural, structural and/or chemical properties of the resulting mineral phases. Therefore, characterizing minerals in extraterrestrial objects and/or on planetary surfaces may provide information about potential modern or past biological activity. In this chapter, we stressed several issues related to such an approach. Firstly, the inference of a biogenic origin of a certain mineralogical property is based on known terrestrial cases of microorganism-mineral interactions. Apart from the issue of extrapolating terrestrial to extraterrestrial cases, which is unavoidable, we still have a limited knowledge about the diversity of these interactions on Earth and we may therefore fail to acknowledge some biogenic mineralogical properties.

Future discoveries in geobiology will likely offer new clues about overlooked mineralogical traces of life. Secondly, it is clear that the characterization of some mineralogical properties affected by the activities of life requires the use of sometimes, large analytical tools such as electron microscopes and/or synchrotron-based microscopies/spectroscopies, which cannot be sent to the surface of another planet on a rover, for example. The return of samples to the Earth will therefore be crucial to achieve the required analyses and complete the in situ characterization required. Thirdly, many properties of biominerals may also be found in abiotic minerals. In several cases, these properties appear under different physical and/or chemical conditions when the catalysing capabilities of life are involved. However, interpreting a biogenic origin for a mineral requires taking into account of all possible abiotic scenarios. Further, this involves determining whether additional environmental constraints may favour a biotic *vs.* an abiotic hypothesis. Last but not least, investigated minerals may have experienced transformations since their formation. Here, we discussed some of these transformations, induced by pressure and temperature changes, which often affect minerals formed at the Earth surface over geological timescales. However, at the surface of planets such as Mars, which is tectonically less active and has a thinner atmosphere, other processes such as irradiation by UV, for example, likely played a bigger role in the ageing of surface minerals and the degradation of potential traces of life. A better evaluation of irradiation-induced transformations in biominerals will therefore be useful to understand what may be expected from future astrobiology-driven missions to Mars.

Acknowledgements Karim Benzerara was supported by funding from the European Research Council under the European Community's Seventh Framework Programme (FP7/2007–2013 Grant Agreement no. 307110—ERC CALCYAN). Sylvain Bernard was supported by funding from the European Research Council under the European Community's Seventh Framework Programme (FP7/2007–2013 Grant Agreement no. 161764—ERC PaleoNanoLife—PI: F. Robert). Jennyfer Miot was supported by funding from the French Agence Nationale de la Recherche (ANR SRB project, grant ANR-14-CE33-0003-01). The SEM facility of the IMPMC is supported by Région Ile de France grant SESAME 2006 N°I-07-593/R, INSU/CNRS, UPMC-Paris 6, and by the Agence Nationale de la Recherche (ANR grant No. ANR-07-BLAN-0124-01. We thank Sebastien Charron and Imene Esteve for their expert support of the SEM at IMPMC. The TEM facility at IMPMC is supported by Région Ile de France grant SESAME 2000 E 1435. We thank Jean-Michel Guigner for his expert support of the TEM at IMPMC.

References

Alleon J, Bernard S, Le Guillou C et al (2016) Early entombment within silica minimizes the molecular degradation of microorganisms during advanced diagenesis. Chem Geol 437:98–108

Aloisi G (2008) The calcium carbonate saturation state in cyanobacterial mats throughout Earth's history. Geochim Cosmochim Acta 72:6037–6060

Amor M, Busigny V, Durand-Dubief M et al (2015) Chemical signature of magnetotactic bacteria. Proc Natl Acad Sci USA 112:1699–1703

Amor M, Busigny V, Louvat P et al (2016) Mass-dependent and -independent signature of Fe isotopes in magnetotactic bacteria. Science 352:705–708

Arp G, Reimer A, Reitner J (2001) Photosynthesis-induced biofilm calcification and calcium concentrations in Phanerozoic oceans. Science 292:1701–1704

Banfield JF (2000) Aggregation-based crystal growth and microstructure development in natural iron oxyhydroxide biomineralization products. Science 289:751–754

Banfield JF, Moreau JW, Chan CS et al (2001) Mineralogical biosignatures and the search for life on Mars. Astrobiology 1:447–465

Barabesi C, Galizzi A, Mastromei G et al (2007) *Bacillus subtilis* gene cluster involved in calcium carbonate biomineralization. J Bacteriol 189:228–235

Barber DJ, Scott ERD (2002) Origin of supposedly biogenic magnetite in the Martian meteorite Allan Hills 84001. Proc Natl Acad Sci USA 99:6556–6561

Bazylinski DA, Frankel RB (2004) Magnetosome formation in prokaryotes. Nat Rev Microbiol 2:217–230

Bell MS (2007) Experimental shock decomposition of siderite and the origin of magnetite in Martian meteorite ALH 84001. Meteorit Planet Sci 42:935–949

Benzerara K, Menguy N (2009) Looking for traces of life in minerals. CR Palevol 8:617–628

Benzerara K, Miot J (2011) Biomineralization mechanisms. In: Gargaud M, López-Garcia P, Martin H (eds) Origins and evolution of life – an astrobiological perspective. Cambridge University Press, Cambridge, pp 450–468

Benzerara K, Menguy N, Guyot F et al (2004) Biologically controlled precipitation of calcium phosphate by *Ramlibacter tataouinensis*. Earth Planet Sci Lett 228:439–449

Benzerara K, Menguy N, López-García P et al (2006) Nanoscale detection of organic signatures in carbonate microbialites. Proc Natl Acad Sci USA 103:9440–9445

Benzerara K, Meibom A, Gautier Q et al (2010) Nanotextures of aragonite in stromatolites from the quasi-marine Satonda crater lake, Indonesia. In: Pedley HM, Rogerson M (eds) Tufas and speleothems: unravelling the microbial and physical controls, vol 336. Geological Society, London, Special Publications, pp 211–224

Benzerara K, Menguy N, Obst M et al (2011) Study of the crystallographic architecture of corals at the nanoscale by scanning transmission x-ray microscopy and transmission electron microscopy. Ultramicroscopy 111:1268–1275

Benzerara K, Skouri-Panet F, Li J et al (2014) Intracellular Ca-carbonate biomineralization is widespread in cyanobacteria. Proc Natl Acad Sci USA 111:10933–10938

Bernard S, Benzerara K, Beyssac O et al (2015) Evolution of the macromolecular structure of sporopollenin during thermal degradation. Heliyon 1:e00034

Braissant O, Decho AW, Dupraz C et al (2007) Exopolymeric substances of sulfate-reducing bacteria: interactions with calcium at alkaline pH and implication for formation of carbonate minerals. Geobiology 5:401–411

Burne RV, Moore LS (1987) Microbialites: organosedimentary deposits of benthic microbial communities. PALAIOS 2:241–254

Burton EA, Walter LM (1987) Relative precipitation rates of aragonite and Mg calcite from seawater-temperature or carbonate ion control. Geology 15:111–114

Butler IB, Rickard D (2000) Framboidal pyrite formation via the oxidation of iron(II) monosulfide by hydrogen sulfide. Geochim Cosmochim Acta 64:2665–2672

Cam N, Benzerara K, Georgelin T et al (2016) Selective uptake of alkaline earth metals by cyanobacteria forming intracellular carbonates. Environ Sci Technol 50:11654–11662

Cam N, Benzerara K, Georgelin T et al (2018) Cyanobacterial formation of intracellular Ca-carbonates in undersaturated solutions. Geobiology 16:49–61

Chan CS (2004) Microbial polysaccharides template assembly of nanocrystal fibers. Science 303:1656–1658

Chan CS, Fakra SC, Edwards DC et al (2009) Iron oxyhydroxide mineralization on microbial extracellular polysaccharides. Geochim Cosmochim Acta 73:3807–3818

Chan CS, Fakra SC, Emerson D et al (2011) Lithotrophic iron-oxidizing bacteria produce organic stalks to control mineral growth: implications for biosignature formation. ISME J 5:717–727
Coker VS, Byrne JM, Telling ND et al (2012) Characterisation of the dissimilatory reduction of Fe (III)-oxyhydroxide at the microbe – mineral interface: the application of STXM-XMCD: STXM-XMCD of microbial Fe(III)-reduction. Geobiology 10:47–354
Cölfen H, Antonietti M (2008) Mesocrystals and nonclassical crystallization. Wiley, Chichester
Cosmidis J, Templeton AS (2016) Self-assembly of biomorphic carbon/sulfur microstructures in sulfidic environments. Nat Commun 7:12812
Cosmidis J, Benzerara K, Gheerbrant E et al (2013) Nanometer-scale characterization of exceptionally preserved bacterial fossils in Paleocene phosphorites from Ouled Abdoun (Morocco). Geobiology 11:139–153
Cosmidis J, Benzerara K, Morin G et al (2014) Biomineralization of iron-phosphates in the water column of Lake Pavin (Massif Central, France). Geochim Cosmochim Acta 126:78–96
Couradeau E, Benzerara K, Gerard E et al (2012) An early-branching microbialite cyanobacterium forms intracellular carbonates. Science 336:459–462
Couradeau E, Benzerara K, Gérard E et al (2013) Cyanobacterial calcification in modern microbialites at the submicrometer scale. Biogeosciences 10:5255–5266
De Yoreo JJ, Vekilov PG (2003) Principles of crystal nucleation and growth. Rev Mineral Geochem 54:57–93
De Yoreo JJ, Gilbert PU, Sommerdijk NA et al (2015) Crystal growth. Crystallization by particle attachment in synthetic, biogenic, and geologic environments. Science 31:aaa6760
Dupraz C, Visscher PT (2005) Microbial lithification in marine stromatolites and hypersaline mats. Trends Microbiol 13:429–438
Dupraz C, Reid RP, Braissant O et al (2009) Processes of carbonate precipitation in modern microbial mats. Earth Sci Rev 96:141–162
Edgar KM, Anagnostou E, Pearson PN et al (2015) Assessing the impact of diagenesis on δ^{11}B, δ^{13}C, δ^{18}O, Sr/Ca and B/Ca values in fossil planktic foraminiferal calcite. Geochim Cosmochim Acta 166:189–209
Emerson D, Fleming EJ, McBeth JM (2010) Iron-oxidizing bacteria: an environmental and genomic perspective. Annu Rev Microbiol 64:561–583
Faivre D, Schüler D (2008) Magnetotactic bacteria and magnetosomes. Chem Rev 108:4875–4898
Folk RL (2005) Nannobacteria and the formation of framboidal pyrite: textural evidence. J Earth Syst Sci 114:369–374
Garcia-Ruiz JM, Hyde ST, Carnerup AM et al (2003) Self-assembled silica-carbonate structures and detection of ancient microfossils. Science 302:1194–1197
Garcia-Ruiz JM, Melero-Garcia E, Hyde ST (2009) Morphogenesis of self-assembled nanocrystalline materials of barium carbonate and silica. Science 323:362–365
Giuffre AJ, Hamm LM, Han N et al (2013) Polysaccharide chemistry regulates kinetics of calcite nucleation through competition of interfacial energies. Proc Natl Acad Sci USA 110:9261–9266
Golden DC, Ming DW, Morris RV et al (2004) Evidence for exclusively inorganic formation of magnetite in Martian meteorite ALH84001. Am Mineral 89:681–695
Grunenwald A, Keyser C, Sautereau AM et al (2014) Adsorption of DNA on biomimetic apatites: toward the understanding of the role of bone and tooth mineral on the preservation of ancient DNA. Appl Surf Sci 292:867–875
Guida BS, Garcia-Pichel F (2016) Extreme cellular adaptations and cell differentiation required by a cyanobacterium for carbonate excavation. Proc Natl Acad Sci USA 113:5712–5717
Hazen RM, Ausubel JH (2016) On the nature and significance of rarity in mineralogy. Am Mineral 101:1245–1251
Hazen RM, Papineau D, Leeker WB et al (2008) Mineral evolution. Am Mineral 93:11–12
Herwartz D, Tütken T, Jochum KP et al (2013) Rare earth element systematics of fossil bone revealed by LA-ICPMS analysis. Geochim Cosmochim Acta 103:161–183
Kalliokoski J, Cathles L (1969) Morphology, mode of formation and diagenetic changes in framboids. Bull Geol Soc Finl 41:125–133

Keenan SW (2016) From bone to fossil: a review of the diagenesis of bioapatite. Am Mineral 101:1943–1951

Kohn MJ, Riciputi LR, Stakes D et al (1998) Sulfur isotope variability in biogenic pyrite: reflections of heterogeneous bacterial colonization? Am Mineral 83:1454–1468

Kopp RE, Kirschvink JL (2008) The identification and biogeochemical interpretation of fossil magnetotactic bacteria. Earth Sci Rev 86:42–61

Krepski ST, Emerson D, Hredzak-Showalter PL et al (2013) Morphology of biogenic iron oxides records microbial physiology and environmental conditions: toward interpreting iron microfossils. Geobiology 11:457–471

Lebon M, Reiche I, Bahain J-J et al (2010) New parameters for the characterization of diagenetic alterations and heat-induced changes of fossil bone mineral using Fourier transform infrared spectrometry. J Archaeol Sci 37:2265–2276

Lefevre CT, Bazylinski DA (2013) Ecology, diversity, and evolution of magnetotactic bacteria. Microbiol Mol Biol Rev 77:497–526

Li J, Benzerara K, Bernard S et al (2013a) The link between biomineralization and fossilization of bacteria: insights from field and experimental studies. Chem Geol 359:49–69

Li YL, Konhauser KO, Kappler A et al (2013b) Experimental low-grade alteration of biogenic magnetite indicates microbial involvement in generation of banded iron formations. Earth Planet Sci Lett 361:229–237

Li J, Bernard S, Benzerara K et al (2014) Impact of biomineralization on the preservation of microorganisms during fossilization: an experimental perspective. Earth Planet Sci Lett 400:113–122

Li J, Margaret Oliver I, Cam N et al (2016) Biomineralization patterns of intracellular carbonatogenesis in cyanobacteria: molecular hypotheses. Fortschr Mineral 6:10

Love LG (1957) Micro-organisms and the presence of syngenetic pyrite. Q J Geol Soc Lond 113:429–440

Lovley DR, Stolz J, Nord GLJ et al (1987) Anaerobic production of magnetite by a dissimilatory iron-reducing microorganism. Nature 330:252–254

Ludwig R, Al-Horani FA, de Beer D et al (2005) Photosynthesis-controlled calcification in a hypersaline microbial mat. Limnol Oceanogr 50:1836–1843

MacFadden BJ, DeSantis LRG, Hochstein JL et al (2010) Physical properties, geochemistry, and diagenesis of xenarthran teeth: prospects for interpreting the paleoecology of extinct species. Palaeogeogr Palaeoclimatol Palaeoecol 291:180–189

MacLean LCW, Tyliszczak T, Gilbert PUPA et al (2008) A high-resolution chemical and structural study of framboidal pyrite formed within a low-temperature bacterial biofilm. Geobiology 6:471–480

Maurice S, Clegg SM, Wiens RC et al (2016) ChemCam activities and discoveries during the nominal mission of the Mars Science Laboratory in Gale crater, Mars. J Anal At Spectrom 31:863–889

Mavromatis V, Bundeleva IA, Shirokova LS et al (2015) The continuous re-equilibration of carbon isotope compositions of hydrous Mg carbonates in the presence of cyanobacteria. Chem Geol 404:41–51

McLoughlin N, Grosch EGA (2015) Hierarchical system for evaluating the biogenicity of metavolcanic- and ultramafic-hosted microalteration textures in the search for extraterrestrial life. Astrobiology 15:901–921

Meldrum FC, Cölfen H (2008) Controlling mineral morphologies and structures in biological and synthetic systems. Chem Rev 108:4332–4432

Millard AR, Hedges REM (1996) A diffusion-adsorption model of uranium uptake by archaeological bone. Geochim Cosmochim Acta 60:2139–2152

Miot J, Etique M (2016) Formation and transformation of iron-bearing minerals by iron(ii)-oxidizing and iron(iii)-reducing bacteria. In: Faivre D (ed) Iron oxides. Wiley-VCH, Weinheim, pp 53–98
Miot J, Benzerara K, Morin G et al (2009a) Iron biomineralization by anaerobic neutrophilic iron-oxidizing bacteria. Geochim Cosmochim Acta 73:696–711
Miot J, Benzerara K, Obst M et al (2009b) Extracellular Iron biomineralization by photoautotrophic iron-oxidizing bacteria. Appl Environ Microbiol 75:5586–5591
Miot J, Maclellan K, Benzerara K et al (2011) Preservation of protein globules and peptidoglycan in the mineralized cell wall of nitrate-reducing, iron(II)-oxidizing bacteria: a cryo-electron microscopy study: persistence of organics in mineralized Fe-oxidizing bacteria. Geobiology 9:459–470
Miot J, Li J, Benzerara K et al (2014a) Formation of single domain magnetite by green rust oxidation promoted by microbial anaerobic nitrate-dependent iron oxidation. Geochim Cosmochim Acta 139:327–343
Miot J, Recham N, Larcher D et al (2014b) Biomineralized α-Fe_2O_3: texture and electrochemical reaction with Li. Energy Environ Sci 7:451–460
Miot J, Jézéquel D, Benzerara K et al (2016) Mineralogical diversity in Lake Pavin: connections with water column chemistry and biomineralization processes. Fortschr Mineral 6:24
Murat D, Quinlan A, Vali H et al (2010) Comprehensive genetic dissection of the magnetosome gene island reveals the step-wise assembly of a prokaryotic organelle. Proc Natl Acad Sci USA 107:5593–5598
Nickel EH, Grice JD (1998) The IMA commission on new minerals and mineral names: procedures and guidelines on mineral nomenclature, 1998. Can Mineral 36:913–926
Oaki Y, Imai H (2005) The hierarchical architecture of nacre and its mimetic material. Angew Chem 44:6571–6575
Ohfuji H, Akai J (2002) Icosahedral domain structure of framboidal pyrite. Am Mineral 87:176–180
Ohfuji H, Rickard D (2005) Experimental syntheses of framboids—a review. Earth Sci Rev 71:147–170
Parker RB, Toots H (1970) Minor elements in fossil bone. Geol Soc Am Bull 81:925
Picard A, Kappler A, Schmid G et al (2015) Experimental diagenesis of organo-mineral structures formed by microaerophilic Fe(II)-oxidizing bacteria. Nat Commun 6:6277
Picard A, Obst M, Schmid G et al (2016) Limited influence of Si on the preservation of Fe mineral-encrusted microbial cells during experimental diagenesis. Geobiology 14:276–292
Pourret O, Davranche M, Gruau G et al (2007) Competition between humic acid and carbonates for rare earth elements complexation. J Colloid Interface Sci 305:25–31
Pucéat E, Reynard B, Lécuyer C (2004) Can crystallinity be used to determine the degree of chemical alteration of biogenic apatites? Chem Geol 205:83–97
Reynard B, Balter V (2014) Trace elements and their isotopes in bones and teeth: diet, environments, diagenesis, and dating of archeological and paleontological samples. Palaeogeogr Palaeoclimatol Palaeoecol 416:4–16
Riechelmann S, Mavromatis V, Buhl D et al (2016) Impact of diagenetic alteration on brachiopod shell magnesium isotope (δ^{26}Mg) signatures: experimental versus field data. Chem Geol 440:191–206
Rogers K, Beckett S, Kuhn S et al (2010) Contrasting the crystallinity indicators of heated and diagenetically altered bone mineral. Palaeogeogr Palaeoclimatol Palaeoecol 296:125–129
Rust GW (1935) Colloidal primary copper ores at Cornwall Mines, southeastern Missouri. J Geol 43:398–426
Saghaï A, Zivanovic Y, Zeyen N et al (2015) Metagenome-based diversity analyses suggest a significant contribution of non-cyanobacterial lineages to carbonate precipitation in modern microbialites. Front Microbiol 6:797

Saghaï A, Zivanovic Y, Moreira D et al (2016) Comparative metagenomics unveils functions and genome features of microbialite-associated communities along a depth gradient. Environ Microbiol 18:4990–5004

Schultze-Lam S, Fortin D, Davis BS et al (1996) Mineralization of bacterial surfaces. Chem Geol 132:171–181

Sethmann I, Putnis A, Grassmann O et al (2005) Observation of nano-clustered calcite growth via a transient phase mediated by organic polyanions: a close match for biomineralization. Am Mineral 90:1213–1217

Shapiro RS, Konhauser KO (2015) Hematite-coated microfossils: primary ecological fingerprint or taphonomic oddity of the Paleoproterozoic? Geobiology 13:209–224

Shiraki R, Brantley SL (1995) Kinetics of near-equilibrium calcite precipitation at 100°C: an evaluation of elementary reaction-based and affinity-based rate laws. Geochim Cosmochim Acta 59:1457–1471

Sievert SM, Wieringa EBA, Wirsen CO et al (2007) Growth and mechanism of filamentous-sulfur formation by Candidatus *Arcobacter sulfidicus* in opposing oxygen-sulfide gradients. Environ Microbiol 9:271–276

Thar R, Kuhl M (2001) Motility of *Marichromatium gracile* in response to light, oxygen, and sulfide. Appl Environ Microbiol 67:5410–5419

Thar R, Kuhl M (2003) Bacteria are not too small for spatial sensing of chemical gradients: an experimental evidence. Proc Natl Acad Sci USA 100:5748–5753

Thomas-Keprta KL, Clemett SJ, Bazylinski DA et al (2001) Truncated hexa-octahedral magnetite crystals in ALH84001: presumptive biosignatures. Proc Natl Acad Sci USA 98:2164–2169

Trueman CN, Behrensmeyer AK, Tuross N et al (2004) Mineralogical and compositional changes in bones exposed on soil surfaces in Amboseli National Park, Kenya: diagenetic mechanisms and the role of sediment pore fluids. J Archaeol Sci 31:721–739

Vali H, Weiss B, Li Y-L et al (2004) Formation of tabular single-domain magnetite induced by *Geobacter metallireducens* GS-15. Proc Natl Acad Sci USA 101:16121–16126

Wacey D, Kilburn MR, Saunders M et al (2015) Uncovering framboidal pyrite biogenicity using nano-scale CNorg mapping. Geology 43:27–30

Weiner S, Dove PM (2003) An overview of biomineralization processes and the problem of the vital effect. Rev Mineral Geochem 54:1–29

Wilkin RT, Barnes HL (1997) Formation process of framboidal pyrite. Geochim Cosmochim Acta 61:323–339

Wilkin RT, Barnes HL, Brantley SL (1996) The size distribution of framboidal pyrite in modern sediments: an indicator of redox conditions. Geochim Cosmochim Acta 60:3897–3912

Yang H, Sun HJ, Downs RT (2011) Hazenite, $KNaMg_2(PO_4)_2.14H_2O$, a new biologically related phosphate mineral, from Mono Lake, California, USA. Am Mineral 96:675–681

Yi H, Balan E, Gervais C et al (2014) Probing atomic scale transformation of fossil dental enamel using Fourier transform infrared and nuclear magnetic resonance spectroscopy: a case study from the Tugen Hills (Rift Gregory, Kenya). Acta Biomater 10:3952–3958

Zazzo A, Saliège J-F (2011) Radiocarbon dating of biological apatites: a review. Palaeogeogr Palaeoclimatol Palaeoecol 310:52–61

Zazzo A, Lécuyer C, Sheppard SMF et al (2004) Diagenesis and the reconstruction of paleoenvironments: a method to restore original δ18O values of carbonate and phosphate from fossil tooth enamel. Geochim Cosmochim Acta 68:2245–2258

Zeyen N, Benzerara K, Li J et al (2015) Microbial formation of low-T hydrated silicates in modern microbialites from Mexico. Front Earth Sci 3:64

Chapter 7
Biosignatures in Deep Time

Frances Westall, Keyron Hickman-Lewis, and Barbara Cavalazzi

Abstract Life on the early Earth inhabited a planet whose environment was vastly different from the Earth of today. An anaerobic and hot early Earth was the birthplace of the first living cells but wide-spread small-scale physico-chemical diversity provided opportunities for a variety of specialists: alkalophiles, acidophiles, halophiles etc. The earliest record of life has been lost due to plate tectonic recycling and the oldest preserved terranes (~3.9–3.7 Ga) are heavily altered by metamorphism, although they may contain traces of fossil life. As of ~3.5 Ga, ancient sediments are so well-preserved that a broad diversity of micro-environments and fossil traces of life can be studied, providing a surprising window into communities of microbes that had already reached the evolutionary stage of photosynthesis. From the wide variety of traces of ancient life that have been reported from the Archaean geological record in Greenland, Canada, South Africa and Western Australia, we examine a few particularly pertinent examples. Biosignatures in the rock record include microfossils, microbial mats, stromatolites, microbially induced sedimentary structures, biominerals, biologically indicative isotopic ratios and fractionations, elemental distributions, organochemical patterns and other geochemical peculiarities best explained by biological mediation. Due to dynamic geological reprocessing over the billions of years since these fossils entered

F. Westall (✉)
Centre de Biophysique Moléculaire, CNRS, Orléans, France
e-mail: frances.westall@cnrs.fr

K. Hickman-Lewis
Centre de Biophysique Moléculaire, CNRS, Orléans, France

Dipartimento di Scienze Biologiche Geologiche e Ambientali, Università di Bologna, Bologna, Italy
e-mail: keyron.hickman-lewis@cnrs-orleans.cnrs.fr

B. Cavalazzi
Dipartimento di Scienze Biologiche Geologiche e Ambientali, Università di Bologna, Bologna, Italy

Department of Geology, University of Johannesburg, Johannesburg, South Africa
e-mail: barbara.cavalazzi@unibo.it

B. Cavalazzi, F. Westall (eds.), *Biosignatures for Astrobiology*, Advances in Astrobiology and Biogeophysics, https://doi.org/10.1007/978-3-319-96175-0_7

the rock record, identifications of very ancient traces of life have been subject to criticism, hence the often complex arguments regarding their biogenicity. We here highlight a range of unambiguously *bona fide* and widely supported examples of fossil biosignatures. Fossil biosignatures have great promise as analogues of life that might be detected on other planets. In this respect, the study of the early Earth is particularly pertinent to the search for life on Mars, given the planetary- and microbial-scale similarities that prevailed on both planets during their early histories, together with the lack of subsequent geological reprocessing on Mars, which may make it an ideal repository for a near-pristine fossil record.

7.1 Introduction

In this chapter, we address purported biosignatures found in the most ancient rocks preserved on Earth. Over the last 50 years there have been many investigations to find the oldest traces of life on Earth, with varying degrees of success. Some notable controversies concern the identification of traces of life in a martian meteorite (e.g. McKay et al. 1996; Golden et al. 2000; Steele et al. 2012), or in a hydrothermal chert vein (e.g. Schopf 1993; Brasier et al. 2002). Over the last few decades, a better understanding of how microorganisms are preserved and an advancement in the techniques used to analyse potential fossil traces have shown themselves to be invaluable, resulting in many excellent investigations of reliable traces of life in rocks from the Archaean Eon. We will showcase a small number of the many examples of early biosignatures to demonstrate the variety in the types of biosignatures and modes of their preservation. From these descriptions, we will consider their significance in terms of the origin and early evolution of life, what they tell us about the early biosphere, and how they may assist the search for life on other planets.

Before describing the biosignatures themselves, it is important to briefly set the scene, since the geological and environmental conditions of the early Earth (until 3.2 to 3.0 Ga before the present) were very different to those from the Mesoarchaean onwards. Indeed, by comparison with the modern Earth these conditions are considered extreme. The microorganisms responsible for the preserved biosignatures from that time therefore originated and evolved in such extreme environmental settings, and the geological settings within in which they are found are an integral factor in the interpretation of their biogenicity and their significance.

Biosignatures from "deep time", in particular, need to be placed in the specific environmental context of the early Earth. By "deep time" we refer to traces of fossil life dating from the oldest, well-preserved rocks on Earth. There is a limit to how far back in time we can go because the most ancient rocks (late Hadean) when life is believed to have emerged, have been destroyed by plate tectonic recycling or are extremely metamorphosed. Some vestiges of these primordial rocks remain in the form of recycled zircon crystals dating back to >4.3 Ga (Mozjsis et al. 2001) and inherited trace element signatures (Kamber 2015). The oldest surviving rocks date back to the Eoarchaean Era (4.0–3.7 Ga) but they are significantly altered by high-grade metamorphism and purported biosignatures from these rocks are either artefacts or highly degraded traces, as will be seen below. In fact, the oldest well-preserved rocks date from 3.5 Ga and are found in two enclaves, in the Barberton Greenstone Belt in eastern South Africa and the East Pilbara terrane in northwest Australia.

These ancient rocks paint a picture of an Earth characterized by such extreme environmental conditions that it can be described as a different planet (Westall et al. 2018). It was not a planet upon which most present day life forms could exist, being characterised by no oxygen or ozone layer, high levels of UV radiation, a mildly reducing atmosphere, much volcanic and hydrothermal activity and, consequently, relatively high temperatures at the rock/water interface where most microbial life existed. The microorganisms that inhabited this environment were therefore anaerobes and extremophiles. The conditions of preservation that existed on the early Earth were special and linked to the high heat flow emanating from the Earth's mantle, which resulted in silica-rich hydrothermal effluent that permeated both volcanic rocks and volcanic sediments, together with their microbial inhabitants. The very rapid silicification of the early terrestrial crust (on the order of days to weeks: Orange et al. 2009; Westall et al. 2015a, b) ensured excellent preservation of the microbial cells. These particular conditions provide a window into an already extraordinary prokaryotic diversity by 3.5 Ga.

7.2 Ancient Biosignatures from the Early-Mid Archaean Eon

Table 7.1 lists the biosignatures described in this chapter. It is not our aim to provide an exhaustive review of biosignatures from the Early to Mid Archaean (including the periods also termed the Eoarchaean, Palaeoarchaean and Mesoarachaean), but rather to show a variety of biosignature types and styles of preservation.

7.2.1 Graphitic Biosignatures from Highly Metamorphosed Sediments from the 3.95 Uivak-Iquluk Gneiss, Labrador and Greenland

7.2.1.1 The 3.95 Uivak-Iqaluk Gneiss, Labrador

Dating the oldest rock remnants is arduous and initially presumed ages often change as a result of improvements in the techniques used. It appears that the 3.95 Uivak-Iqaluk Gneiss from Labrador may be the one of the oldest known crustal remnants. Careful mapping and analysis has documented the existence of marine and possibly hydrothermal sediments within which are graphitic crystallites up to several hundreds of micrometres in size located along bedding planes in the metasediments, and along mineral boundaries, as well as within minerals, such as quartz, garnet and biotite. Tashiro et al. (2017) demonstrated the syngenicity of the graphite, which has

Table 7.1 List of the purported Early to Mid Archaean bioisgnatures described in this chapter

Age—Formation—Location	Biosignature	Context	Reference
3.95 Ga Saglek Block, Uivak-Iqaluk Gneiss Labrador, Canada	Graphitic crystallites with a restricted range of negative carbon isotope signatures (−28.2‰)	Highly metamorphosed (amphibolite-granulite facies) marine and possibly hydrothermal sediments	(1)
> 3.7 Ga Isua greenstone belt, Greenland	Graphitic crystallites with a restricted range of negative carbon isotope signatures (−19‰); graphite with a tubular and polygonal nanostructure; kerogen with diverse functional groups preserved in garnet	Highly metamorphosed (greenschist-amphibolite facies) marine sediments and turbidite sequences, with volcanic tholeiites and pillow lavas	(2–4)
3.48 Ga Dresser Formation, Warrawoona Group, Western Australia	Stromatolites, microbial mats and microbially induced sedimentary structures (MISS), putative microfossils	Either tidal platform/lagoonal setting or volcanic caldera	(5–6)
3.472 Ga Middle Marker horizon, Hooggenoeg Formation, South Africa	Diverse microbial mats	Shallow-water basin with regular and voluminous komatiitic-basaltic-felsic volcanic-volcaniclastic input, weak hydrothermal activity	(7)
3.46 Ga stratiform Apex chert, Warrawoona Group, Western Australia	Microbially induced sedimentary structures, filament-like laminations	Shallow-water basin setting, strongly influenced by hydrothermal activity and mafic-felsic volcanic activity	(8–9)
~3.45 Ga upper Hooggenoeg Formation cherts, South Africa	Microbial mats, microfossils	Shallow to deep water volcanically influenced setting associated with periods of mafic (komatiitic-basaltic) and felsic (dacitic) volcanism.	(10–11)
3.446 Ga Kitty's Gap Chert, Warrawoona Group, Western Australia	Monolayer coccoidal colonics coating volcanic particles, putative microbial etching features	Shallow water volcanic-volcaniclastic sediments, more distal hydrothermal influence, mafic and felsic volcanic inputs	(12–13)
3.43 Ga Strelley Pool Chert, various greenstone belts, Western Australia	Diverse stromatolites, microbial mats, diverse microfossils associated with biologically indicative sulphur and carbon isotope signatures, organic ambient inclusion trails	Hydrothermally influenced carbonate platform with sandstone units, broadly elucidated as a rocky shoreline	(14–21)

(continued)

Table 7.1 (continued)

Age—Formation—Location	Biosignature	Context	Reference
3.42 Ga Buck Reef Chert, Kromberg Formation, South Africa	Diverse microbial mats, microfossils, putative stromatolites	Shallow to deep basins with volcanic and volcaniclastic input; presence of hydrothermal activity remains controversial	(10, 22–24)
3.33 Ga Josefsdal Chert, Kromberg Formation, South Africa	Microbial mats and biofilms, microfossils, putative chemotrophic biomass	Shallow-water volcaniclastic basin deposits with widespread hydrothermal influence, and evidence for periodic sub-aerial exposure	(25–28)
3.22 Ga Moodies Group, South Africa	Diverse microbial mats, microfossils, 'acritarchs'	Siliciclastic, terrestrial, fluvially dominated environment	(29–32)

Note that this is not an exhaustive list

(1) Tashiro et al. (2017); (2) Rosing (1999); (3) Ohtomo et al. (2013); (4) Hassenkam et al. (2017): (5) Buick et al. (1981); (6) Ueno et al. (2001); (7) Hickman-Lewis et al. (2018); (8, 9) Hickman-Lewis et al. (2016, 2017); (10) Walsh (1992); (11) Walsh and Lowe (1999); (12, 13) Westall et al. (2006a, 2011a); (14) Hofmann et al. (1999); (15, 16) Allwood et al. (2006, 2009); (17) Wacey (2010); (18, 19) Wacey et al. (2006, 2011); (20) Bontognali et al. (2012); (21) Sugitani et al. (2015); (22, 23) Tice and Lowe (2004, 2006); (24) Tice (2009); (25–28) Westall et al. (2006b, 2011b, 2015a, b); (29) Noffke et al. (2006); (30) Heubeck (2009); (31) Javaux et al. (2010); (32, 33) Homann et al. (2015, 2016)

an average $\delta^{13}C$ isotope value of −28.2‰, a value consistent with the fractionation of carbon by microbial metabolisms (Schidlowki 2001). The relatively restricted range of $\delta^{13}C$ values associated with the graphite suggest that they may indeed have been produced by living metabolic processes, although it should be noted that the isotopic signatures of abiotic organic carbon of meteoritic origin overlap with those produced by living processes (cf. Pearson et al. 2006).

This observation is important because it supports the theory that life had already appeared on Earth during the Hadean, when the environmental conditions were probably somewhat similar to those of the Early Archean Earth (Westall et al. 2018). Unfortunately, the degree of metamorphic degradation undergone by the Uivak-Iqaluk sediments makes it impossible to say more. What kinds of life forms produced the isotopic fractionation and in what kinds of habitats did the microbes live? Based on what we understand about microbial communities in the natural environment comprising normally a number of species (Tan et al. 2017), the isotopic biosignature probably represents the combined fractionation signature of a number of co-existing microorganisms.

7.2.1.2 Greenland

Isua Greenstone Belt, 3.7 Ga Despite its greenschist to amphibolite grade metamorphism, *bona fide* sediments do occur in the Isua Greenstone Belt (IGB) on the mainland of Greenland, dated at >3.7 Ga (Moorbath 2009; Rosing 1999; Ohtomo et al. 2013). Graphite with a ^{13}C-depleted isotopic signature was detected in several metamorphosed sedimentary formations in the IGB. The Garbenschiefer Formation, described as layered turbiditic sediments derived from volcanic tholeites and associated with pillow lavas (Rosing 1999) contains graphitic crystallites (2–4 μm in size) enclosed in re-crystallised sediment forming biotite and garnet porphyroblasts that have an average $\delta^{13}C$ isotopic composition of −19‰ (bulk sample measurements). On this basis, Rosing (1999) suggested that the carbon originated as possibly photoautotrophic microbial plankton that sedimented onto the seafloor. Similar values were measured in graphite crystallites from a less deformed portion of the IGB that was mapped as schist with geochemical signatures of clastic marine sediments (Ohtomo et al. 2013). In addition to the isotopic measurements, high-resolution transmission electron micrographs of the tubular and polygonal nanostructure of the graphite suggest that it originates from a precursor comprising heterogeneous organic compounds, i.e. is of biogenic origin rather than abiotic origin.

In these studies, despite the heavy metamorphism, the geological context is consistent with a marine sedimentary setting and the syngenicity of a reduced carbon precursor of the graphite has been established. Nonetheless, as with the metasediments of Labrador, it is necessary to take into account the possibility of the input of meteoritic carbon, although the relatively limited range in $\delta^{13}C$ suggests a biogenic origin. On the other hand, the so-called "biogenic" nanostructures, the tubular and polygonal shapes of the graphite have also been observed in carbonaceous matter from meteorites, such as carbonaceous chondrites (Le Guillou et al. 2012).

An interesting recent study of metasediments in the low-strain area of the IGB has documented the presence of kerogen in inclusions trapped in metamorphic garnet crystals (Hassenkam et al. 2017). The sediments are finely laminated, iron-rich, amphibolite-grade metapelites that still retain evidence of original sub-millimetric carbonaceous layering, now graphitized. These layers are contiguous with kerogenous (liquid) inclusions in millimetre-sized garnets that later overgrew the layers. In situ analyses documented a number of functional groups in the kerogen including Si–O, P–O, C–O, C–N, C–O, C–P, and C≡N, all suggestive of a biogenic precursor. This study is particularly unique because, despite the high-grade metamorphism, a time capsule of preserved kerogen within the garnet crystals permits the kind of organo-chemical study normally impossible in such rocks where carbon is in the form of graphite.

7.2.2 *Biosignatures from the 3.5–3.0 Ga Pilbara Greenstone Belt*

7.2.2.1 The 3.43 Ga Strelley Pool Formation

Seven morphotypes of well-preserved stromatolites have been described from the 3.43 Ga Strelley Pool Chert (Hofmann et al. 1999; Allwood et al. 2006; Fig. 7.1a), each of which can be related to specific locations on a peritidal carbonate platform.

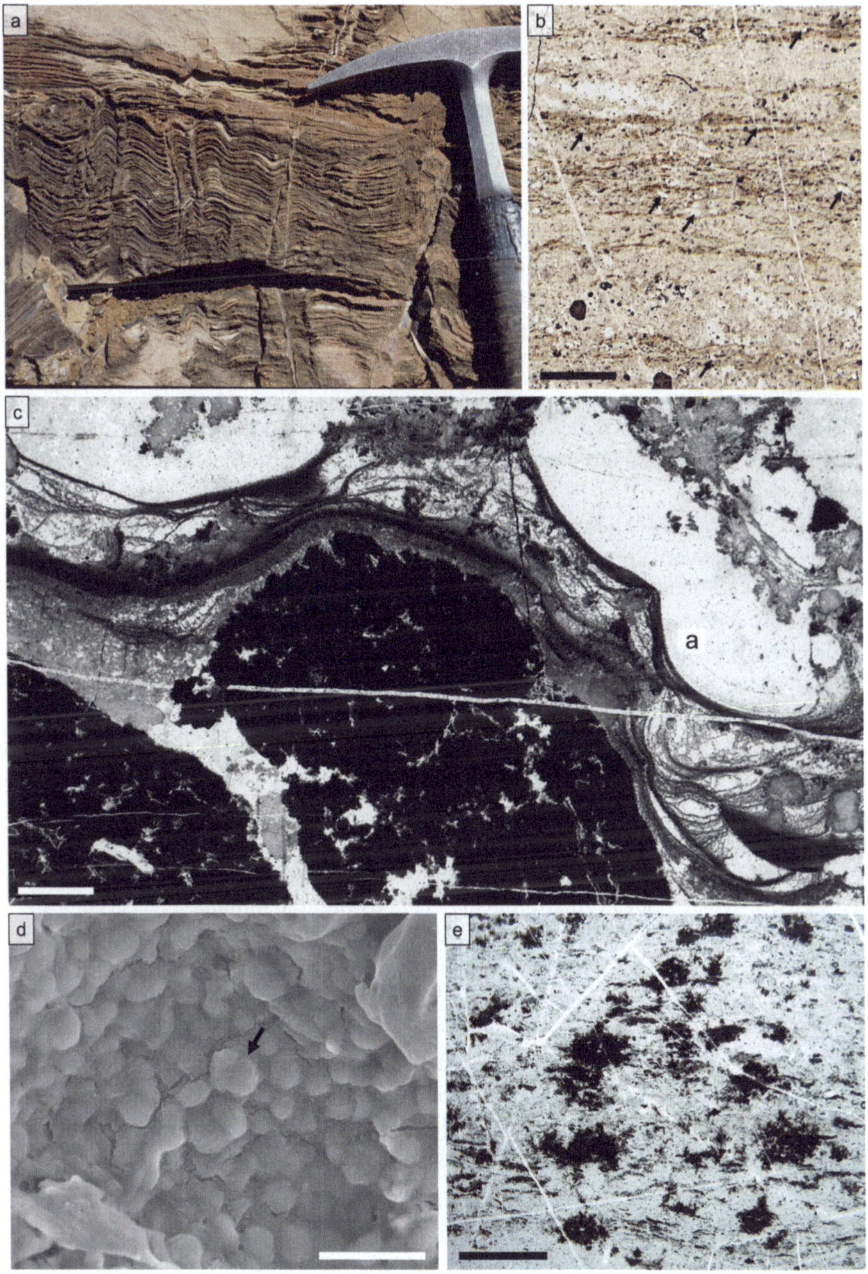

Fig. 7.1 Palaeoarchaean biosignatures. (**a**) Coniform-pseudocolumnar stromatolites from the Strelley Pool Formation. Hammer (30 cm) for scale. Source: F Westall. (**b**) Persistent, wrinkly, filament-like laminations interpreted as microbially induced sedimentary structures (MISS) from the stratiform “Apex chert”; arrows indicate grains that have been trapped and bound, probably due to sediment bio-stabilisation efforts. Scale bar = 1 mm. Source: K Hickman-Lewis. (**c**) Mesh-like, diaphanous microbial mat from the Buck Reef Chert, consisting of bundles of aligned carbonaceous strands; abundant open lenses are present near (**a**). Note the draping of large carbonaceous particles

The geological setting is well constrained with sediments being deposited on a rocky coastline under a regionally transgressive regime (i.e. gradually deepening water with time), culminating in a spate of volcanic and hydrothermal activity. Lindsay et al. (2005) further suggested that some carbonate units within the sequence—specifically one rich in coniform stromatolites—may have been deposited out of hydrothermal solutions, although the hydrothermal setting is now more generally applied to the unit (e.g. Sugitani et al. 2015).

The wide range of morphologies of the stromatolites (encrusting domical morphologies, 'egg-carton' forms, laminites, large complex cones, cuspate swales, small crested and conical types), the restriction of each to a specific microenvironment on the peritidal platform, and the characteristics of their internal laminae suggest that these were features produced by sticky, cohesive films, such as those produced by phototrophic microorganisms, and not by abiogenic phenomena (cf. Pope et al. 2000). Wacey (2010) provides a comprehensive outline of the evidence that—from the macro-scale to the nano-scale—is consistent with the biological interpretation of these stromatolites.

Wacey et al. (2011) described spheroidal carbonaceous objects that are not carbon-coated quartz grains, and which are constructed of thermally mature but disordered carbonaceous material within which bio-essential nitrogen and sulphur are present.

There is a clear spatial relationship between microfossil cell walls and micron-scale pyrite grains, which Wacey et al. (2011) conclude to be the result of sulphate-reducing or sulphur-disproportionating microbes. Sugitani et al. (2009, 2015) described morphologically diverse spherical and lenticular structures, which they interpreted as microfossils based initially on morphology and colonial occurrence, and later through high-resolution in situ isotopic studies and NanoSIMS ion mapping (Oehler et al. 2010; Delarue et al. 2017).

Apart from the more obvious stromatolites and microfossils, carbonaceous material in the Strelley Pool Formation also occurs as organic laminae preserved within the stromatolitic layers (Allwood et al. 2009), and as clasts and clots finely disseminated throughout a chert matrix (Marshall et al. 2007). The kerogen consists of highly aromatic molecules and its bulk $\delta^{13}C$ values range from −28.3 to −35.8‰. Derenne et al. (2008) noted an odd-over-even carbon number pattern associated with this kerogen, generally considered consistent with a biological origin, although this pattern is also found in meteoritic carbon (R Summons personal communication 2013). The Strelley Pool organic matter also contains sulphur and analysis of the $\delta^{33}S$ and $\delta^{34}S$ signatures indicated (1) that sulphur has been incorporated into the

Fig. 7.1 (continued) by the mat. Scale bar = 500 μm. Source: Adapted from Tice and Lowe (2006). (**d**) Monolayer colony of coccoidal organisms coating a volcanic particle from the Kitty's Gap Chert. The central arrowed organism was silicified during the process of cell division. Scale bar = 2 μm. Source: F Westall. (**e**) Carbonaceous clots within a hydrothermally influenced, silica gel-like chemical sediment from the Josefsdal Chert, interpreted as the degraded remnants of chemosynthetic biomass. Scale bar = 500 μm. Source: K Hickman-Lewis

organic matter during degradation of the organics by H_2S produced in the mat pore-waters due to microbes respiring sulphur, and that (2) the positive $\delta^{33}S$ anomalies indicate a metabolism using disproportionation of elemental sulphur (Bontognali et al. 2012).

In contrast to many older indications of life, which are based on isotopic and inconclusive morphological evidence, stromatolites of the Strelley Pool Chert incorporate both morphological and organic biosignatures, including molecular structure, and carbon and sulphur isotopes that are clearly consistent with microorganisms within a geological context that could have hosted photosynthetic mats and stromatolites.

7.2.2.2 The 3.46 Ga Apex Chert: A Contentious Case

Other examples of interpreted microfossil or mat-like remains occur in the Pilbara region of which the 3.46 Ga Apex Chert is an important and controversial example. Filamentous carbonaceous objects superficially resembling cyanobacteria have been described from a chert in the Apex Formation (Schopf 1993), which later geological context studies showed to be a hydrothermal feeder vein (Brasier et al. 2002). Continued investigation of the carbon has shown that it had an isotopic signature consistent with a biological origin (Schopf et al. 2018), while, at the same time, multi-technique analysis demonstrated that this carbon is merely condensed onto stacked phyllosilicate grains, thus imparting the 'sheath-like' appearance (Brasier et al. 2015; Wacey et al. 2015). It can be concluded that hydrothermal recycling can result in the condensation of carbon originally of biogenic origin into bacteriomorph structures that are essentially abiotic.

Nevertheless, the adjacent, shallow water stratiform sediments of the Apex chert, deposited in the vicinity of the hydrothermal vein cherts, contain detrital fragments of silicified, carbonaceous laminated structures of apparently biogenic origin. These include (1) laminated grains built of non-isopachous carbonaceous laminations, (2) carbonaceous roll-up structures both within the laminated grains and scattered within the matrix, (3) elongate, tapering flaky grains, and (4) persistent, filament-like, wrinkled laminations that entrain relict sedimentary grains. Multiple techniques (confocal laser scanning microscopy, Raman spectroscopy, nanoSIMS and X-ray micro-CT) have demonstrated that these structures pass many criteria for biogenicity in three dimensions, and have documented the presence of the bio-essential elements C, N, and S associated with the carbonaceous laminae (Hickman-Lewis et al. 2016), as well as the plastic behaviour of the laminated grains and filament-like carbonaceous horizons (Fig. 7.1b), and the fact that they exhibit particle-trapping behaviour (Hickman-Lewis et al. 2016, 2017) typical of that found in sediment-stabilising epibenthic microbial mats, and are thus directly comparable to a type of microbially induced sedimentary structure (MISS, cf. Noffke 2009). The non-isopachous morphology, particle-trapping behaviour and elemental composition of the carbonaceous films attest to their biogenicity and the fact that they occur in particles within a sedimentary horizon is evidence for their syngenicity.

7.2.2.3 The Kitty's Gap Microfossils: Chemotrophic Fossil Colonies

The 3.446 Ga Kitty's Gap Chert is part of the Warrawoona Group in the Pilbara. Unlike the stromatolites of the Strelley Pool, Dresser (Buick et al. 1981) and Mendon (Byerly et al. 1986) Formations, there is no macroscopically visible structure suggestive of microbial activity. The Kitty's Gap Chert comprises volcaniclastic sediments deposited in a shallow water, coastal environment (de Vries et al. 2010) influenced to varying degrees by hydrothermal activity (van den Boorn et al. 2007), the sediments and any microbial colonies being very rapidly silicified (Westall et al. 2006a; de Vries et al. 2010). Sediments from an infilling tidal channel were selected for detailed investigation of possible biosignatures (Westall et al. 2006a, 2011a, 2015a). The finely layered sediments revealed a variety of carbonaceous bacteriomorph structures, including coccoids, rods and filaments, generally associated with a film-like substance of either smooth or ropy appearance. Clusters of bimodal (~0.4 μm and ~0.8 μm) coccoids form monolayer colony-like associations on the surfaces of volcanic clasts (Fig. 7.1d), the carbonaceous material of which Raman spectroscopy shows to be consistent with the regional prehnite-pumpellyite grade metamorphism. Some of the volcanic grains exhibit tunnels filled with a mucus-like substance extending into a volcanic clast from the outer surface, interpreted as fossilised extracellular polymeric substance (Foucher et al. 2010).

The exceptional state of morphological preservation of some of these microfossils due to rapid silicification has retained in vivo details, such as cell division and cell lysis, as well as rugosity of cell surfaces. Bulk carbon isotope analyses of individual mm-thick layers are in the range of −25.9 to −27.8‰, consistent with microbial fractionation.

Westall et al. (2006a, 2011a, 2015a) interpreted the bacteriomorph structures as microbial remains on the basis of their carbonaceous composition, isotopic signature (which overlaps with abiotic carbon and perhaps represents a mixed signature), cellular morphology, surface-specific behaviour around volcanic clasts, and variability in diversity and state of preservation. The close association of monolayer clusters of coccoids with the surfaces of the volcanic grains and the evidence of tunnelling into the grains suggests that the microorganisms could have used a metabolism based on lithotrophy.

This example is unusual in that it is rare to observe individual microfossils in rocks of this age. We will see below that, generally, the microbial colonies were thoroughly degraded by later colonisers, the heterotrophs, thus destroying individual cell morphology.

7.2.3 *Biosignatures of the 3.5–3.2 Ga Barberton Greenstone Belt*

7.2.3.1 The 3.42 Ga Buck Reef Chert: Photosynthetic Fossil Biofilms

Biofilms and mats are a recurrent feature throughout almost the entirety of the stratigraphy of the BGB, from its oldest chert horizon, the Middle Marker (Hickman-Lewis et al. 2018), throughout its stratigraphy (Walsh 1992; Walsh and Lowe 1999) to its uppermost unit, the Moodies Group (Noffke et al. 2006; Heubeck 2009; Homann et al. 2016). The best-studied of these are the diverse mat systems of the Buck Reef Chert (Tice and Lowe 2004, 2006; Tice 2009). The Buck Reef Chert is a volcano-sedimentary complex with many similarities to the Kitty's Gap Chert in the Pilbara (de Vries et al. 2010). Both sedimentary successions were deposited in shallow coastal waters, alternating between lacustrine and littoral marine facies for the Buck Reef Chert and predominantly subaqueous with tidal influence and a regressive sequential trend for Kitty's Gap Chert. Evidence for hydrothermal influence comes from macroscopic chert veins, as well as often Fe-rich, high-energy breccia pods interpreted as the result of explosive hydrothermal activity and Fe-rich layers strongly influenced by hydrothermal fluids. de Vries and Touret (2007) further described hydrothermal and lower-salinity aqueous fluids from fluid inclusions, yielding similarities to modern shallow-water hydrothermal systems.

From the Buck Reef Chert, Tice (2009) described three photosynthetic mat morphotypes—alpha, beta and gamma—and their relationships with the evolution of the shallow water sedimentary succession, hypothesising that the mats were probably constructed by a dominantly photosynthetic community given their sedimentological setting (Fig. 7.1c). Furthermore, an estimation of the fluid dynamics of the depositional environment in terms of shear stress at the mat-water boundary and the Reynolds number (i.e., turbidity) of the water led Tice (2009) to suggest that morphologically more complex mats, in this case alpha laminations, can be the product of increased current dynamics relative to more simple mats. This is consistent with both the modelled growth of biofilms and observation of growing biofilms: some systems have increased strength, density and complexity when grown under enhanced shear stress (Lewandowski and Walser 1991; Stoodley et al. 2001). In this case, syngenicity of the biofilms and mats is clearly documented by complex intergrowth with the surrounding clastic particles and delicate, diaphanous webs caught in silica cement. The complexity and diversity of the biofilms testifies to their biogenicity.

7.2.3.2 The 3.33 Ga Josefsdal Chert: Chemotrophic Fossil Biofilms

The Josefsdal Chert is a volcaniclastic sedimentary horizon deposited in a shallow marine, littoral environment very similar to the Kitty's Gap Chert and the Buck Reef Chert. The abundant, sunlit surfaces were colonised by phototrophic biofilms and mats, some of which were episodically exposed (Westall et al. 2006b, 2011b). Although the phototrophic biofilms form characteristic features that clearly demonstrate direct interaction with their immediate environment and plastic deformation and are readily identifiable, interpreted chemotrophic colonies produce far more subtle biosignatures, as in the Kitty's Gap Chert described above. In contrast to the Kitty's Gap Chert, individually identifiable morphological fossils have not yet been identified in the Josefsdal Chert, even though Raman spectroscopy shows that volcanic particles are often coated with carbon (Westall et al. 2015a, b). In the vicinity of hydrothermal vents, these coats are quite irregular with protuberances up to 30 μm in thickness that suggest a non-isopachous growth inconsistent with simple carbon coating of volcanic grains by abiotic means. In addition, certain deposits of hydrothermal silica contain layers of carbon having a clotted texture. The clots, ranging from ~50 to 500 μm, appear to be free floating in a silica matrix. They may be rounded but sometimes they exhibit a spiky morphology that indicates in situ formation (Fig. 7.1e). The clotted layers are often associated with and inter-grown with photosynthetic biofilms. Westall et al. (2015a, b) interpreted these features as chemotrophic colonies, the irregular coatings around the volcanic clasts representing initial chemolithotrophic communities later degraded by chemoorganotrophs. Likewise the free-floating/free-growing clots were interpreted as chemotrophic colonies. In both cases, the clotted, carbon-rich biofilms occur in the vicinity of hydrothermal effluent, which would have served as a source of nutrients for the microorganisms. The semi-bulk (carbon-rich horizons selected) carbon isotope signature of the sediments is −22 to −27.8‰.

7.2.3.3 The 3.2 Ga Moodies Group

Microbial Mats and MISS To describe the variety of features produced by microbial mat interaction with sediments, Noffke et al. (2001) coined the term "microbially induced sedimentary structures" (MISS) of which several examples have been provided above. Such features were subsequently identified in the rock record back to the Mesoarchaean (Noffke et al. 2006) and even the Palaeoarchaean (Noffke et al. 2013; Hickman-Lewis et al. 2017). MISS in the Moodies Group are produced by photosynthetic microbial mats—which are also preserved in some cases—interacting with the sediment substrate in the dynamic tidal environment, where the mats are exposed to destruction by erosion and burial by sediment deposition. Responses to these environmental challenges are the production of large amounts of EPS to stabilize the underlying sediments, as well as the baffling, trapping, and binding of sediment particles (Noffke et al. 2003). This microbial interaction leads to the formation of distinctive sedimentary structures collectively termed MISS, similar to microbial mats formed on modern tidal flat sediments (Noffke et al. 2003). In cross-section, these mats and mat-like features exhibit a variety of textures including pinnacles, tufts, convex-upward domes, or anvil-shaped protrusions (Heubeck 2009; Noffke et al. 2006; Noffke 2009; Homann et al. 2015). These studies come to the conclusion that these features were indeed produced by photosynthetic microbial mats on the basis of the intimate interaction between the carbonaceous films, the underlying sediments and trapping/baffling behaviour. Carbon isotope values of −21‰ for the Moodies Group examples is consistent with microbial fractionation. Lithification of the sandy tidal environments was through silicification owing to high silica concentrations in the seawater (Heubeck 2009) and accounts for the exceptional preservation of biosignatures therein.

Coelobiontic (Cavity-Dwelling) Microorganisms The beautifully preserved Moodies Group sediments contain bedding parallel cavities formed in a peritidal environment filled with downward-growing microstromatolitic columns composed of kerogenous laminae whose in situ carbon isotope values ranging from −32.3‰ to −21.3‰ are consistent with a biogenic origin (Homann et al. 2016). The cavities, subsequently filled by silica, contain chains of 2 μm long, <0.5 μm wide, rod-shaped moulds forming non-branching filaments up to tens of microns in length and which resemble microbial filaments (Fig. 7.2a).

Homann et al. (2016) suggest that the isotopic data and the observed microfossils are consistent with a chemotrophic or photosynthetic community especially since there are phototrophic mats abound in the direct vicinity of the cavity dwellers (coelobionts). This is one of the first descriptions of cavity-dwelling behaviour and clearly documents the hold life was taking out of permanent contact with water. It is also the oldest known example of microbial moulds as a biosignature.

Organic-Walled Cells The microbial fossils described earlier in this chapter concern silicified microbial structures, albeit with associated carbon in the form of kerogen. The Mesoarchaean Moodies Group sediments document new varieties of biosignatures—the microbial moulds described above from vadose cavities in sandy facies, as well as non-silicified, organic-walled structures from finer grained, silty to muddy sediments. Javaux et al. (2010) have identified large (up to about 300 μm in diameter) carbonaceous spheroidal microstructures (Fig. 7.2b) that have carbon

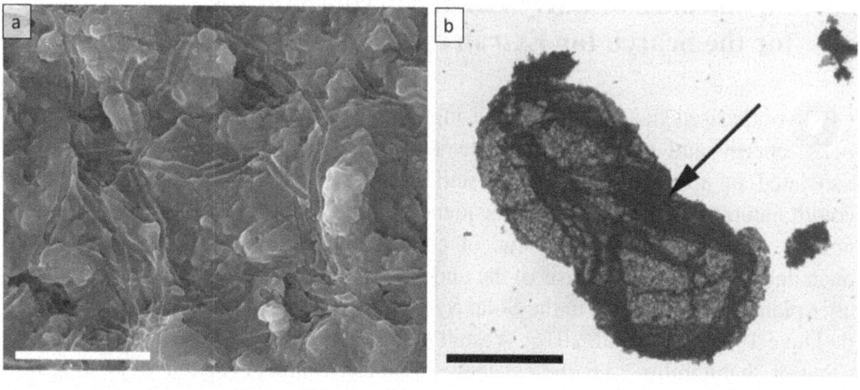

Fig. 7.2 Mesoarchaean biosignatures. (**a**) Moulds of fossil coelobiontic (cavity-dwelling) bacteria, from microbial mat-rich horizons in the Moodies Group. Scale bar = 5 μm. Source: Adapted from Homann et al. (2016). (**b**) Acritarch from the Moodies Goup, showing folding and collapsing deformation (dark arrowed region). Scale bar = 100 μm. Source: Adapted from Javaux et al. (2010)

isotope values ranging from −16.4 to −28.3‰ and present Raman spectra of the carbonaceous material consistent with the metamorphic grade of the country rock. The spherical cell-like objects occur in clusters in the silty-muddy sediments, range in size from 31 to 298 μm with a modal value of 50–75 μm, and exhibit soft-wall deformation, appearing torn and wrinkled. Javaux et al. (2010) eliminated possible abiotic origins from coalescing extraterrestrial carbon to form vesicles (vesicles <100 nm in diameter and processes not known in nature) or from hydrothermal carbon, the latter not documented in this environment. The large size of the organisms in such an environment suggests the requirement of oxygen, although there is no evidence for oxygen in the Moodies Group sediments. Javaux et al. (2010) suggest that their large size, taphonomy, and habitat could indicate either early evolution of a compartmentalized eukaryotic cytoplasm or colonial envelopes of cyanobacteria, or extinct prokaryotes with an unknown metabolism.

This is not the only description of large microfossils from the Palaeo-Mesoarchaean. Other carbonaceous, spherical to lenticular flange-shaped structures, 20–70 μm in size, occurring isolated or in apparent chains and with carbon isotope signatures consistent with life have been described from the 3.43 Ga Strelley Pool Formation by Sugitani et al. (2015) and from the 3.3–3.4 Ga Kromberg Formation by Walsh (1992) and Oehler et al. (2017). Even larger organic walled features have been discovered in the younger, 3.0 Ga siliciclastic sediments in the Mount Goldsworthy-Mount Grant region of the Pilbara (Sugitani et al. 2009). The large, organic-walled spheres of the Moodies Group are therefore not alone. Given the overwhelming evidence for non-oxygenic conditions at the time that these sediments were being deposited, the nature and origin of these large organic structures, whose carbon isotope signature is consistent with a microbial origin, is enigmatic.

7.3 Significance of Early Earth Microorganisms for the Search for Extraterrestrial Life

It is hypothesised that, wherever the ingredients of carbon molecules, water and rocks, energy and other essential elements (HNOPS and transition metals) are associated in a suitable geological setting for a sufficient amount of time, life would naturally appear as a consequence of prebiotic chemistry merging into biology. This is a simplistic view, of course, and the details of the process are much debated, but it is the basis of the understanding that life could have emerged on other planets and satellites in the Solar System, as well as elsewhere in the Universe (de Duve 1995; Dass et al. 2016; Westall et al. 2018). One of the criteria often cited is that of "habitability", i.e. the existence of an environment capable of hosting life. However, the kind of habitable conditions necessary for the emergence of life are not of the same order of magnitude for hosting well-established cellular life or slow-growing or dormant life. These differences have an impact on what kinds of life forms might exist or have existed on extraterrestrial bodies, where they might be (or have been), and how we could detect them.

From this point of view, the microbial communities of the early Earth are relevant for the search for life elsewhere in the Solar System for several reasons. On microbial scales, the anaerobic environment of the early Earth was similar to that of the other telluric (rocky) planets, Venus and Mars, in the early phases of their development. Venus was most similar to Earth, likely having a moderate amount of water although possibly not a global ocean (Way et al. 2016), while Mars has always been a landlocked planet characterised by isolated pockets of habitable areas of variable sizes. The environments of these early planets were similar: anaerobic, CO_2 atmosphere, volcanically and hydrothermally active, energy from heat sources, chemical reactions, radiation and light, carbon from extraterrestrial (carbonaceous chondrites, micrometeorites) and endogenous (atmosphere, subsurface hydrothermal reactions) sources. The implication is that life could have emerged on any of these planets. Way et al. (2016) consider that planets with a moderate amount of water, like early Venus, could be more habitable than ocean planets, like the early Earth. If Venus was habitable for about 2 Ga, and if life emerged in the planet, by comparison with the evolution of life on Earth, Venusian life forms could have reached quite a sophisticated level of evolution with the appearance of photosynthesis and, even perhaps oxygenic photosynthesis. However, its closer proximity to the Sun resulted in what is termed a run-away greenhouse effect (Ingersoll 1969) that resulted in Venus becoming uninhabitable to such an extent that temperatures >350 °C at the surface, an atmosphere rich in sulphuric acid, together with volcanic resurfacing have most likely erased any trace of potential past life.

Mars, for different reasons, also became uninhabitable at the surface after about 3.8 Ga. With a decrease of its internal heat flow and cessation of the dynamo that created its magnetic field, this planet half the size of the Earth lost its protection against the solar wind, which stripped the atmosphere leaving the unprotected planet almost airless and frozen (Jakosky et al. 2015). As a result of its limited habitability, i.e. consisting of habitats that were not necessarily co-located in time or space—termed "punctuated habitability" by Westall et al. (2015a), or "uninhabited habitats" by Cockell et al. (2012)—life, if it emerged, is likely to have remained in a primitive, anaerobic prokaryote stage and, potentially, without the evolution of even anaerobic photosynthesis. As the planet became a frozen desert during the Hesperian Period (after about 3.8 Ga), its pockets of habitability became increasingly restricted and finally disappeared from the surface. Nevertheless, at any time since the disappearance of major occurrences of surface water, pockets of water could have appeared either through impacts tapping into the subsurface cryosphere, or through melting of subsurface water or ice at the poles during changes in the obliquity of the planet (Mars' axis of rotation is not stabilised by a large moon, as is the case on Earth). It is even possible that there is extant life in the subsurface of the planet, possibly as dormant or slow-living cells (Cockell 2014). Thus, viable cells preserved in the cryosphere could rapidly colonise short-lived habitats, if they could be transported to them. Transport can be envisaged *via* impacted rocks with cells in fractures or through groundwater transport.

The early terrestrial anaerobic chemotrophic colonies represent a precious record of the kinds of life forms that could have lived on Mars and early Venus. Their habitats and modes of preservation are a useful indication of how microbial life on Mars could have been preserved, although the analogy is nevertheless limited because life on Mars could have inhabited a wider range of environmental ecosystems including cold to frozen deserts and ice, hyperarid, hypersaline environments, and the subsurface. Ice was not present on the early Earth, although saline evaporative conditions (but not salt desert) are recorded at the coast (Lowe and Fisher Worrell 1999; Westall et al. 2006b). Moreover, rapid silicification of the water/rock and sediment interface during the Palaeoarchaean would have effectively sealed off the subsurface environments (cf. Chap. 4). These are environments found on the Earth today and therefore, despite the presence of oxygen, the extreme biomes are relevant to the search for life on environmentally diverse Mars.

Nevertheless, early Earth is the only truly anaerobic analogue for life on Mars. As described above, the chemotrophic cells that inhabited it occupied mainly nutrient-rich hydrothermal settings, but they also emphemerally colonized the surfaces of volcanic particles in oligotrophic environments, co-existing with phosynthetic biofilms and mats in both. Unfortunately, much sample preparation and sometimes highly sophisticated instrumentation is necessary to identify the purported microfossils and to verify their biogenicity and syngenicity. From this point of view, the methods used to study the early microbes provide a useful indication of what would be required to search for traces of fossil life in samples returned from Mars (e.g. the Mars 2020 mission). They also demonstrate the difficulty of finding and identifying fossil microbes in situ on Mars (Westall et al. 2015a; Vago et al. 2017).

7.4 Conclusions

The early record of life on Earth starts late, when life had already evolved to the photosynthetic stage, the earlier steps being lost due to recycling of Hadean and Eoarchaean crust and high-grade metamorphism of the remnants. Despite metamorphism, certain isotopic and kerogenous biosignatures can still be detected that testify to the likely presence of life back to ~3.9 Ga. Good preservation of crustal rocks started after ~3.5 Ga and document mainly shallow-water environments inhabited by a great diversity of microbial consortia ranging from chemotrophic colonies to photsynthetic microbial biofilms and mats, as well as larger, enigmatic cell-like structures of unclear origin.

In terms of the search for extraterrestrial life, it is our closest neighbour, Mars, which has the most likely chance of hosting past and possibly extant traces of life. Early life on Earth is the most representative anaerobic analogue for Martian life: in a shallow-water hydrothermal setting. Although biomes on early Mars were similar on a microbial scale to those of the early Earth, throughout its geological history Mars had many more diverse environments, including icy and hot deserts, as well as the subsurface realm.

Acknowledgements FW and KHL acknowledge support from The French Space Agency (CNES) and the Mars Analogues for Space Exploration (MASE) project supported by the FP7/2007-2013 (grant no. 607297). BC acknowledges support from the FP7-PEOPLE-2013-CIG/INACMa (grant no. 618657).

References

Allwood AC, Walter MR, Kamber BS et al (2006) Stromatolite reef from the early Archaean era of Australia. Nature 441:714–719

Allwood AC, Grotzinger JP, Knoll AH et al (2009) Controls on development and diversity of Early Archean stromatolites. Proc Natl Acad Sci USA 106:9548–9555

Bontognali TRR, Sessions AL, Allwood AC et al (2012) Sulfur isotopes of organic matter preserved in 3.45-billion-year-old stromatolites reveal microbial metabolism. Proc Natl Acad Sci USA 109:15146–15151

Brasier MD, Green OR, Jephcoat AP et al (2002) Questioning the evidence for Earth's oldest fossils. Nature 416:76–81

Brasier MD, Antcliffe J, Saunders M et al (2015) Changing the picture of Earth's earliest fossils (3.5–1.9 Ga) with new approaches and new discoveries. Proc Natl Acad Sci USA 112:4859–4864

Buick R, Dunlop J, Groves D (1981) Stromatolite recognition in ancient rocks: an appraisal of irregularly laminated structures in an Early Archean chert-barite unit from North Pole, Western Australia. Alcheringa 5:161–181

Byerly GR, Lowe DR, Walsh MM (1986) Stromatolites from the 3300–3500 Myr Swaziland Supergroup, Barberton Mountain Land, South Africa. Nature 319:489–491

Cockell CS (2014) The subsurface habitability of terrestrial rocky planets: Mars. In: Kellmeyer J, Wagner D (eds) Microbial life of the deep biosphere. De Gruyter, Boston, pp 225–259

Cockell CS, Balme M, Bridges JC et al (2012) Uninhabited habitats on Mars. Icarus 217:184–193

Dass AV, Hickman-Lewis K, Brack A et al (2016) Stochastic prebiotic chemistry within realistic geological systems. ChemistrySelect 1:4906–4926

de Duve C (1995) Vital dust. BasicBooks, New York

de Vries ST, Touret JLR (2007) Early Archaean hydrothermal fluids; a study of inclusions from the ~3.4 Ga Buck Ridge Chert, Barberton Greenstone Belt, South Africa. Chem Geol 237:289–302

de Vries ST, Nijman W, de Boer PL (2010) Sedimentary geology of the Palaeoarchaean Buck Ridge (South Africa) and Kittys Gap (Western Australia) volcano-sedimentary complexes. Precambrian Res 183:749–769

Delarue F, Robert F, Sugitani K et al (2017) Investigation of the geochemical preservation of ca. 3.0 Ga permineralized and encapsulated microfossils by nanoscale secondary ion mass spectrometry. Astrobiology 17:1192–1202

Derenne C, Robert F, Skrzypczak-Bonduelle A et al (2008) Molecular evidence for life in the 3.5 billion year old Warrawoona chert. Earth Planet Sci Lett 272:476–448

Foucher F, Westall F, Brandstatter F et al (2010) Testing the survival of microfossils in artificial Martian sedimentary meteorites during entry into Earth's atmosphere: the STONE 6 experiment. Icarus 207:616–630

Golden DC, Ming DW, Schwandt CS et al (2000) An experimental study on kinetically-driven precipitation of Ca-Mg-Fe carbonates from solution: implications for the low temperature formation of carbonates in Martian meteorite ALH84001. Meteorit Planet Sci 35:457–465

Hassenkam T, Andersson MP, Dalby KN et al (2017) Elements of Eoarchean life trapped in mineral inclusions. Nature 548:78–81

Heubeck C (2009) An early ecosystem of Archean tidal microbial mats (Moodies Group, South Africa, ca. 3.2 Ga). Geology 37:931–934

Hickman-Lewis K, Garwood RJ, Brasier MD et al (2016) Carbonaceous microstructures from sedimentary laminated chert within the 3.46 Ga Apex Basalt, Chinaman Creek locality, Pilbara, Western Australia. Precambrian Res 278:161–178

Hickman-Lewis K, Garwood RJ, Withers PJ et al (2017) X-ray microtomography as a tool for investigating the petrological context of Precambrian cellular Remains. In: Brasier AT, McIlroy D, McLoughlin N (eds) Earth system evolution and early life: a celebration of the work of Martin Brasier, vol 448. Geological Society, London, Special Publications, pp 33–56

Hickman-Lewis K, Cavalazzi B, Foucher F et al (2018) Most ancient evidence for life in the Barberton Greenstone Belt: microbial mats and biofabrics of the ~3.47 Ga Middle Marker Horizon. Precambrian Res 312:45–67

Hofmann HJ, Grey K, Hickman A et al (1999) Origin of 3.45 Ga coniform stromatolites in Warrawoona Group, Western Australia. Geol Soc Am Bull 111:1256–1262

Homann M, Heubeck C, Airo A et al (2015) Morphological adaptations of 3.22 Ga-old tufted microbial mats to Archean coastal habitats (Moodies Group, Barberton Greenstone Belt, South Africa). Precambrian Res 266:47–64

Homann M, Heubeck C, Bontognali TRR et al (2016) Evidence for cavity-dwelling microbial life in 3.22 Ga tidal deposits. Geology 44:51–54

Ingersoll AP (1969) The runaway greenhouse: a history of water on Venus. J Atmos Sci 26:1191–1198

Jakosky BM, Grebowsky JM, Luhmann JG et al (2015) Initial results from the MAVEN mission to Mars. Geophys Res Lett 42:8791–8802

Javaux EJ, Marshall CP, Bekker A (2010) Organic-walled microfossils in 3.2-billion-year-old shallow-marine siliciclastic deposits. Nature 463:934–938

Kamber BS (2015) The evolving nature of terrestrial crust from the Hadean, through the Archaean, into the Proterozoic. Precambrian Res 258:48–82

Le Guillou C, Rouzaud JN, Bonal L et al (2012) High resolution TEM of chondritic carbonaceous matter: metamorphic evolution and heterogeneity. Meteorit Planet Sci 47:345–362

Lewandowski Z, Walser G (1991) Influence of hydrodynamics on biofilm accumulation. In: Krenkel PA (ed) Environmental Engineering Proceedings. American Society of Civil Engineers, New York, pp 619–624

Lindsay JF, Brasier MD, McLoughlin N et al (2005) The problem of deep carbon – an Archean paradox. Precambrian Res 143:1–22

Lowe DR, Fisher Worrell G (1999) Sedimentology, mineralogy and implications of silicified evaporites in the Kromberg Formation, Barberton Greenstone Belt, South Africa. In: Lowe DR, Byerly GR (eds) Geolgic evolution of the Barberton Greenstone Belt, South Africa, vol 329. Geological Society of America, Special Paper, pp 167–180

Marshall C, Love GD, Snape CE et al (2007) Structural characterization of kerogen in 3.4 Ga Archaean cherts from the Pilbara Craton, Western Australia. Precambrian Res 155:1–23

McKay DS, Gibson EK Jr, Thomas-Keprta KL et al (1996) Search for past life on Mars: possible relic biogenic activity in Martian meteorite ALH84001. Science 273:924–930

Moorbath S (2009) The discovery of the Earth's oldest rocks. Notes Rec R Soc 63:381–392

Mozjsis SJ, Harrison TM, Pidgeon RT (2001) Oxygen-isotope evidence from ancient zircons for liquid water at the Earth's surface 4,300 Myr ago. Nature 409:178–181

Noffke N, Gerdes G, Klenke T, Krumbein WE (2001) Microbially induced sedimentary structures—a new category within the classification of primary sedimentary structures. J Sediment Res 71:649–656

Noffke N (2009) The criteria for the biogeneicity of microbially induced sedimentary structures (MISS) in Archean and younger, sandy deposits. Earth Sci Rev 96:173–180

Noffke N, Hazen RM, Nhleko N (2003) Earth's earliest microbial mats in a siliciclastic marine environment (2.9 Ga Mozaan group, South Africa). Geology 31:673–676

Noffke N, Eriksson KA, Hazen RM et al (2006) A new window into Archean life: microbial mats in Earth's oldest siliciclastic tidal deposits (3.2 Ga Moodies Group, South Africa). Geology 34:253–256

Noffke N, Christian D, Wacey D et al (2013) Microbially induced sedimentary structures recording an ancient ecosystem in the ca. 3.48 billion year-old Dresser Formation, Pilbara, Western Australia. Astrobiology 13:1103–1124
Oehler DZ, Robert F, Walter MR et al (2010) Diversity in the Archean Biosphere: new insights from NanoSIMS. Astrobiology 10:413–424
Oehler DZ, Walsh MM, Sugitani K et al (2017) Large and robust lenticular microorganisms on the young Earth. Precambrian Res 296:112–119
Ohtomo Y, Kakegawa T, Ishida A et al (2013) Evidence for biogenic graphite in early Archaean Isua metasedimentary rocks. Nat Geosci 7:25–28
Orange F, Westall F, Disnar JR et al (2009) Experimental silicification of the extremophilic Archaea *Pyroccus abyssi* and *Methanocaldococcus jannaschii*. Applications in the search for evidence of life in early Earth and extraterrestrial rocks. Geobiology 7:403–418
Pearson VK, Sephton MA, Franchi IA et al (2006) Carbon and nitrogen in carbonaceous chondrites: elemental abundances and stable isotopic compositions. Meteorit Planet Sci 41:1899–1918
Pope MC, Grotzinger JP, Schreiber BC (2000) Evaporitic subtidal stromatolites produced by in situ precipitation: textures, facies associations, and temporal significance. J Sediment Res 70:1139–1151
Rosing MT (1999) 13C-depleted carbon microparticles in >3700-Ma seafloor sedimentary rocks from West Greenland. Science 283:674–676
Schidlowki M (2001) Carbon isotopes as biogeochemical recorders of life over 3.8 Ga of Earth history: evolution of a concept. Precambrian Res 106:117–134
Schopf JW (1993) Microfossils of the Early Archean Apex Chert: new evidence of the antiquity of life. Science 260:640–646
Schopf JW, Kitajima K, Spicuzza MJ et al (2018) SIMS analyses of the oldest known assemblage of microfossils document their taxon-correlated carbon isotope compositions. Proc Natl Acad Sci USA 115:53–58
Steele A, McGubbin FM, Agee C et al (2012) A reduced organic carbon component to Martian Basalts. Science 337:212–215
Stoodley P, Jacobsen A, Dunsmore BC, Purevdorj B, Wilson S, Lapin-Scott HM, Costerton JW (2001) The influence of fluid shear and alcl3 on the material properties of pseudomonas aeruginosa pao1 and desulfovibrio sp. ex265 biofilms. Water Sci and Technol 43:113–120
Sugitani K, Grey K, Nagaoka T et al (2009) Taxonomy and biogenicity of Archaean spheroidal microfossils (ca. 3.0 Ga) from the Mount Goldsworthy-Mount Grant area in the northeastern Pilbara Craton, Western Australia. Precambrian Res 173:50–59
Sugitani K, Mimura K, Takeuchi M et al (2015) A Paleoarchean coastal hydrothermal field inhabited by diverse microbial communities: the Strelley Pool Formation, Pilbara Craton, Western Australia. Geobiology 13:522–545
Tan CH, Lee KWK, Burmølle M et al (2017) All together now: experimental multispecies biofilm model systems. Environ Microbiol 19:42–53
Tashiro T, Ishida A, Hori M et al (2017) Early trace of life from 3.95 Ga sedimentary rocks in Labrador, Canada. Nature 549:516–518
Tice MM (2009) Environmental controls on photosynthetic microbial mat distribution and morphogenesis on a 3.42 Ga clastic-starved platform. Astrobiology 9:989–1000
Tice M, Lowe DR (2004) Photosynthetic microbial mats in the 3,416-Myr-old ocean. Nature 431:549–552
Tice M, Lowe DR (2006) The origin of carbonaceous matter in pre-3.0 Ga greenstone terrains: a review and new evidence from the 3.42 Ga Buck Reef Chert. Earth Sci Rev 76:259–300
Ueno Y, Isozaki Y, Yurimoto H, Maruyama S (2001) Carbon isotopic signatures of individual Archean microfossils (?) From Western Australia. Int Geol Rev 43:196–212
Vago JL, Westall F, Pasteur Instrument Teams et al (2017) Habitability on Early Mars and the search for biosignatures with the ExoMars Rover. Astrobiology 17:471–510

van den Boorn SHJM, van Bergen MJ, Nijman W et al (2007) Dual role of seawater and hydrothermal fluids in Early Archean chert formation: evidence from silicon isotopes. Geology 10:939–942

Wacey D, McLoughlin N, Green OR, Parnell J, Stoakes CA, Brasier MD (2006) The *3.4 billion-year-old strelley pool sandstone: a new window into early life on earth. Int J Astrobiol 5:333–342

Wacey D (2010) Stromatolites in the ~3400 Ma Strelley Pool Formation, Western Australia: examining biogenicity from the macro- to the nano-scale. Astrobiology 10:381–395

Wacey D, Kilburn MR, Saunders M et al (2011) Microfossils of sulphur-metabolizing cells in 3.4-billion-year-old rocks of Western Australia. Nat Geosci 4:698–670

Wacey D, Saunders M, Kong C et al (2015) 3.46 Ga Apex chert 'microfossils' reinterpreted as mineral artefacts produced during phyllosilicate exfoliation. Gondwana Res 36:296–313

Walsh MM (1992) Microfossils and possible microfossils from the Early Archean Onverwacht Group, Barberton Mountain Land, South Africa. Precambrian Res 54:271–293

Walsh MM, Lowe DR (1999) Modes of accumulation of carbonaceous matter in the early Archean: a petrographic and geochemical study of the carbonaceous cherts of the Swaziland Supergroup. In: Lowe DR, Byerly GR (eds) Geological evolution of the Barberton Greenstone Belt, South Africa, vol 329. Geological Society of America, Special Publications, pp 115–132

Way MJ, Del Genio AD, Kiang NY et al (2016) Was Venus the first habitable world of our solar system? Geophys Res Lett 3:8376–8383

Westall F, de Vries ST, Nijman W et al (2006a) The 3.466 Ga Kitty's Gap Chert, an Early Archaean microbial ecosystem. In Reimold WU, Gibson R (eds) Processes on the Early Earth, vol 405. Geological Society of American, Special Publications, pp 105–131

Westall F, de Ronde CEJ, Southam G et al (2006b) Implications of a 3.472–3.333 Ga-old subaerial microbial mat from the Barberton greenstone belt, South Africa for the UV environmental conditions on the early Earth. Philos Trans R Soc Lond B 361:1857–1875

Westall F, Foucher F, Cavalazzi B et al (2011a) Early life on Earth and Mars: a case study from ~3.5 Ga-old rocks from the Pilbara, Australia. Planet Space Sci 59:1093–1106

Westall F, Cavalazzi B, Lemelle L et al (2011b) Implications of in situ calcification for photosynthesis in a ~3.3 Ga-old microbial biofilm from the Barberton greenstone belt, South Africa. Earth Planet Sci Lett 310:468–479

Westall F, Foucher F, Bost N et al (2015a) Biosignatures on Mars: what, where and how? Implications for the search for Martian life. Astrobiology 15:998–1029

Westall F, Campbell KA, Bréhéret JG et al (2015b) Archean (3.33 Ga) microbe-sediment systems were diverse and flourished in a hydrothermal context. Geology 43:615–618

Westall F, Hickman-Lewis K, Hinman N et al (2018) A hydrothermal sedimentary context for the origin of life. Astrobiology 18:259–293

Part II
Biosignatures in Space

Chapter 8
The Search for Biosignatures in Martian Meteorite Allan Hills 84001

Harry Y. McSween Jr.

Abstract Proposed biosignatures in the ancient Allan Hills 84001 martian meteorite are most plausibly explained as abiotic features. The purported evidence of biological activity on Mars included biogenic minerals (magnetite and sulphide formed by magnetotactic and sulphate-respiring microorganisms), organic matter resulting from the decay of such organisms, microfossils, and biofilms, all physically associated with biologically mediated carbonates. The zoned carbonate globules formed by inorganic precipitation from an aqueous fluid or evaporative brine circulating within fractures in this igneous rock. A subsequent shock event partially volatilized Fe-carbonate, and its decomposition produced nanophase magnetite crystals with unusual morphologies, structures, and compositions consistent with vapour condensation. Sulphur isotopes in sulphide are unlike those in terrestrial biogenic sulphides. The organic compounds identified in ALH 84001 include polycyclic aromatic hydrocarbons, complex macromolecules, graphite, and amino acids, most of which are terrestrial, based on their carbon isotopes and stereochemistry. A small amount of the organic matter may be martian, but even that likely had an exogenic (chondritic) source. The putative microfossils were identified only by morphology, without any other supporting observations. These forms are apparently too small to represent viable organisms, which has engendered controversy about the plausibility of nanobacteria. Observations of possible fossilized biofilms are compromised by infiltration of the meteorite by terrestrial microorganisms in the Antarctic environment from which the meteorite was recovered. The controversial hypothesis that ALH 84001 contains evidence of extraterrestrial biology has mostly subsided, but it has fuelled a Mars exploration program focused on the search for life and has helped refine the criteria for the recognition of biosignatures.

H. Y. McSween Jr. (✉)
University of Tennessee, Knoxville, TN, USA
e-mail: mcsween@utk.edu

B. Cavalazzi, F. Westall (eds.), *Biosignatures for Astrobiology*, Advances in Astrobiology and Biogeophysics, https://doi.org/10.1007/978-3-319-96175-0_8

8.1 Introduction

The intriguing proposal that the martian meteorite Allan Hills (ALH) 84001 contains biochemical markers, biogenic minerals, and microfossils (McKay et al. 1996) engendered great public interest and a flurry of claims supporting or refuting the hypothesis (e.g., McSween 1997; Gibson et al. 2001; Golden et al. 2001; Treiman 2003; and many more scientific squabbles). Comparable features were also reported in other martian meteorites (Nakhla and Shergotty), leading Gibson et al. (2001) to suppose that all these samples might contain evidence for extraterrestrial life, although that hypothesis has been largely ignored. While the scientific controversy about life in a martian meteorite has now mostly subsided, this work has spawned an international Mars exploration program focused on the search for life.

ALH 84001 was collected on the Antarctic ice in 1984 but, owing to its distinctive petrography relative to other martian meteorites, it was not recognized as a Mars rock until a decade later (Mittlefehldt 1994). Its martian provenance has been confirmed by its diagnostic oxygen isotopic composition (Clayton and Mayeda 1996) and by noble gas data that indicate a trapped martian atmospheric component (Swindle et al. 1995; Goswami et al. 1997; Bogard and Garrison 1998). Based on its cosmic-ray exposure, an impact launched the meteorite from Mars ~14 million years ago (Eugster et al. 1997). The crystallization age of ALH 84001 has been radiometrically dated at 4.091 $\pm$ 0.030 billion years (Lapen et al. 2010). This ancient formation age corresponds to the Noachian Period, which is commonly thought to have been more hospitable to life than more recent periods.

ALH 84001 is an ultramafic igneous rock, seemingly an unlikely sample to contain biosignatures. However, the meteorite is cut by fracture zones in which carbonates of martian origin (Jull et al. 1995) were deposited. The age of the carbonates is slightly younger, ~3.90–4.04 billion years old (Borg et al. 1999), than the host rock. Associated with the carbonate grains are unusual microparticles and organic matter that together constitute the evidence for life cited by McKay et al. (1996).

Gram for gram, ALH 84001 is arguably the most intensely studied rock in history. Here I review and evaluate the extensive research published in, and in response to, the original proposal that it contains indications of martian life, and conclude (as have others) that the evidence is best explained by abiotic processes. Nonetheless, this controversy has prompted useful examination of the limits of life and has helped sharpen the tools employed in the search for biosignatures in extraterrestrial samples.

Fig. 8.1 Portion of the ALH 84001 meteorite, ~8 cm across, showing dark fusion crust. Source: Image courtesy of the Smithsonian Institution

8.2 Petrologic Context for Claims of Biologic Activity

ALH 84001 (originally a 1.93 kg find, Score and MacPherson 1985) is a grey stone covered by dark fusion crust (Fig. 8.1). It consists primarily of orthopyroxene crystals, with minor plagioclase and chromite, accumulated from basaltic parent magma (Mittlefehldt 1994; Treiman 1995; Gleason et al. 1997). Minor igneous minerals include olivine, alkali feldspar, augite, sulphide, and phosphate. Trace element analyses of the meteorite and its constituent minerals (Warren and Kallemeyn 1996; Wadhwa and Crozaz 1998) are consistent with its origin as a magmatic cumulate.

The meteorite has been heavily shock metamorphosed and sheared. Multiple shock events bracket the formation of carbonates (Treiman 1998; Greenwood and McSween 2001; Corrigan and Harvey 2004). The earliest shock produced granular "crush zones" of orthopyroxene and other minerals that transect the original igneous fabric (Fig. 8.2). These bands may have originally been cataclastic zones or melt dikes; however, their former textures have been annealed and coarsened by recrystallization at high temperature (Treiman 1998). This shock event also formed feldspathic impact melts (Greenwood and McSween 2001). Although Scott et al. (1997) suggested that the carbonates were deposited from carbonate impact melt, other evidence described below demonstrates their precipitation from aqueous fluids. Following carbonate precipitation, additional shock(s) produced microfaults that offset the carbonate grains and granular bands, and melted and mobilized silica.

There is some disagreement about the total number of shock metamorphic events recorded in ALH 84001: Treiman (1998) argued for at least four, whereas Greenwood and McSween (2001) required only two. In any case, the purported biosignatures are associated with carbonates within the granular bands, and may be related to significant shock effects after carbonate deposition.

Fig. 8.2 Granular "crush zones" in ALH 84001, viewed in thin section under crossed nicols. Image is ~2 cm across. Source: Image courtesy of DS Lauretta

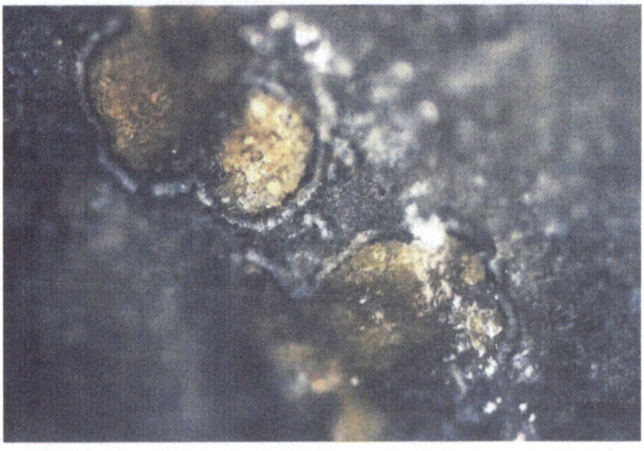

Fig. 8.3 Carbonate globules, showing orange cores and alternating iron-rich (black) and magnesium-rich (white) rims. Globules are ~150 μm across. Source: Image courtesy of M Grady

8.3 Biogenic Carbonates?

The carbonates in ALH 84001 occur mostly as flattened (discoidal) globules or "rosettes", 20–200 μm in diameter and 10–50 μm thick. They have semi-circular cross-sections, are orange in visible light, and exhibit alternating dark and light rims (Fig. 8.3). Each rosette is composed of needle-shaped crystallites that radiate outward from the center (Treiman and Romanek 1998). The rosettes are compositionally zoned (Fig. 8.4), ranging from calcium-rich (calcite $CaCO_3$ or ankerite $Ca(Mg,Fe)(CO_3)_2$) compositions in the interior to rims alternating between magnesium-rich

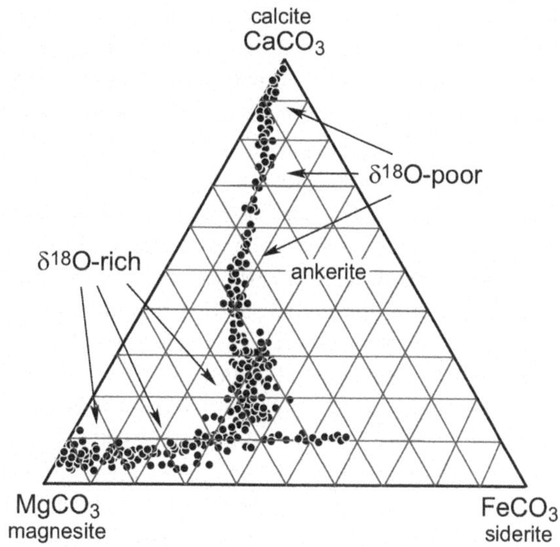

Fig. 8.4 Compositional zoning in carbonates. Oxygen isotopic compositions of carbonates (Eiler et al. 2002) vary with calcium contents: $\delta^{18}O$-rich = +14.5 ± 6.5‰, $\delta^{18}O$-poor = +5.5 ± 2.2‰. Source: Adapted from Corrigan and Harvey (2004) and Treiman (2003)

(magnesite $MgCO_3$) and iron-rich (siderite $FeCO_3$) compositions (e.g., Harvey and McSween 1996; Corrigan and Harvey 2004).

The same compositional zoning occurs along the length of each crystallite needle, so it clearly formed during crystallization and likely reflects a changing chemical or thermal environment during formation. The oxygen isotopic compositions of carbonates (e.g., Eiler et al. 2002) vary systematically with their calcium contents (Fig. 8.4). In addition to the rosettes, other less common textural forms of carbonate have also been noted. These include slabs with zoning that duplicates that in rosettes, lacy carbonate associated with impact glass, and magnesite that fills voids. Corrigan and Harvey (2004) presented textural evidence that these forms represent multiple, distinct generations of carbonate formation.

All of the cited evidence for life is associated with the carbonates, and McKay et al. (1996) suggested that the carbonate globules themselves might be bacterially induced precipitates. Consequently, their environment and conditions (especially temperature) of formation are critical to the argument for life. The carbonates are clearly martian in origin, based on their carbon isotopic composition (Jull et al. 1995). Based on the oxygen isotopes, Romanek et al. (1994) first suggested that the carbonate formed by reaction with a hydrothermal fluid at low temperature (0–80 °C). Harvey and McSween (1996), on the other hand, favoured formation at high temperatures (>500 °C) by reaction of the rock with hot CO_2-rich fluid during an impact event, based on carbonate mineral thermometry and the absence of phyllosilicates in ALH 84001. Finally, Scott et al. (1998) proposed that the carbonates crystallized from carbonate shock melts. However, the geothermometry assumes that the carbonate phases are in equilibrium, which is probably not correct. Subsequent oxygen isotope measurements of the carbonates have yielded temperatures ranging from 20 to 375 °C (Valley et al. 1997; Leshin et al. 1998; Treiman and Romanek 1998; Eiler et al. 2002), and clumped isotope analyses fix the temperature of carbonate precipitation at 18 ± 4 °C (Halevy et al. 2011). Paleomagnetic measurements of ALH 84001 indirectly support a low formation temperature for the carbonates (Kirschvink et al. 1997). The calcium-rich cores of rosettes give higher temperatures than the rims, based on both oxygen isotope and Sr/Ca thermometry (Eiler et al. 2002).

Treiman et al. (2002) favoured precipitation of the carbonates from hydrothermal fluids, and described an analogous terrestrial occurrence of zoned carbonate globules in basalts from Spitsbergen. A hydrothermal model would likely imply carbonate formation temperatures of 100 °C or more. Golden et al. (2001) experimentally produced chemically zoned carbonates with compositions like those in ALH 84001 under non-equilibrium, hydrothermal (150 °C) conditions. Noting that the absence of phyllosilicates in ALH 84001 was inconsistent with hydrothermal alteration, McSween and Harvey (1998) and Warren (1998) instead suggested the evaporation of brines, implying temperatures below 100 °C. A ponded, evaporating brine in an impact crater could have infiltrated fractures in the crater floor and directly precipitated carbonates. Either model can be reconciled with the observed carbonate compositional zoning and extreme oxygen isotope fractionations.

Following the precipitation of carbonates, likely during multiple events (Eiler et al. 2002; Corrigan and Harvey 2004), they experienced partial thermal decomposition by shock (Scott et al. 1997; Golden et al. 2001), as described below. This seems to be in disagreement with a maximum limiting temperature of 40 °C estimated from paleomagnetic measurements (Weiss et al. 2000). Although Scott et al. (1997) suggested that carbonate (and plagioclase) had experienced shock melting, other observations indicate that the carbonate in ALH 84001 replaced maskelynite (Kring et al. 1998). Some carbonates are also intruded by impact-generated silica glass (Eiler et al. 2002).

8.4 Biogenic Magnetite and Sulphide?

Nanophase magnetite (Fe_3O_4) grains within the ALH 84001 carbonate (Fig. 8.5) were argued to be morphologically similar to those produced by terrestrial magnetotactic bacteria (McKay et al. 1996). The grains occur in a variety of shapes, including equant (octahedral or rhombic dodecahedral), elongated (hexa-octahedral), whiskers, blades, and plates (Bradley et al. 1996; Thomas-Keprta et al. 2000; Barber and Scott 2002; Treiman 2003). The morphologies of bacterially produced magnetites are species-specific (Buseck et al. 2001), so the variations in magnetite shapes in this meteorite would imply a diverse community of organisms. The elongated grains, in particular, were cited as having the sizes and shapes of biogenic magnetite (Thomas-Keprta et al. 2000, 2001).

Magnetotactic bacteria string crystallographically oriented magnetite grains together into chains (magnetosomes) that increase the net magnetic moment, and that are used for sensing the Earth's geomagnetic field. Friedmann et al. (2001) examined broken surfaces of the meteorite and identified aligned magnetite crystals

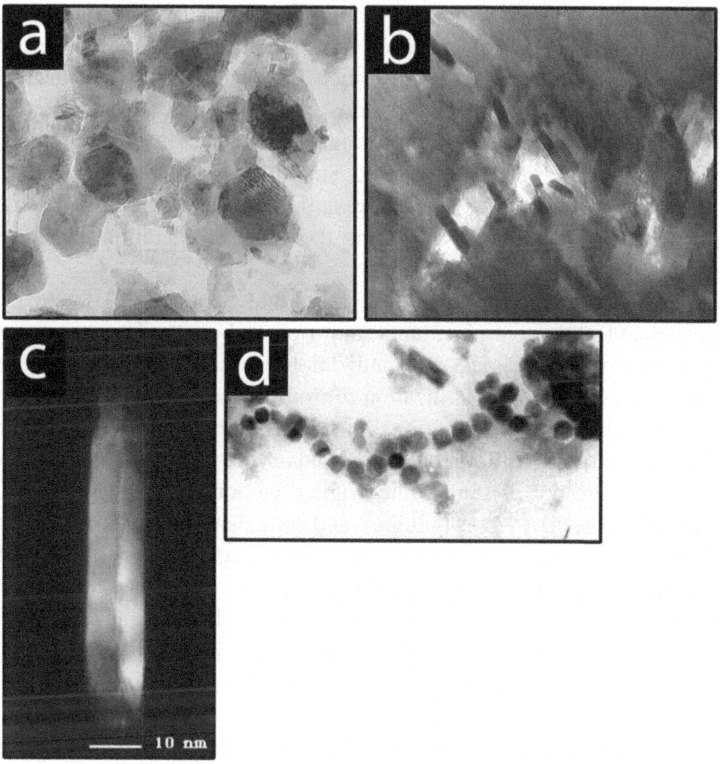

Fig. 8.5 TEM images of nanophase magnetite. (**a**) Lattice-fringe image of hexa-octahedra, measuring up to 40 nm across, in ALH 84001. (**b**) Bright-field image of magnetite whiskers in ALH 84001; several touching crystals are expitaxially intergrown whiskers of 50–100 nm length. (**c**) Dark-field image of magnetite whisker with axial screw dislocation. (**d**) Chain of crystallographically oriented nanophase magnetites (each ~100 nm in size) in terrestrial *Magnetococcus bacterium*. Source: Images courtesy of NASA and JP Bradley

that they interpreted as magnetosomes. However, Buseck et al. (2001) questioned their identification of magnetite and noted that intact magnetite chains were unlikely to survive in rocks.

Although most of the nanophase magnetites in ALH 84001 provide no specific constraints on their growth mode, Bradley et al. (1996) found elongated rods and platelets with internal screw dislocations, consistent with spiral growth from a hot vapour or supercritical fluid. However, biogenic magnetite whiskers with possible screw locations have also been reported (Taylor et al. 2001). The elongated magnetites in the meteorite occur in stacks of single-domain crystals epitaxially intergrown with one another (Bradley et al. 1998), providing a further indication of high-temperature origin and evidence against intracellular precipitation by organisms.

Barber and Scott (2002) recognized two distinct types of nanophase magnetites in ALH 84001. Euhedral magnetite nanocrystals like those described by Bradley et al. (1996) are topotactically oriented with respect to the lattice of the enclosing carbonate, whereas magnetites that are irregular in shape and more varied in size are not well oriented. Topotaxy is evidence for solid-state exsolution. The euhedral, oriented grains are enclosed within iron-rich carbonate zones, and the irregular, unoriented grains are found in the outermost iron-rich rims. Thomas-Keprta et al. (2000) countered that only the elongated grains in the outer rims of carbonate grains, where epitaxy is not observed, are biogenic. Support for a low-temperature origin for magnetite was provided by Melwani Daswani et al. (2016), who calculated that magnetite would be stable at 20 °C during the aqueous alteration that formed the carbonate.

In addition to nanophase magnetite, Barber and Scott (2002) found periclase (MgO) associated with voids in magnesium-rich carbonate. They interpreted the periclase as evidence of shock devolatilization of magnesite, and similarly the magnetites as products of devolatilization of iron-rich carbonate (siderite). Treiman (2003) also explained the compositional purity of the magnetites by formation via siderite decomposition, as they could not contain elements (titanium, chromium, aluminum) that were not present in the original carbonate. He synthesized all the information about the sizes, morphologies, and compositions of ALH 84001 magnetites, and provided detailed explanations for all these features as the results of thermal decomposition of sideritic carbonate following shock. Based on an observation of reduced carbon in ALH 84001 (Steele et al. 2007), Treiman and Essene (2011) calculated the magnetite formation pathway ($FeCO_3 = Fe_3O_4 + CO_2 + C$).

The hypothesis that fine-grained sulphide coexisting with nanophase magnetite in ALH 84001 carbonate could be the product of sulphate-respiring bacteria was also offered by McKay et al. (1996). These sulphides were tentatively identified as pyrrhotite ($Fe_{1-x}S$) and greigite (Fe_3S_4), distinct from the coarser-grained pyrite (FeS_2) occurring in the crushed zones of the meteorite. Sulphur isotope ratios are fractionated by terrestrial bacteria depending on the concentration and species of the sulphur source, but biogenic sulphate reduction typically results in significant enrichment in the light sulphur isotope ^{32}S. The measured heavy sulphur isotope ratio in ALH 84001 sulphides (Shearer et al. 1996), and especially in a pyrrhotite-bearing zone within a carbonate globule ($\delta^{34}S$ = +6‰, Greenwood et al. 1997), are indistinguishable from that of the pyrite in the meteorite. However, in a later study Greenwood et al. (2000) found pyrite with ($\delta^{34}S$ = −9.7‰), which they attributed to precipitation from a fluid derived by impact of the regolith.

8.5 Biological Organic Matter?

McKay et al. (1996) found fused carbon rings (polycyclic aromatic hydrocarbons, or PAHs) associated with the carbonate globules, and suggested that they represented the decomposed remains of microorganisms. Although PAHs play no known role in biochemistry, they can form from the chemical transformation of decayed organisms. In response, Anders (1996) and Oró (1998) pointed out that, given enough time and elevated temperatures, PAHs form readily from any organic materials, whether biological or not. Anders (1996) further observed that the specific PAHs from C_{14} to

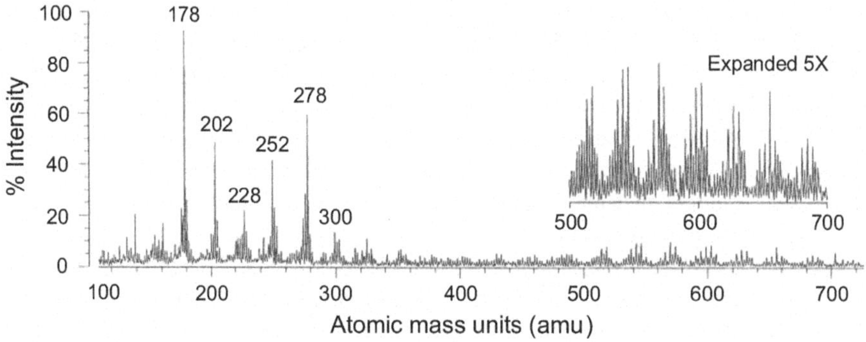

Fig. 8.6 PAHs detected in ALH 84001 carbonate globules. Peaks at 178, 202, 228, 252, 276, 278, and 300 amu are assigned to $C_{14}H_{10}$, $C_{16}H_{10}$, $C_{18}H_{12}$, $C_{20}H_{12}$, $C_{22}H_{12}$, $C_{22}H_{14}$, and $C_{24}H_{12}$. A high-mass envelope, shown expanded 5 times, is consistent with a chondritic, kerogen-like component. Source: Modified from Becker et al. (1997, 1999)

C_{22} (Fig. 8.6), thought by McKay et al. (1996) to require a selective (i.e., biologic) source, comprise all homologues in this carbon number range and, thus, argued that the interpretation of these molecules as biomarkers was not justified.

Various abiotic origins for the PAHs in ALH 84001 have been subsequently proposed. Zolotov and Shock (2000) modelled the formation of PAHs in ALH 84001 by condensation from an impact-generated gas liberated from devolatilized carbonate. Becker et al. (1997, 1999) noted that the PAHs could have been introduced as exogenic (meteoritic) debris, as their carbon isotopic composition (δ^{13} C = −15‰) is similar to PAHs found in the kerogens of carbonaceous chondrites.

Even the claim of McKay et al. (1996) that the PAHs were present in the meteorite before its fall was challenged by Becker et al. (1997), who instead proposed that the PAHs were adsorbed onto carbonates from Antarctic seasonal melt water. They noted similarities in PAH abundances in melted Antarctic ice and performed experiments to demonstrate their adsorption onto carbonate. Clemett et al. (1998) responded with contradictory measurements of PAHs in Antarctic ice and differing adsorption experiment results, and observed that other Antarctic meteorites were not similarly contaminated with PAHs. They also indicated that the PAHs in ALH 84001 differed from those in chondrites by the presence or absence of some of the attached side chains. In addition to PAHs, the only other analysed organic components in ALH 84001 are amino acids. Bada et al. (1998) extracted trace amounts of glycine, serine, and alanine from the carbonate and from the rest of the meteorite.

PAHs and amino acids together constitute only a small fraction of the organic matter in ALH 84001. In fact, most of the carbon in the meteorite occurs as a refractory macromolecular phase (Becker et al. 1999; Sephton et al. 2002; Steele et al. 2007) and as graphite (Steele et al. 2012), both interpreted to be abiotic. The δ^{13} C values for organic matter extracted from the carbonate average −26‰, similar to terrestrial C_4 plants (Becker et al. 1999). Analyses of ^{14}C in the organic matter of

ALH 84001 also indicate that at least 80% of the organic carbon is terrestrial (Jull et al. 1998), although an acid-resistant component did not contain modern-day atmospheric ^{14}C. Grady et al. (1994) and Jull et al. (1998) interpreted the carbon component lacking ^{14}C to be martian. In any case, most if not all of the amino acids in ALH 84001 carbonate are clearly of terrestrial origin, based on the identical proportions of stereoisomers in the meteorite and Antarctic ice (Bada et al. 1998), so there is little doubt that the meteorite has been contaminated on Earth by water-soluble organic compounds.

The whole issue of the characterization and origin of organic matter in ALH 84001 is further confounded by the detection of contaminating terrestrial microorganisms on the meteorite (Steele et al. 2000). Fibrous structures resembling mycelia are similar to those in crypto-endolithic communities, implying that the organism was introduced during the long residence of the meteorite in Antarctica. These organisms could have penetrated the interior *via* fractures.

8.6 Microfossils and Related Biogenic Structures?

Certainly the most intriguing evidence cited by McKay et al. (1996) was an assortment of elongated and tubular forms observed in fractures in the carbonates, which they argued were bacterial microfossils (Fig. 8.7a, b). One form illustrated in their paper even appeared to be segmented. However, morphology alone is not sufficient to identify these structures as microfossils—Precambrian microfossil experts generally look for evidence of cell walls, growth, reproduction, and colonies as necessary supporting observations, none of which have been reliably described in ALH 84001. Similar elongated forms in the meteorite appear to be magnetites (Fig. 8.7c).

Although the forms in meteoritic carbonates appear morphologically similar to some terrestrial bacteria, they are much smaller, only 20–100 nm in size. This size differential has prompted a debate about how small free-living (non-parasitic) organisms can be and still have sufficient internal volumes to carry the required genetic information and molecules necessary for metabolism. The generally accepted lower size limit is ~200–250 nm diameter (Maniloff 1997; Nealson 1997; Knoll and Osborne 1999), significantly larger than the putative microfossils in ALH 84001. The putative nanobacteria in ALH 84001 are the same size as an average ribosome and, if they contain a cell wall and membrane, would have no internal volume and thus contain no molecules necessary for life functions. Nevertheless, a few microbiologists (Kajander and Çiftçioğlu 1998; Uwins et al. 1998) have claimed finding viable cells 80–100 nm in size. The prevailing perception of the biological community caused the McKay group to recant a biological interpretation for some of the structures in ALH 84001 (McKay et al. 1997; Thomas-Keprta et al. 1998), and other structures were suggested to be fossilized bacterial appendages rather than whole organisms.

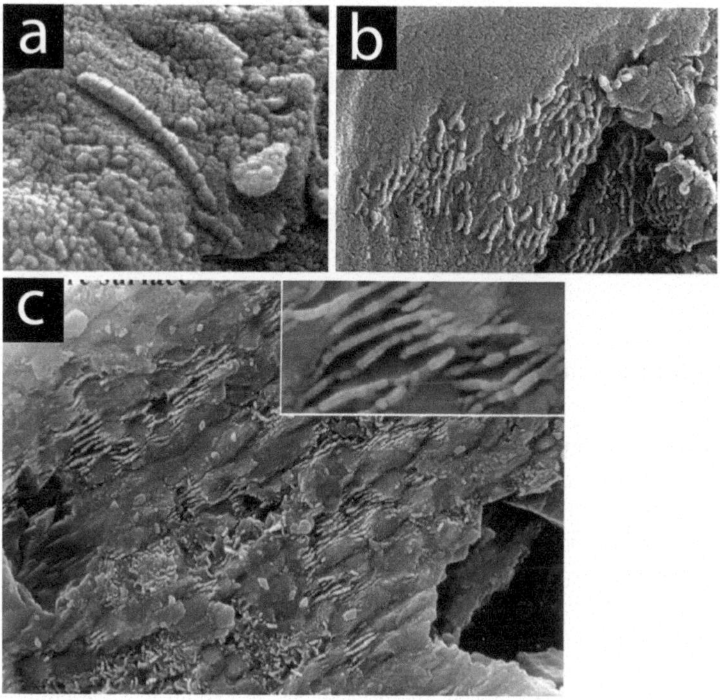

Fig. 8.7 SEM images of purported microfossils in ALH 84001. (**a**) Segmented form, measuring ~100 nm in length. (**b**) Other forms suggested to be microfossils, ~100 nm in length (McKay et al. 1996). (**c**) Similar elongated forms, measuring 200–300 nm and identified as crystal growth steps with metallic coatings (Bradley et al. 1997). Source: Images courtesy of JP Bradley, EK Gibson and NASA

Taking another tact, Folk and Taylor (2002) examined igneous pyroxenes in ALH 84001 for evidence of microfossils. They observed clusters (described as "colonies") of spherical, ovoid, and caterpillar-shaped bodies 30–100 nm in size, which they interpreted as "nannobacteria" (note the different spelling). They also found a few occurrences of larger ellipsoidal bodies, of the size of terrestrial bacteria. Arguing that the shapes of the forms they interpreted as nannobacteria were definitive for biology, not mineralogy, they concluded that the meteorite had been invaded with microorganisms, either on Mars or Earth. This conclusion suffers from being based on morphology alone, and the occurrence of these forms on igneous crystals not associated with crush zones and carbonate is problematic.

Folk (1997) has previously argued for the existence of fossilized nannobacteria in terrestrial sedimentary rocks; however, similar ovoid features in calcite have been demonstrated to be inorganic or artifacts of sample preparation (Kirkland et al. 1999). Bradley et al. (1997) examined the microfossil forms in ALH 84001 carbonates, and argued that the majority of these structures were growth steps on mineral surfaces, accentuated by the Au/Pd coating required for scanning electron microscopy. However, metal lumps produced during coating are smaller (5–10 nm) than some of the putative microfossils, so the larger structures are not entirely artifacts (Steele et al. 1998). Bacterial colonies modify their environment by producing secretions of polysaccharides ("biofilms"). Fine films coating fractures in ALH84001 have been suggested to be fossilized biofilms (Gibson et al. 2001). In addition, oval hollows on the fracture surface of a carbonate grain also resemble bacterial moulds in terrestrial rocks (Westall 1999). However, the biofilm evidence may be compromised by infiltration of terrestrial microorganisms, and the formation mechanism of the hollows is unknown.

8.7 Summary and Lessons Learned

ALH 84001 is an igneous rock, a sample of the ancient martian crust that has experienced hydrothermal alteration and a complex impact history. The zoned carbonate globules in fracture zones precipitated from a circulating hydrothermal fluid or evaporative brine, possibly within the floor of a large impact crater. Subsequent shock metamorphism plausibly accounts for much of the evidence previously cited for biological activity. Putatively biogenic magnetite formed from the shock decomposition of Fe-carbonate, and the small portion of the organic matter in the meteorite that is not terrestrial may be exogenic (chondritic) material modified at high temperatures by shock. Proposed microfossils are too small to have contained the molecules necessary for life functions, and may simply be surface features in shocked carbonate, possibly enhanced by metallic films during sample preparation. Contamination of the meteorite by terrestrial microorganisms in the Antarctic environment renders other evidence suspect.

Despite how it has been resolved, the "life in a martian meteorite" controversy can be seen as a trial run for the examination of returned Mars samples for biosignatures. As already pointed out (Treiman 2003), some of the ambiguities encountered in the meteorite will likely recur when rocks collected on Mars for their biological potential are returned to Earth. Demonstrating a plausible geological context for entrapping and preserving evidence of life is essential, as unravelling the origin of carbonates in ALH 84001 demonstrates. Biogenic minerals are rarely unique, and a combination of morphological, chemical, and isotopic data may be required to recognize them unambiguously (cf. Chap. 6). Thermally processed organic matter can have any number of precursors, and the likelihood of terrestrial organic contamination is a constant concern. Furthermore, the identification of fossilized (especially microscopic) forms requires more than a morphological similarity to life. The scientific community has gained considerable expertise and insights from the analysis of ALH 84001, and our ability to interrogate possible evidence for life is better for the experience.

Acknowledgement John W. Valley and M. Melwani Daswani provided helpful reviews.

References

Anders E (1996) Technical comment. Science 274:2119–2120
Bada JL, Glavin DP, McDonald GD et al (1998) A search for endogenous amino acids in Martian meteorite ALH 84001. Science 2789:362–365
Barber DJ, Scott ERD (2002) Origin of supposedly biogenic magnetite in the Martian meteorite Allan Hills 84001. Proc Natl Acad Sci USA 99:6551–6561
Becker L, Glavin DP, Bada JL (1997) Polycyclic aromatic hydrocarbons (PAHs) in Antarctic Martian meteorites, carbonaceous chondrites, and polar ice. Geochim Cosmochim Acta 61:475–481
Becker L, Popp B, Rust T et al (1999) The origin of organic matter in the Martian meteorite ALH 84001. Earth Planet Sci Lett 167:71–79
Bogard DD, Garrison DH (1998) Relative abundances of argon, krypton, and xenon in the Martian atmosphere as measured in Martian meteorites. Geochim Cosmochim Acta 62:1829–1835
Borg LE, Connelly JN, Nyquist LE et al (1999) The age of the carbonates in Martian meteorite ALH 84001. Science 286:90–94
Bradley JP, Harvey RP, McSween HY (1996) Magnetite whiskers and platelets in ALH 84001 Martian meteorite: evidence for vapor phase growth. Geochim Cosmochim Acta 60:5149–5155
Bradley JP, Harvey RP, McSween HY (1997) No 'nanofossils' in Martian meteorite. Nature 390:454–455
Bradley JP, McSween HY, Harvey RP (1998) Epitaxial growth of nanophase magnetite in Martian meteorite ALH 84001: implications for biogenic mineralization. Meteorit Planet Sci 33:765–773
Buseck PR, Dunin-Borkowski RE, Devouard B et al (2001) Magnetite morphology and life on Mars. Proc Natl Acad Sci USA 98:13490–13495
Clayton RN, Mayeda TK (1996) Oxygen isotope studies of achondrites. Geochim Cosmochim Acta 60:1999–2017
Clemett SJ, Dulay MT, Gilette JS et al (1998) Evidence for the extraterrestrial origin of polycyclic aromatic hydrocarbons (PAHs) in the Martian meteorite ALH 84001. Farady Discuss R Soc Chem 109:417–436
Corrigan CM, Harvey RP (2004) Multi-generational carbonate assemblages in Martian meteorite Allan Hills 84001: implications for nucleation, growth, and alteration. Meteorit Planet Sci 39:17–30
Eiler JM, Valley JW, Graham CM et al (2002) Two populations of carbonate in ALH 84001: geochemical evidence for discrimination and genesis. Geochim Cosmochim Acta 66:1285–1303
Eugster O, Weigel A, Polnau E (1997) Ejection times of Martian meteorites. Geochim Cosmochim Acta 61:2749–2757
Folk RL (1997) The possible role of nanobacteria (dwarf bacteria) in clay mineral diagenesis and the importance of careful sample preparation in high magnification SEM study. J Sediment Res 67:583–589
Folk RL, Taylor LA (2002) Nannobacterial alteration of pyroxenes in Martian meteorite Allan Hills 84001. Meteorit Planet Sci 37:1057–1069
Friedmann EI, Wierzchos J, Ascaso C et al (2001) Chains of magnetite crystals in the meteorite ALH 84001: evidence of biologic origin. Proc Natl Acad Sci USA 98:2176–2181
Gibson EK, McKay DS, Thomas-Keprta KL et al (2001) Life on Mars: evaluation of the evidence within Martian meteorites ALH 84001, Nakhla, and Shergotty. Precambrian Res 106:15–34
Gleason JD, Kring DA, Hill DH et al (1997) Petrography and bulk chemistry of Martian orthopyroxenite ALH 84001: implications for the origin of secondary carbonates. Geochim Cosmochim Acta 61:3503–3512
Golden DC, Ming DW, Schwandt CS et al (2001) A simple inorganic process for formation of carbonates, magnetite, and sulfides in Martian meteorite ALH 84001. Am Mineral 86:370–375

Goswami JN, Sinha N, Murty SVS et al (1997) Nuclear tracks and light noble gases in Allan Hills 84001: preatmospheric size, fall characteristics, cosmic-ray exposure duration and formation age. Meteorit Planet Sci 32:91–96

Grady MM, Wright IP, Douglas C et al (1994) Carbon and nitrogen in ALH84001. Meteoritics 29:469

Greenwood JP, McSween HY (2001) Petrogenesis of Allan Hills 84001: constraints from impact-melted feldspathic and silica glasses. Meteorit Planet Sci 36:43–61

Greenwood JP, Riciputi LR, McSween HY (1997) Sulfide isotopic compositions in shergottites and ALH 84001, and possible implications for life on Mars. Geochim Cosmochim Acta 61:4449–4453

Greenwood JP, Mojzsis SJ, Coath CD (2000) Sulfur isotopic compositions of individual sulfides in Martian meteorite ALH 84001 and Nakhla: implications for crust-regolith exchange on Mars. Earth Planet Sci Lett 184:23–35

Halevy I, Fischer WW, Eiler JM (2011) Carbonates in the Martian meteorite Allan Hills 84001 formed at 18+4°C in a near-surface aqueous environment. Proc Natl Acad Sci USA 108:16895–16899

Harvey RP, McSween HY (1996) A possible high-temperature origin for the carbonates in the Martian meteorite ALH84001. Nature 382:49–51

Jull AJT, Eastoe CJ, Xue S et al (1995) Isotopic composition of carbonates in the SNC meteorites and Allan Hills 84001 and Nakhla. Meteoritics 30:311–318

Jull AJT, Courtney C, Jeffrey DA et al (1998) Isotopic evidence for a terrestrial source of organic compounds found in Martian meteorites Allan Hills 84001 and Elephant Moraine 79001. Science 279:366–369

Kajander EO, Çiftçioğlu N (1998) Nanobacteria: an alternative mechanism for pathogenic intra- and extracellular calcification and stone formation. Proc Natl Acad Sci USA 95:8274–8279

Kirkland BL, Lynch FL, Rahnis MA et al (1999) Alternative origins for nanobacteria-like objects in calcite. Geology 27:347–350

Kirschvink JL, Maine AT, Vali H (1997) Paleomagnetic evidence of a low-temperature origin of carbonate in the Martian meteorite ALH84001. Science 275:1629–1632

Knoll A, Osborne MJ (1999) Size limits of very small microorganisms. National Research Council, National Academy Press, Washington, DC

Kring DA, Swindle TD, Gleason JD et al (1998) Formation and relative ages of maskelynite and carbonate in ALH 84001. Geochim Cosmochim Acta 62:2155–2166

Lapen TJ, Righter M, Brandon AD et al (2010) A younger age for ALH 84001 and its geochemical link to shergottite sources in Mars. Science 328:347–351

Leshin LA, McKeegan KD, Carpenter PK et al (1998) Oxygen isotopic constraints on the genesis of carbonates from Martian meteorite ALH84001. Geochim Cosmochim Acta 62:3–13

Maniloff J (1997) Nannobacteria: size limits and evidence. Science 276:1776

McKay DS, Gibson EK, Thomas-Keprta KL et al (1996) Search for past life on Mars: possible relic biogenic activity in Martian meteorite ALH 84001. Science 273:924–930

McKay DS, Gibson EK, Thomas-Keprta KL et al (1997) No 'nanofossils' in Martian meteorite: reply. Nature 390:455–456

McSween HY (1997) Evidence for life in a Martian meteorite? GSA Today 7:1–6

McSween HY, Harvey RP (1998) An evaporation model for formation of carbonates in the ALH84001 Martian meteorite. Int Geol Rev 40:774–783

Melwani Daswani M, Schwenzer SP et al (2016) Alteration minerals, fluids, and gases on early Mars: predictions from 1-D flow geochemical modeling of mineral assemblages in meteorite ALH 84001. Meteorit Planet Sci 51:2154–2174

Mittlefehldt DW (1994) ALH 84001, a cumulate orthopyroxenite nember of the Martian meteorite clan. Meteoritics 29:214–221

Nealson KH (1997) The limits of life on Earth and searching for life on Mars. J Geophys Res 102:23675–23686

Oró J (1998) The case for life on Mars, part 1: An "open" skeptical view. BioAstron News 10:1–6

Romanek CS, Grady MM, Wright IP et al (1994) Record of fluid-rock interactions on Mars from the meteorite ALH 84001. Nature 372:655–657

Score R, MacPherson G (1985) Macroscopic and thin section description of ALH 84001. Antarct Meteorit Newsl JSC Curator Office 8:5

Scott ERD, Yamaguchi A, Krot AN (1997) Petrological evidence for shock melting of carbonates in the Martian meteorite ALH84001. Science 387:377–379

Scott ERD, Krot AN, Yamaguchi A (1998) Carbonates in fractures of Martian meteorite Allan Hills 84001: petrologic evidence for impact origin. Meteorit Planet Sci 33:709–719

Sephton MA, Wright IP, Gilmour I et al (2002) High molecular weight organic matter in Martian meteorites. Planet Space Sci 50:711–716

Shearer CK, Layne GD, Papike JJ et al (1996) Sulfur isotopic systematics in altearation assemblages in Martian meteorite Allan Hills 84001. Geochim Cosmochim Acta 60:2921–2926

Steele A, Goddard D, Beech IB et al (1998) Atomic force microscopy imaging of fragments from the Martian meteorite ALH 84001. J Microsc 189:2–7

Steele A, Goddard DT, Stapleton D et al (2000) Investigations into an unknown organism on the Martian meteorite Allan Hills 84001. Meteorit Planet Sci 35:237–241

Steele A, Fries MD, Amundsen HEF et al (2007) Comprehensive imaging and Raman spectroscopy of carbonate globules from Martian meteorite ALH 84001 and a terrestrial analogue from Svalbard. Meteorit Planet Sci 42:1549–1566

Steele A, McCubbin FM, Fries MD et al (2012) Graphite in the Martian meteorite Allan Hills 84001. Am Mineral 97:1256–1259

Swindle TD, Grier JA, Burkland MK (1995) Noble gases in orthopyroxenite ALH 84001: a different kind of Martian meteorite with an atmospheric signature. Geochim Cosmochim Acta 59:793–801

Taylor AP, Barry JC, Webb RI (2001) Structural and morphological anomalies in magnetosomes: possible biogenic origin for magnetite in ALH 84001. J Microsc 201:84–106

Thomas-Keprta KL, McKay DS, Wentworth SJ et al (1998) Bacterial mineralization patterns in basaltic aquifers: implications for possible life in Martian meteorite ALH 84001. Geology 26:1031–1035

Thomas-Keprta KL, Bazylinski DA, Kirschvink JL et al (2000) Elongated prismatic magnetite crystals in ALH 84001 carbonate globules: potential Martian magnetofossils. Geochim Cosmochim Acta 64:4049–4081

Thomas-Keprta KL, Clemett SJ, Bazylinski DA et al (2001) Truncated hexa-octahedral magnetite crystals in ALH 84001: presumptive biosignatures. Proc Natl Acad Sci USA 98:2164–2169

Treiman AH (1995) A petrographic history of Martian meteorite ALH 84001: two shocks and an ancient age. Meteoritics 30:294–302

Treiman AH (1998) The history of Allan Hills 84001 revised: multiple shock events. Meteorit Planet Sci 33:753–764

Treiman AH (2003) Submicron magnetite grains and carbon compounds in Martian meteorite ALH 84001: inorganic, abiotic formation by shock and thermal metamorphism. Astrobiology 3:369–392

Treiman AH, Essene EJ (2011) Chemical composition of magnetite in Martian meteorite ALH 84001: revised appraisal from thermochemistry of phases in Fe-Mg-C-O. Geochim Cosmochim Acta 75:5324–5335

Treiman AH, Romanek CS (1998) Bulk and stable isotopic compositions of carbonate minerals in Martian meteorite Allan Hills 84001: no proof of high formation temperature. Meteorit Planet Sci 33:737–742

Treiman AH, Amundsen HEF, Blake DF et al (2002) Hydrothermal origin for carbonate globules in Martian meteorite ALH84001: a terrestrial analogue from Spitsbergen (Norway). Earth Planet Sci Lett 204:323–332

Uwins P, Webb RI, Taylor AP (1998) Novel nano-organisms from Australian sandstones. Am Mineral 83:1541–1550

Valley JW, Eiler JM, Graham CM et al (1997) Low-temperature carbonate concretions in the Martian meteorite ALH84001: evidence from stable isotopes and mineralogy. Science 275:1633–1638

Wadhwa M, Crozaz G (1998) The igneous crystallization history of an ancient Martian meteorite from rare earth element microdistributions. Meteorit Planet Sci 33:685–692

Warren PH (1998) Petrologic evidence for low-temperature, possibly flood evaporatic origin of carbonates in the ALH84001 meteorite. J Geophys Res 103:16759–16773

Warren PH, Kallemeyn GW (1996) Siderophile trace elements in ALH84001, other SNC meteorites and eucrites: evidence of heterogeneity, possibly time-linked, in the mantle of Mars. Meteorit Planet Sci 31:97–105

Weiss BP, Kirschvink JL, Baudenbacher FJ et al (2000) A low temperature transfer of ALH 84001 from Mars to Earth. Science 290:791–794

Westall F (1999) The nature of fossil bacteria: a guide to the search for extraterrestrial life. J Geophys Res 104:E7

Zolotov MY, Shock EL (2000) An abiotic origin for hydrocarbons in the Allan Hills 84001 Martian meteorite through cooling of magmatic and impact-generated gases. Meteorit Planet Sci 35:629–638

Chapter 9
Role of Mineral Surfaces in Prebiotic Processes and Space-Like Conditions

John Robert Brucato and Teresa Fornaro

Abstract The study of the interactions between organic molecules and minerals is fundamental to unravel the prebiotic processes that led to the emergence of life on Earth or possibly on other planets. Mineral surfaces may act as adsorbents, templates and catalysts driving the abiotic evolution of chemical systems on early Earth and in space towards increasing molecular complexity. Investigations about molecule-mineral interactions provide also important scientific support to space missions devoted to the search of past or present signs of life in the form of molecular biomarkers that can be included inside rock samples. Such studies are essential for establishing habitability of other planets, selection of sampling sites, identification of potential biomarkers, correct interpretation of data collected during mission operative periods, development of suitable life detection methods and technologies for in situ analysis.

In this chapter, the possible roles of minerals have been examined both from the standpoint of prebiotic chemistry and life detection investigations focusing mainly on Mars exploration.

9.1 Introduction

A fundamental question in astrobiology concerns the transition from the geochemical world to the biochemical one. The so-called abiogenesis, the set of processes that led to emergence of life from inorganic matter on early Earth and potentially on other planets, is still matter of debate within the scientific community. To face such a complex issue, several aspects need to be addressed. First, it is necessary to investigate the plausible conditions and the possible chemical pathways for the synthesis of

J. R. Brucato (✉)
INAF – Astrophysical Observatory of Arcetri, Florence, Italy
e-mail: john.brucato@inaf.it

T. Fornaro
Carnegie Science, Geophysical Laboratory, Washington, DC, USA
e-mail: tfornaro@carnegiescience.edu

B. Cavalazzi, F. Westall (eds.), *Biosignatures for Astrobiology*, Advances in Astrobiology and Biogeophysics, https://doi.org/10.1007/978-3-319-96175-0_9

the monomeric components of the biomolecules ("building blocks of life"). The subsequent steps involve the study of the interactions among such abiotically-formed organics in plausible environmental conditions, investigating the transformations induced by astrobiologically relevant external agents, towards the study of the co-evolution of compartmentalization, metabolism and informational polymers, up to the emergence of the first simple living systems (Miller 1953; Wächtershäuser 1990).

A variety of chemical routes, both endogenous and exogenous, have been found for the abiotic formation of key organic molecules and more complex biomolecules in many plausible environmental conditions.

The presence of complex organic compounds in meteorites and comets (Brownlee et al. 2006; Sandford et al. 2006) is a proof of the occurrence of abiotic processes in space. Several amino acids, such as alanine, valine, isovaline, proline, leucine, isoleucine, threonine, serine, aspartic acid and glutamic acid, occur in meteorites. Some of these amino acids showed enantiomeric excess. There is also evidence for the presence of extraterrestrial hydrocarbons, carboxylic acids, urea, ketones, alcohols, aldehydes, glycolaldehyde, sugars, purines (e.g. xanthine and adenine) and pyrimidines (e.g. uracil) (Pizzarello and Cronin 2000; Martins et al. 2006; Callahan et al. 2011). Many experiments have been carried out to synthesize complex organic compounds in laboratory space simulating conditions (Dworkin et al. 2001; Brucato et al. 2006; Saladino et al. 2015). In the interstellar medium (ISM) about 200 organic molecules, i.e. saturated or unsaturated molecules containing carbon with more than five atoms, have been found so far (Herbst and Van Dishoeck 2009). Some examples of organic interstellar molecules are acetone, dimethyl ether, ethylene glycol, formamide, acetamide, acetaldehyde, simple sugars like glycolaldehyde, and other species like cyanopolyynes, cumulenes, benzene (Herbst and Van Dishoeck 2009). The presence of the simplest amino acid, glycine, in the ISM is still a matter of controversy, even if it was detected in several carbonaceous meteorites and in cometary dust from the Stardust mission (Elsila et al. 2009). A significant amount of such abiotically-formed organics was delivered to early Earth and other terrestrial planets as a consequence of impacts by interplanetary dust particles, asteroids and comets during the Late Heavy Bombardment period occurred approximately 4.1 to 3.8 billion years ago (Gomes et al. 2005).

On the other hand, the endogenous generation of organic molecules likely occurred through many physico-chemical processes promoted by the energy sources available on the prebiotic Earth, such as electric discharges in the atmosphere, volcanic activity, high energy radiation, hydrothermal fluxes, and by mineral catalysts (Miller 1953; Wächtershäuser 1990; Chyba and Sagan 1992; Botta and Bada 2002; Ehrenfreund et al. 2002; Hazen 2005). Indeed, several plausible reactive pathways were found for the synthesis of the building blocks of the biological macromolecules occurring in the Archaean atmosphere (Dobson et al. 2000), or in mid-ocean hydrothermal systems (Martin et al. 2008; Braakman 2013), and on surface and shallow subsurface environments (Miller and Urey 1959; Oró 1961). Abiotic routes have been discovered also to account for homochirality, which is one of the most distinctive biochemical signatures of life and a prerequisite for life.

Examples are deracemization induced by ultraviolet (UV), circular polarized light (Meinert et al. 2015), partial sublimation and homochiral self-organization (Tarasevych et al. 2015), enantio-selective adsorption on mineral surfaces (Hazen et al. 2001; Yun and Gellman 2015), and asymmetric auto-catalysis (Sato et al. 2003; Blanco et al. 2013). In support of the prebiotic origin of homochirality, as already mentioned, enantiomeric excesses in chiral organic compounds have been detected in meteorites (Pizzarello and Cronin 2000; Pizzarello et al. 2008; Glavin and Dworkin 2009).

Given the prebiotic availability of biomolecule precursors, the crucial transition from inanimate matter to biological systems probably occurred through the selection, concentration and organization of the organic precursors that led to the essential macromolecules of life. In this context, mineral surfaces might have played a very important role in such prebiotic processes, potentially acting as sorbents, templates, and catalysts (Hazen 2006; Hazen and Sverjensky 2010).

Therefore, study of molecule-mineral interactions under plausible prebiotic and space-like conditions is fundamental to understand the possible role of extraterrestrial bodies such as asteroids and comets in the delivery of biomolecules in space (e.g. Gomes et al. 2005) and shed light on the emergence of molecular complexity in prebiotic conditions.

Beyond prebiotic chemistry, the study of molecule-mineral interactions finds broad applications in the space mission context providing an essential support to both remote-sensing and in situ exploration missions of objects of the Solar System. Indeed, knowledge of the adsorption properties on minerals and the nature of the interactions between molecules and mineral surfaces and study of stability and reactivity of molecule-mineral complexes under space-like conditions are essential for detecting organic compounds on the surface of planets, moons, comets and asteroids in the Solar System. Such investigations help, for example, the selection of sampling sites, identification of potential biomarkers, correct interpretation of data collected on the ground or obtained by remote sensing, development of suitable life detection methods and technologies for in situ analysis.

Life detection investigations in space exploration missions have always been a great challenge. Interesting questions concern the kind of organic molecules to be expected, the techniques that are most suitable for the detection of molecular biomarkers, taking into account also the possible effect of the interactions with mineral matrices on the detectable molecular features.

The definition of life itself is controversial (Ruiz-Mirazo et al. 2004) and leads to serious problems in the design of instruments dedicated to life detection. All the instruments developed so far for detecting molecular biomarkers on extraterrestrial objects were designed to look for Earth-like biomolecules. The main reason is that life on Earth is the only example we currently have, and all diverse life forms on Earth have the same basic biochemistry. It is also important to note that all the rocky planets of the Solar System, i.e. Earth, Mars, Venus and Mercury, should have received the same early organic inventory during the Late Heavy Bombardment period. From a deterministic point of view, the same abiotically-formed molecules may evolve into the same biochemistry under quite similar environmental conditions. Another interesting theory, named lithopanspermia, argues that impact-expelled rocks from a planet's surface serve as transfer vehicles for spreading biological material (Mileikowsky et al. 2000). In this regard, it has been demonstrated that microorganisms are able to survive to extreme conditions during interplanetary cruise travelling through space and landing on another planet (Horneck et al. 2002). Within these hypotheses, life on different planets might have had a common ancestor and consequently a similar biochemistry, thus, Earth-like biomarkers might be suitable for detecting extraterrestrial life on those planets or moons that are habitable, i.e. able to sustain environments supporting the activity of at least one known organism (see Cockell et al. 2016 for a review on habitability).

The main common assumptions about possible extraterrestrial life are: (1) life forms based on carbon, hydrogen, nitrogen, oxygen, phosphorus, sulphur and bio-essential metals; (2) a requirement for liquid water as a solvent; (3) self-contained, cell-like entities; (4) metabolic characteristics determined by the same physical, chemical and thermodynamic factors found on Earth, and employing complex organic molecules in biochemical roles (Committee on an Astrobiology Strategy for the Exploration of Mars, 2007). As a consequence, the potential biosignatures can be classified in: (1) structural evidence, such as fossilized organisms; (2) isotopic fractionation between reservoirs; (3) organic molecules of biotic origin; (4) chiral excess of appropriate organic molecules; and (5) metabolic activity.

Commonly, molecular biomarkers are divided in biomarkers of ancient/extinct life and recent/extant life. Biomarkers of extinct life include fossil derivatives, also called geo-molecules, because they represent the products of diagenesis, degradation, alteration and preservation of ancient life. They are membrane molecules that are very stable in harsh environmental conditions, such as hopanes, isoprenoids, steranes, porphyrins, straight-chain hydrocarbons, long-chain fatty acids, as well as compounds such as amino acids and proteins that are not resistant for long periods in the fossil record on Earth due to microbial degradation, diagenesis and metamorphism. On Mars, metamorphic effects would be less important because of the smaller size and lower gravity of the planet and, thus, some resistant biomarker molecules could be preserved. The biomarkers of extant life, instead, are short-lived products of present life such as nucleobases, chiral amino acids, carbohydrates, fatty acids and their polymers. They can be classified depending on their function in energy storage and electron transfer compounds (e.g., ATP, phosphoenolpyruvate, cyclic AMP, Coenzime A, ATP synthase, quinines, NAD, carotenoids), informational macromolecules (e.g., DNA, RNA) and structural molecules (e.g., phytane, squalene, teichoic acids, lipopolysaccharides).

It is important to consider also the presence of abiotic organic molecules deriving from meteoritic in-fall and preservation or diagenesis of the early inventory of organics. For instance, the influx of meteorites on Mars, estimated at approximately 240 tons per year (Flynn 1996), corresponds to a great influx of organic matter—such as polycyclic aromatic hydrocarbons (PAHs), amino acids, aromatic carboxylic acids, sugar-related compounds, etc.—to the surface of the planet. However, the degradation mechanisms on the surface of Mars probably do not make the carbon budget increase. Distinguishing between abiogenic and biogenic materials is a very important challenge.

Abiotic compounds are formed by physico-chemical processes, generally driven by thermodynamic control dependent on the environmental conditions without the intervention of enzymes promoting selective recurring patterns of biosynthesis. They are, therefore, characterized by wide equilibrium distribution of structures (structural isomers), predominance of branched-chain compounds (thermodynamically more stable) and decrease in abundance with increase in carbon number—the extreme conditions of the interstellar medium in which they formed do not allow the growth of large molecules. Biogenic compounds, instead, are characterized by diastereoisomeric and structural isomer preference, repeating structural subunits, uneven distribution patterns and homochirality, even if a slight chiral excess of abiotic origin has been observed in meteorites as well (Glavin and Dworkin 2009).

Establishing the nature of life is essential for the development of life detection methods and technologies for in situ analysis. However, versatility is also required if diverse life forms developed on extraterrestrial objects.

In this chapter, the study of molecule-mineral interactions is addressed both from the standpoint of prebiotic chemistry and life detection investigations. Firstly, the possible roles of minerals as adsorbents, templates and catalysts in prebiotic processes are inspected. Then, the space mission instruments and techniques developed for detection of biomarkers are described, particularly focusing on Mars exploration. Finally, a brief overview of the laboratory studies performed so far to support life detection investigations is provided.

9.2 Role of Minerals in the Prebiotic Context

9.2.1 Adsorption Processes

It is presumable that any extraterrestrial water body, as well as the Hadean ocean on early Earth, is a highly diluting aqueous environment. In dilute solutions, the probability of molecular reactions is very low and, in general, the formation of macromolecules is inhibited because hydrolysis prevails over condensation, which is a fundamental reaction towards polymerization. Therefore, a mechanism of molecular concentration is needed for prebiotic processes to occur. Interfaces between aqueous solutions and mineral matrices may play a key role in concentrating organics on a local scale. Indeed, mineral surfaces are highly energetic environments characterized by non-saturated asymmetric unbalanced forces, which can be partially or totally compensated due to molecular adsorption. For this reason, adsorption processes are thermodynamically favoured and always characterized by a decrease in the total free energy. The binding mechanism depends on the balance of the great variety of possible specific and non-specific interactions that may be established between the molecules adsorbed, the so-called adsorbates, and the surface. Such different kinds of interactions may determine selective adsorption of specific molecules in particular conditions.

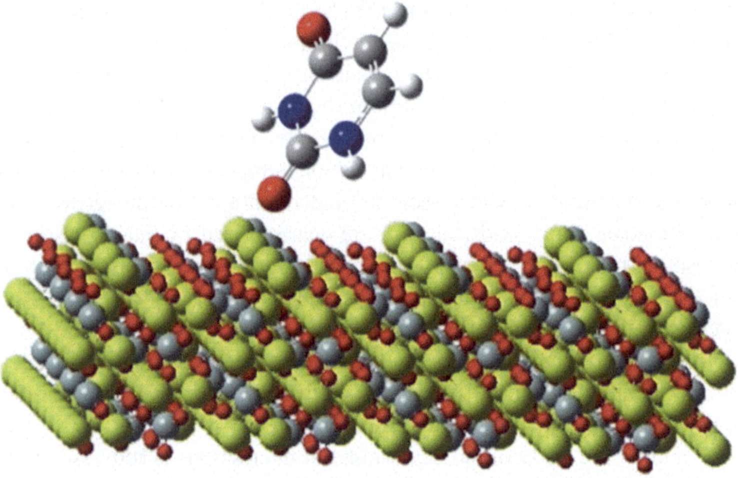

Fig. 9.1 Example of interaction between an important "building block of life", namely uracil, with the surface of forsterite mineral, an abundant mineral in space. Source: JR Brucato

Several studies (Sowerby and Heckl 1998; Fornaro et al. 2013a) have provided evidence that organic molecules, which are "building blocks of life", adsorb spontaneously at solid-liquid interfaces (Fig. 9.1). This potentially favours further prebiotic synthesis and concentration of these molecules, even from low yielding reactions.

The key components of modern biochemistry might be the result of concentration and specific selection by minerals in prebiotic processes. For instance, it has been argued that minerals might have been involved in the emergence of homochirality in prebiotic conditions, through mechanisms of chiral selection on specific surfaces (Downs and Hazen 2004).

The term "sorption" indicates processes of mass exchange among the different phases in a system. Depending on the depth reached by solute molecules in the solid phase, it is possible to distinguish between absorption (where molecules penetrate several nanometers into the surface of the solid) and adsorption (where molecules attach just at the interface). In the majority of the cases adsorption occurs. The kind of physico-chemical interactions established between molecules and mineral surfaces depends on many factors: (1) the characteristics of the mineral, such as chemical composition, surface charge, surface area, grain size, porosity, crystal structure; (2) characteristics of the molecules, including solubility, protonation constants, nucleophilicity or electrophilicity, electronic structure; and (3) experimental conditions, such as molecule-mineral ratio, pH, temperature, background electrolytes, ionic strength. For example, pH determines the protonation status of the molecules, which in turn has a great influence on the adsorption process. Generally, the adsorption of organic molecules onto mineral surfaces is thermodynamically complex resulting in a molecular configuration that is the most energetically favourable depending on the synergy of the different types of specific and non-specific

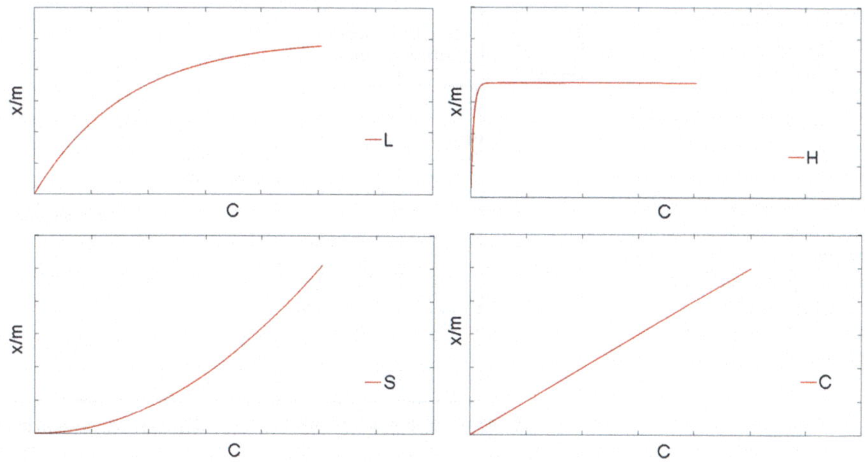

Fig. 9.2 Classification of adsorption isotherms: x/m = $f(C)_T$, where x is the amount of solute adsorbed on the solid mass m, C is the solute concentration in the aqueous phase at equilibrium, f (C) indicates a function of the concentration C at the temperature T. Source: JR Brucato

interactions, ranging from physisorption (van der Waals, electrostatic interactions) to chemisorption (covalent, hydrogen bonding, coordination bonding).

Models of equilibrium adsorption are based on a hypothesis known as Local Equilibrium Assumption (Valocchi 1985), which assumes that, at a microscopic scale, the adsorption process reaches equilibrium instantaneously and, if the concentration in the fluid phase does not change, the adsorbed concentration remains constant. This is possible when, at the microscopic level, the adsorption and desorption rates are much higher than the flow rate of the liquid phase.

Methods used to study equilibrium adsorption processes include those that identify a dependency between the adsorbed solute concentration and the solute concentration in the liquid phase. At a given temperature, the relationship between the degree of surface coverage and the equilibrium solute concentration is the adsorption isotherm: x/m = $f(C)_T$, where x is the amount of solute adsorbed on the solid mass m, C is the solute concentration in the aqueous phase at equilibrium, f(C) indicates a function of the concentration C at the temperature T. According to the initial trend of the experimental isotherms, as shown in Fig. 9.2, they can be classified in: (1) *L* (or *Langmuir*) which features a downward curvature because the number of available surface sites is constant and decreases as adsorption occurs, and is typical of systems with solute-surface interactions stronger than solute-solute and solute-solvent interactions; (2) *S* (*S*-shape isotherm) which is typical of unfavourable cooperative adsorption where solute-solute or solute-solvent interactions are stronger than interactions with the surface; (3) *H* (or *high affinity*) is observed when solutes have high affinity for the surface and the initial part of the isotherm is nearly vertical; (4) *C* (or *constant partition*) is observed when the solute-surface interactions are stronger than those of the solvent with the solid and the intrasolid ones, allowing

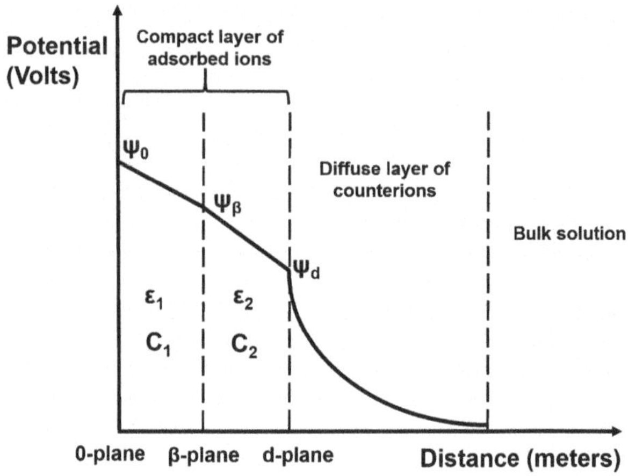

Fig. 9.3 Schematic representation of the relationship between electrical potential Ψ and distance x from the mineral surface in the triple-layer model. The three planes of charge are associated with three planes of potential and treated as a series of pairs of parallel-plate capacitors of capacitances C_1 and C_2. Source: Adpted from Westall (1987)

molecules to penetrate the solid, which determines a constant partition of the solute between solid and liquid phases.

Mechanistic models of equilibrium adsorption, instead, describe specific surface-molecule interactions and treat the energies involved quantitatively. The processes that can be studied with these models are the ionic exchange and the surface complexation, which strongly depend on the mineral characteristics, including pH, surface charge, ionic strength, solute aqueous equilibrium etc. The ionic exchange depends primarily on electrostatic interactions, while the formation of surface complexes may involve both weak "outer-sphere" non-specific interactions (molecules keep their solvation shell) and more specific "inner-sphere" interactions (molecules directly interact with the surface sites).

One of the most successful models to describe surface complexation is the extended triple-layer model (ETLM) which allows the prediction of surface speciation as a function of environmental conditions, such as pH, ionic strength, ligand-to-solid ratios (Fukushi and Sverjensky 2007; Jonsson et al. 2009). According to the triple-layer model, there are three planes in the interfacial region, schematically represented in Fig. 9.3: the mineral surface or 0-plane, where protonation/deprotonation reactions occur; the inner Helmholtz or β-plane, where electrolyte ions adsorb, and which is defined by centers of specifically adsorbed anions and cations; and the outer Helmholtz or d-plane, corresponding to the closest distance of approach of the diffuse swarm of counterions induced near the surface to balance the charge of the 0- and β-planes of the mineral/water interface.

The three planes of charge are associated with three planes of potential and treated as a series of pairs of parallel-plate capacitors. The ETLM not only accounts for specific electrolyte adsorption, i.e. the equilibrium constant for the adsorption of a particular electrolyte ion, but also specifically accounts for the electrical work associated with desorption of chemisorbed molecule during inner-sphere surface complexation given by $\Delta\Psi = -n_{H2O}\ (\Psi_0\text{-}\Psi_\beta)$, where n_{H2O} is the number of water molecules released from the surface when inner-sphere bonds are formed, Ψ_0 is the electric potential at the 0-plane and Ψ_β is the electric potential at the β-plane. As a consequence, it indicates the number of inner-sphere linkages for the adsorbate, as well as the number of surface sites involved in the reaction stoichiometry, significantly constraining the likely mode of surface attachment. This approach has been validated with respect to spectroscopic data and quantum chemical computations, providing reliable predictions for the attachment of aqueous organic compounds to the surface of oxide minerals (Jonsson et al. 2009; Parikh et al. 2011; Lee et al. 2012). Such surface complexation modelling provides reasonable inferences for the possible surface complexes determining the number of inner/outer-sphere linkages for the adsorbates and the number of surface sites involved in the reaction stoichiometry. However, to distinguish the specific functional groups, which constitute the points of attachment to the surface, further quantum mechanical simulations on the energetics of the possible complexes can be performed.

Computational studies have been performed to study the adsorption of several amino acids on different substrates, specifically: cometary dust analogs, such as forsterite, in the presence of amorphous water ice (Escamilla-Roa and Moreno 2012), other silicates, including clays (Newman et al. 2002), silicon (Carnimeo et al. 2011), silica (Zhao et al. 2011), quartz (Gambino et al. 2006), metal and metal oxide surfaces (Parikh et al. 2011; Monti et al. 2012), carbonates like calcite (Asthagiri and Hazen 2007), phosphates like hydroxyapatite (Jiménez-Serra et al. 2012), and sulphides, such as pyrite (Pollet et al. 2006). Also, the adsorption of nucleobases has been widely computationally studied on the surfaces of clays (Mignon and Sodupe 2013), graphite (Sowerby et al. 1998), graphene (Panigrahi et al. 2012) and metals (Kong et al. 2014).

Over the last few years, computer simulations have been very useful in aiding understanding of various aspects of astrobiology (Rimola et al. 2012), ranging from the chemistry of simple molecules in the early Universe (Galli and Palla 1998), the characterization of the properties and reactivity of key prebiotic molecules in various plausible environmental conditions (Barone et al. 2015), the evolution of planetary atmospheres (Smith et al. 2014), up to the evolution of biomolecules (Gallori and Branciamore 2012). Therefore, theoretical tools have become essential for the interpretation of experimental/observational data helping to unravel intricate issues in astrobiology.

9.2.2 Catalysis

Beyond their sorption capabilities, minerals can act as templates promoting molecular self-organization by specific molecule-mineral interactions (Hazen 2006). Minerals can also act as catalysts, accelerating chemical reactions by lowering the corresponding activation energy or promoting the formation of specific products (Hazen and Sverjensky 2010). This can happen in different ways. Mineral surfaces may be responsible for inducing the orientation of reacting groups of neighbouring molecules through specific molecule-mineral interactions. In this case, the spatial approach of reactants adsorbed on a surface can be ascribed to catalysis. In other cases, there may be an "electronic"-type of catalysis, i.e. the mineral surface may act as a means to transfer electrons between reactants, or the changes in the electronic structure of adsorbates induced by interaction with the surface may favour specific reactions (Schoonen et al. 2004). In the presence of radiation, minerals can also be involved in photo-processes, such as photolysis, photocatalysis or photosynthesis. Molecular photolysis may be enhanced by adsorption onto a mineral surface because the interactions of a molecule with a mineral can weaken intramolecular bonds, thus, facilitating their breakage upon exposure to electromagnetic radiation. In other cases, minerals absorb photons, determining the formation of an electron/hole pair inside the solid, which can be followed by redox processes of the adsorbate due to electron transfer between molecules and the solid (Schoonen et al. 2004). On the other hand, some minerals can protect molecules against electromagnetic radiation because of the presence of optimal interlayer sites where molecules can be adsorbed and shielded (Scappini et al. 2004; Biondi et al. 2007). Therefore, the physico-chemical interactions between molecules and minerals might affect the lifetime of molecules under space conditions in different ways: photocatalytic minerals may promote rapid degradation of the adsorbed molecules or catalyse reactions towards more complex species (Brucato et al. 2006; Fornaro et al. 2013b; Saladino et al. 2015), while other mineral matrices are able to protect molecules against photo-degradation (Scappini et al. 2004; Biondi et al. 2007; Poch et al. 2015; Fornaro et al. 2018).

All this suggests a pivotal role of minerals in the prebiotic evolution of complex chemical systems. From this perspective, the study of molecule-mineral interactions under plausible prebiotic and space conditions may be a step forward in resolving unanswered questions about the origins of life on Earth and beyond.

9.3 Role of Minerals in the Context of Space Missions

9.3.1 Mission Instruments to Detect Biomarkers in Space

Mars is the most plausible place to look for life elsewhere in the Solar System due to its similarities with Earth in terms of its geological environment, availability of liquid water throughout time, and potentiality to support microbial life, especially during the early history of the planet (Westall et al. 2013; Grotzinger et al. 2014).

The first space mission to Mars with the specific aim of searching for signs of life on the red planet were the 1976 Viking landers, equipped with a Gas Chromatograph Mass Spectrometer (GCMS), a device that uses a gas chromatograph to chemically separate volatile components released as the Martian soil is heated to different temperatures, and then measure the molecular weight of each chemical through a mass spectrometer to identify a great variety of compounds. This instrument detected no organic matter at the surface and at a depth of 10 cm, at parts per billion (ppb) level for complex molecules and parts per million (ppm) level for simpler compounds (Biemann 1979). In fact, present conditions on Mars' surface are inhospitable to life, especially due to intense ultraviolet (UV) irradiation (Patel et al. 2003; Hassler et al. 2014) and highly oxidizing and acidic nature of Martian soil (Glavin et al. 2013) causing sterilization and degradation of organic compounds (Benner et al. 2000). The presence of perchlorates in the Martian regolith was confirmed by instruments on board Mars Phoenix Lander 9 in 2008 and on Mars Science Laboratory's (MSL) Curiosity rover in 2013. Complex models showed a decrease in oxidizing agents with depth, becoming zero between 2 to 4 m (Patel et al. 2003). Therefore, subsurface locations represent more plausible places to look for preserved organics on Mars. In this regard, the ExoMars2020 rover mission is designed to study the Martian subsurface environment using a drill that will collect samples down to a maximum depth of 2 m.

Nevertheless, it is worth mentioning that the Sample Analysis at Mars (SAM) instrument on board MSL's Curiosity rover recently detected chlorinated hydrocarbons in drilled samples of the Sheepbed mudstone (Glavin et al. 2013; Freissinet et al. 2015) at Gale crater, the landing site for Curiosity, formed from sediments deposited in an ancient lake ~3.6 billions of years ago. SAM comprises a mass spectrometer, a gas chromatograph, and a laser spectrometer. The mass spectrometer separates elements and compounds by mass for identification. The gas chromatograph heats soil and rock samples until they vaporize, and then separates the resulting gases into various components for analysis. The laser spectrometer measures the abundance of various isotopes of carbon, hydrogen, and oxygen in atmospheric gases such as methane, water vapour, and carbon dioxide. The accuracy of these measurements is within 10 parts per thousand. SAM has been used also to characterize the mineral composition of the Sheepbed mudstone, together with the Chemistry and Mineralogy (CheMin) instrument also present in the MSL's Analytical Laboratory, which identifies and measures the abundances of various minerals. Interestingly, drilled samples of mudstone from the Sheepbed unit at Yellowknife Bay analyzed by SAM and CheMin contain 20% smectite clays (Grotzinger et al. 2014), which are relevant minerals in the prebiotic context for their capabilities of concentrating and protecting organic compounds. The organic molecules found by the team have chlorine atoms, and include chlorobenzene and several dichloroalkanes, such as dichloroethane, dichloropropane and dichlorobutane. Chlorobenzene is the most abundant with concentrations between 150 and 300 ppb. It is likely that chlorinated organics formed from reactions inside the SAM instrument as the sample was heated for analysis (Freissinet et al. 2015). In 1976, the GCMS instrument on NASA's Viking landers detected two simple chlorinated hydrocarbons after heating Martian soils for analysis (chloromethane and dichloromethane) (Navarro-González et al. 2010). However it was not possible to rule out that the compounds were derived from the instrument itself. Also, within the SAM instrument, there are potential sources of chlorinated hydrocarbons, but the amount of chlorobenzene that can be produced in this way does not exceed 22 ppb, which is far below the amounts detected in the mudstone sample. This indicates that the organic molecules detected on Mars are not contaminants. SAM's GCMS oven used to volatilize organic compounds is much more powerful than the one used by Viking. The oven on the Viking landers only went up to 500 °C, while SAM reaches 850 °C allowing the release of very refractory materials from Martian soil. Additional sulphur-containing organics were discovered that could have been either released from the sample or formed inside SAM by combination of sulphur from SO_2 and organics released from the sample during high-temperature pyrolysis (Freissinet et al. 2016). On Earth, sulphurization of organic matter is a key process that aids preservation over geological time-scales. This is because it reduces reactive functional groups and adds cross-links between small unstable molecules, thereby converting them into recalcitrant macromolecules. Sulphurization of organic materials prior to deposition and during early diagenesis might have been a key mechanism responsible for organic matter preservation in the Murray formation mudstones. Sulphur-bearing organics have also been observed in carbonaceous meteorites and there is indication of their presence in the Tissint Martian meteorite (Eigenbrode et al. 2016).

These latest discoveries indicate that some preserved organics can be found also on the surface, possibly associated with specific minerals, in regions where recent impact events or erosion processes might have unearthed the underlying terrain.

The choice of the landing sites is the key for success of mission searching signs of past and/or present life on Mars. Considering that the most important requirement for life is the availability of liquid water, life detection investigations have been based on a "follow the water" principle and all the potential landing sites currently under discussion for the next rover missions to Mars, such as NASA Mars 2020 and ESA ExoMars2020, show evidence of having been influenced by water in the past. Indication of aqueous alteration is generally given by the presence of specific minerals formed in the presence of water, such as phyllosilicates, carbonates, sulphates, etc. Compositional investigations of Mars with various remote sensing instruments and in situ payloads have revealed changes in the mineralogy at the surface of Mars corresponding to specific geological Eras and related to specific environmental conditions (Bibring et al. 2006). Three types of hydrated minerals have been identified on Mars so far: phyllosilicates and sulphates (Bibring et al. 2006), and more recently opals (Milliken et al. 2008). Most of the phyllosilicate minerals are iron and magnesium rich (such as chamosite and nontronite). Aluminium rich phyllosilicates (such as montmorillonite) appear to be dominant in ancient Noachian-aged locations. Phyllosilicates are found in the oldest terrains because they formed by aqueous alteration of basaltic crust very early in the history of Mars (the "phyllocian" Era). This corresponds to the Noachian Eon (~4.1 to 3.7 billion years ago), characterized by a warm, wet and nonacidic chemistry indicating the presence of liquid water on the surface, which resulted in high rates of weathering and the formation of valley networks. Thus, these minerals are present in rocks buried by more recent deposits that are locally exposed because of impact, faulting, or erosion (Bibring et al. 2006). Sulphates, including magnesium sulphates (such as kieserite) and calcium sulphates (such as gypsum), constitute the second major class of hydrated minerals mapped by Mars Express OMEGA instrument and detected by NASA rovers (Squyres 2004; Bibring et al. 2006). These kinds of minerals formed from volcanic outgassing during an acid wet phase, the "theiikian" era corresponding to the Hesperian period (~3.7 to 3 billion years ago). Sulphur-bearing phases have also been detected by SAM instrument at Gale crater (McAdam et al. 2014). Interestingly, the Mars Reconnaissance Orbiter has recently discovered across large regions of Mars the presence of hydrated silica, commonly known as opal (Milliken et al. 2008). Opal represents the youngest hydrated mineral ever detected on Mars, and indicates that water might have existed on Mars as recently as 2 billion years ago likely playing an important role in shaping the planet's surface and possibly hosting life (Viviano-Beck et al. 2014). There is evidence that acidic water remained on the Martian surface for a long time, because iron sulphate minerals have been detected in some locations around dry river channels together with opaline silica (Viviano-Beck et al. 2014). Exploration of opaline silica deposits can be particularly useful to assess habitability on Mars, especially in younger terrains. Opaline deposits at Home Plate in Gusev Crater have been interpreted as having a hydrothermal origin (Ruff and Farmer 2016).

Another primary class of minerals detected on Mars is anhydrous ferric oxides, formed by a slow, superficial, water-free weathering process beginning about 3.5 billion years ago, during the last era called the "siderikian" (Bibring et al. 2006), which corresponds to the Amazonian Eon (~3 billion years ago to the present day).

Regarding mafic minerals, OMEGA and previous instruments have shown that pyroxene and olivine are still present at the surface in the older terrains, included within sand dunes associated with ancient (early) Noachian crustal rocks and early Hesperian volcanism (Ody et al. 2012). However, much of the younger surface, particularly within the large lowlands of the northern hemisphere, does not exhibit mafic spectral signatures. The exposed original materials have been heavily chemically altered or covered by heavily altered dust (Bibring et al. 2006; Ody et al. 2012). In order to understand the phenomena acting on organics in the harsh Martian environment and to identify the most suitable conditions for their preservation, it is essential to evaluate the catalytic or protecting effect of minerals. In this regard, laboratory simulations allow investigating the adsorption properties of biomarkers on minerals and the nature of the interactions between molecules and mineral surfaces. It is also possible to examine in the laboratory the molecular degradation under simulated space conditions in order to evaluate the residence time of biomarkers on minerals and develop models for biomarker degradation on geological time-scales (ten Kate et al. 2005, 2006; Fornaro et al. 2013a, b, c; dos Santos et al. 2016). Such activities provide essential information to maximize the chances of finding biomarkers and, thus, aid selection of sampling sites for future Mars missions or other Solar System objects.

9.3.2 Future Exploration of Mars

The Mars Science Laboratory's (MSL) Curiosity rover was designed to assess the habitability of the planet Mars by investigating the geological setting, atmosphere, environmental conditions and potential biosignatures at Gale Crater with a relevant suite of instruments. The MSL mission has shown that Mars hosted a chemical environment that could have sustained life. Interestingly, the detection of organic compounds on the surface of the planet has demonstrated that the "building blocks of life" may be preserved even in the harsh Martian environment. Given these latest discoveries, the next rover missions to Mars will mainly focus on looking for signs of past or present life, concentrating specifically on whether or not the molecules detected on Mars are biogenic in origin.

The NASA Mars2020 rover mission will look for signs of habitable conditions on Mars in the ancient past and for associated biosignatures. A drill will be used to collect interesting core samples of Martian rocks and set them aside in a "cache" on the surface of Mars. A future mission could potentially return these samples to Earth. An important instrument on board this rover will be the Scanning Habitable Environments with Raman & Luminescence for Organics & Chemicals (SHERLOC), mounted on the rover's robotic arm (Beegle and Bhartia 2016). SHERLOC will use spectrometers, a laser and a camera to search for organics and minerals that have been altered by aqueous environments and may be signs of past microbial life.

Analogously, the goals of ESA ExoMars2020 mission pertain to the search for signs of past and present life on Mars, investigating how the water and geochemical environment change as a function of depth in the shallow subsurface and examining Martian atmospheric trace gases and their sources. The key novel feature of this rover will be a drill capable of collecting samples from a depth of 2 m below the surface, where the amount of oxidants should be quite low.

The Mars Organic Molecule Analyzer (MOMA) will be the largest instrument in the ExoMars rover, and the one directly targeting biomarkers (Goetz et al. 2016). MOMA will provide insights into the potential origin, evolution and distribution of life on Mars by detecting organic compounds at very low concentrations in Martian soil samples collected by the rover's drill, discriminating between biotic or abiotic origin by molecular identification in terms of chirality, and analysing gases in the Martian atmosphere. Specifically, MOMA has two complementary operational modes: Gas Chromatograph-Mass Spectrometry (MOMA GC-MS) and Laser Desorption-Mass Spectrometry (MOMA LD-MS). The GC-MS mode will be used to identify and analyse volatile molecules found in Martian soil samples. The powdered sample material dispensed by the Rover's drill will be used to fill one of twenty one-time-use, small ovens. All volatile materials will be evaporated at a high temperature (850 °C) and routed to the gas chromatograph, where they will be separated and identified. The volatile molecules will be ionized and then analysed individually with the mass spectrometer. Some ovens will be filled with a chemical product, the derivatisation agent, used to make the samples suitable for chiral analysis. In the LD-MS mode, the powdered sample material will be carefully positioned under a high-power laser head, which will be then pulsed to liberate ions. The ions will be guided into the mass spectrometer where they will be analysed. This mode will allow desorbing more refractory compounds from rock samples.

Another important instrument on the ExoMars rover is MicrOmega, which consists of a visible light microscope and a near-infrared imaging spectrometer designed to identify, at grain scale, the mineralogical and the molecular composition of the Martian samples collected by the drill. The study of the mineralogical composition of the Martian soil is key to determine the physical and chemical conditions under which the minerals formed inspecting the geological processes, climate and environment of Mars. These measurements will help to assess habitability and possible preservation of biosignatures in the Martian soil. In addition, through infrared imaging, MicrOmega will be able to identify spectral features typical of organics with high sensitivity even at very low densities, including small amounts of organics on mineral grains. This can be used to locate grains of particular interest (for example, grains that might have been processed by liquid water, or grains which present typical spectral features of organics) in a totally non-destructive manner, thus identifying the targets for Raman and MOMA measurements.

Another non-destructive analysis, complementary to MicrOmega measurements, will be performed by the Raman Laser Spectrometer instrument (RLS) using the Raman spectroscopy technique, which consists in collecting and analysing the light scattered from a sample illuminated by a laser. The scattering process is, indeed, determined by the composition of the material analysed. RLS will provide a rapid preliminary analysis of Martian samples that will be then further inspected by the other instruments on board the ExoMars rover. RLS will help to identify minerals and biomarkers, specifically characterizing mineral phases derived by aqueous alteration, igneous minerals, the products of their alteration thus searching for biosignatures (Rull et al. 2017).

9.3.3 Laboratory Studies to Support Life Detection Investigations

Laboratory studies are fundamental to support in situ life detection investigations. It is necessary to simulate extraterrestrial environments in order to obtain information about potential biosignatures and to develop suitable life detection technologies. Laboratory simulations provide an essential aid for correct interpretation of observational data as well.

Spectroscopy is extensively used for detection of organics in space. However, the analysis and interpretation of spectral data can be rather intricate. Even in the case of small molecules in the gas-phase, a variety of intramolecular interactions can occur that strongly affect the spectroscopic features (Espinoza et al. 2010). Additionally, an important source of complexity originates from environmental effects, such as the molecule-solvent and molecule-mineral interactions. Indeed, the formation of supramolecular complexes significantly alters the spectroscopic features with respect to the gas-phase scenario, due to the introduction of new intermolecular interactions as well as the perturbation of the intramolecular ones, causing changes in the spectroscopic selection rules (Tolstoy et al. 2003). This casts doubts on the interpretation of astronomical data because assignments of the spectroscopic features based on gas-phase data may be misleading. In such circumstances, in silico modelling represents a useful tool in order to assist the interpretation of the experimental/observational measurements, and several computational methods have been developed so far for simulating the spectroscopic features of organic molecules such as the "building blocks of life" in heterogeneous environments (Fornaro et al. 2014, 2015a, b, 2016; Fornaro and Carnimeo 2014).

Another important aspect to consider is that the molecules might have been subjected to deep transformations due to processing by a variety of possible external agents. For instance, extensive degradation of organic compounds occurs on Mars owing to the lack of an ozone layer, the thinness of the atmosphere and the absence of a strong magnetic field, which allow ionizing radiation, i.e. solar light down to 190 nanometres (far UV), energetic solar protons and galactic cosmic rays, to reach the surface of the planet (Patel et al. 2003). In particular, UV photons represent the most dangerous type of radiation leading to the formation of highly reactive radical species by photochemical processes (Benner et al. 2000). Nevertheless, UV photons can penetrate only few millimetres in the Martian soil and in the absence of aeolian transport any organics in the subsurface of Mars are shielded. However, UV-driven chemical reactions between molecules from the atmosphere and the soil cause the formation of very reactive species in the Martian soil that act as strong oxidants capable of penetrating down to 2–3 m into the subsurface (Benner et al. 2000).

The likelihood of preservation of possible biomarkers in such conditions can be evaluated only by simulating the harsh conditions occurring on the surface of Mars. Several experiments have been conducted using UV sources and oxidants comparing the different capabilities of mineral matrices to preserve or degrade organics. Garry et al. (2006) investigated the native amino acid composition of two analogs of Martian soil, JSC Mars-1 and Salten Skov, using a Mars simulation chamber to expose samples of these analogs to temperature and UV irradiation conditions similar to those found at low latitudes on the Martian surface. As a consequence of UV irradiation, a slight increase in the concentration of amino acids, such as L-aspartic acid, L-leucine, L-glutamic acid and L-alanine, was observed probably due to the degradation of microorganisms contained in the soils. On the other hand, low temperatures caused the condensation of water onto the soil, which enhanced the destruction of amino acids. These results support the idea that water in the Martian soil may be responsible for the production of reactive species, which are involved in the degradation of organics.

Recent studies (dos Santos et al. 2016) confirm that clay minerals preserve amino acids due to their high surface areas and small pore sizes where molecules can be adsorbed. Sulphates also protect amino acids, probably because these minerals do not absorb UV radiation or because they can trap amino acids during recrystallization after their partial dissolution. Minerals containing ferrous iron instead are photocatalysts. dos Santos and co-workers (2016) observed also that D- and L-amino acids are degraded at equal rates, and that there is a certain correlation between preservation/degradation of amino acids and their molecular structure. Specifically, the amino acids with higher stability under UV radiation present alkyl substitution in the α-carbon.

The photostability of another important class of "building blocks of life", the nucleobases, has also been investigated. Fornaro et al. (2013b) examined the degradation kinetics of the nucleobases adenine, cytosine, uracil and hypoxanthine adsorbed on minerals, including magnesium oxide and forsterite, through in situ UV irradiation experiments. They observed that the half-lifetimes of degradation decrease and the degradation cross-sections increase in the presence of the minerals, which act as catalysts.

Recently Ertem et al. (2017) observed some protection capability of the minerals calcite, calcium sulphate, kaolinite and clay-bearing Atacama desert soil for the biomolecules purine, pyrimidine and uracil against UV photolysis in the presence of sodium perchlorate. They observed that, in the absence of these minerals, organic compounds were completely degraded when subjected directly to UV photolysis equivalent to only 5 Martian day's exposure. However, in the presence of the minerals, only 1–2% loss of organics was obtained. Mixtures of purine and uracil with calcium carbonate exposed to gamma radiation of 3 Gray, which corresponds to approximately 15,000 days on Mars, resulted in up to 10% loss of organics. By contrast, these organic compounds completely decomposed upon mixing with iron oxide (Fe_2O_3) before UV irradiation.

From these examples it appears clear that establishing the conditions for preservation of organics on Mars is not a trivial issue given all the possible combinations of biomarkers-Martian minerals and the variety of environmental factors, which may affect molecular stability.

The study of the evolution of organic matter subjected to space conditions, and more specifically to Solar photons in the vacuum ultraviolet range (120–200 nm) has been undertaken also in low-Earth orbit since the 1990s, and implemented on various space platforms (Cottin et al. 2015; see also Chap. 10).

All these results provide essential information for selecting sampling sites for future life detection missions on the surface of Mars or other objects of the Solar System.

9.4 Conclusions

This review on the role of minerals in prebiotic processes and in the context of life detection underlines the importance of studying the adsorption properties of biomolecules on mineral surfaces under plausible prebiotic and space-like conditions for a variety of reasons. Such studies allow investigation of the physico-chemical mechanisms leading to the synthesis of complex chemical compounds in space in order to understand the possible role of extraterrestrial bodies in the delivery of biomolecules in space.

From the prebiotic point of view, mineral surfaces might have played key roles in the processes that led to the emergence of life on early Earth, acting as adsorbents, templates and catalysts. Minerals might have been involved in the selection of specific molecular characteristics, which were subsequently reflected in modern biochemistry. Moreover, knowledge of the adsorption properties of extraterrestrial analogues and molecular stability under harsh planetary conditions is essential for assessing the likelihood of preservation of possible biomarkers and supporting the scientific activity and technology development of instruments for life detection.

Acknowledgements This research was supported by INAF-Astrophysical Observatory of Arcetri through the Italian Space Agency (ASI) grant agreement ASI/INAF nr. 2015-002-R.0 and by the Carnegie Institution for Science.

References

Asthagiri A, Hazen RM (2007) An ab initio study of adsorption of alanine on the chiral calcite surface. Mol Simul 33:343–351

Barone V, Biczysko M, Puzzarini C (2015) Quantum chemistry meets spectroscopy for astrochemistry: increasing complexity toward prebiotic molecules. Acc Chem Res 48: $32#1413–1422

Beegle L, Bhartia R (2016) SHERLOC: an investigation for Mars 2020. Geophys Res Abstr EGU Gen Assem 18:EGU2016–11215

Benner S, Devine KG, Matveeva LN et al (2000) The missing organic molecules on Mars. Proc Natl Acad Sci USA 97:2425–2430

Bibring JP, Langevin Y, Mustard J (2006) Global mineralogical and aqueous Mars history derived from OMEGA/Mars Express Data. Science 312:400–404

Biemann K (1979) The implications and limitations of the findings of the Viking organic analysis experiment. J Mol Evol 14:65–70

Biondi E, Branciamore S, Maurel MC et al (2007) Montmorillonite protection of an UV-irradiated hairpin ribozyme: evolution of the RNA world in a mineral environment. BMC Evol Biol 7:S2

Blanco C, Ribó JM, Crusats J et al (2013) Mirror symmetry breaking with limited enantioselective autocatalysis and temperature gradients: a stability survey. Phys Chem Chem Phys 15:1546–1556

Botta O, Bada J (2002) Extraterrestrial organic compounds in meteorites. Surv Geophys 23: $32#411–467

Braakman R (2013) Mapping metabolism onto the prebiotic organic chemistry of hydrothermal vents. Proc Natl Acad Sci USA 110:13236–13237

Brownlee D, Tsou P, Aléon J et al (2006) Comet 81P/Wild 2 under a microscope. Science 314: $32#1711–1716

Brucato JR, Strazzulla G, Baratta GA et al (2006) Cryogenic synthesis of molecules of astrobiological interest: catalytic role of cosmic dust analogues. Orig Life Evol Biosph 36:451–457

Callahan MP, Smith KE, Cleaves HJ et al (2011) Carbonaceous meteorites contain a wide range of extraterrestrial nucleobases. Proc Natl Acad Sci USA 108:13995–13998

Carnimeo I, Biczysko M, Bloino J et al (2011) Reliable structural, thermodynamic, and spectroscopic properties of organic molecules adsorbed on silicon surfaces from computational modeling: the case of glycine@Si(100). Phys Chem Chem Phys 13:16713–16727

Chyba C, Sagan C (1992) Endogenous production, exogenous delivery and impact-shock synthesis of organic molecules: an inventory for the origins of life. Nature 355:125–132

Cockell CS, Bush T, Bryce C et al (2016) Habitability: a review. Astrobiology 16:89–117

Cottin H, Saiagh K, Guan YY et al (2015) The AMINO experiment: a laboratory for astrochemistry and astrobiology on the EXPOSE-R facility of the International Space Station. Int J Astrobiol 14:67–77

Dobson CM, Ellison GB, Tuck AF et al (2000) Atmospheric aerosols as prebiotic chemical reactors. Proc Natl Acad Sci USA 97:11864–11868

dos Santos R, Patel M, Cuadros J et al (2016) Influence of mineralogy on the preservation of amino acids under simulated Mars conditions. Icarus 277:342–353

Downs RT, Hazen RM (2004) Chiral indices of crystalline surfaces as a measure of enantioselective potential. J Mol Catal A Chem 216:273–285
Dworkin JP, Deamer DW, Sandford SA, Allamandola LJ (2001) Self-assembling amphiphilic molecules: synthesis in simulated interstellar/precometary ices. Proc Natl Acad Sci 98 (3):815–819
Ehrenfreund P, Irvine W, Becker L et al (2002) Astrophysical and astrochemical insights into the origin of life. Rep Prog Phys 65:1427
Eigenbrode JL, Steele A, Summons RE et al (2016) Preservation of organic matter on Mars by sulfur. AGU Fall Meet, P21D-08
Elsila JE, Glavin DP, Dworkin JP (2009) Cometary glycine detected in samples returned by Stardust. Meteorit Planet Sci 44:1323–1330
Ertem G, Ertem MC, McKay CP et al (2017) Shielding biomolecules from effects of radiation by Mars analogue minerals and soils. Int J Astrobiol 16:280–285
Escamilla-Roa E, Moreno F (2012) Adsorption of glycine by cometary dust: astrobiological implications. Planet Space Sci 70:1–9
Espinoza C, Szczepanski J, Vala M et al (2010) Glycine and its hydrated complexes: a matrix isolation infrared study. J Phys Chem A 114:5919–5927
Flynn GJ (1996) The delivery of organic matter from asteroids and comets to the early surface of Mars. Earth Moon Planets 72:469–474
Fornaro T, Carnimeo I (2014) Computer simulations of prebiotic systems. In: Reference module in chemistry, molecular sciences and chemical engineering. Elsevier
Fornaro T, Brucato JR, Branciamore S et al (2013a) Adsorption of nucleic acid bases on magnesium oxide (MgO). Int J Astrobiol 12:78–86
Fornaro T, Brucato JR, Pace E et al (2013b) Infrared spectral investigations of UV irradiated nucleobases adsorbed on mineral surfaces. Icarus 226:1068–1085
Fornaro T, Brucato JR, Pucci A et al (2013c) Development of extraction protocols for life detection biosensor-based instruments. Planet Space Sci 86:75–79
Fornaro T, Biczysko M, Monti S et al (2014) Dispersion corrected DFT approaches for anharmonic vibrational frequency calculations: nucleobases and their dimers. Phys Chem Chem Phys 16: $32#10112–10128
Fornaro T, Burini D, Biczysko M et al (2015a) Hydrogen-bonding effects on infrared spectra from anharmonic computations: uracil-water complexes and uracil dimers. J Phys Chem A 119:4224–4236
Fornaro T, Carnimeo I, Biczysko M (2015b) Toward feasible and comprehensive computational protocol for simulation of the spectroscopic properties of large molecular systems: the anharmonic infrared spectrum of uracil in the solid state by the reduced dimensionality/hybrid VPT2 approach. J Phys Chem A 119:5313–5326
Fornaro T, Biczysko M, Bloino J et al (2016) Reliable vibrational wavenumbers for C=O and N-H stretchings of isolated and hydrogen-bonded nucleic acid bases. Phys Chem Chem Phys 18: $32#8479–8490
Fornaro T, Boosman A, Brucato JR et al (2018) UV irradiation of biomarkers adsorbed on minerals under Martian-like conditions: hints for life detection on Mars. Icarus 313:38–60
Freissinet C, Glavin DP, Mahaffy PR et al (2015) Organic molecules in the sheepbed mudstone, Gale crater, Mars. J Geophys Res Planets 120:495–514
Freissinet C, Glavin DP, Buch A et al (2016) First detection of non-chlorinated organic molecules indigenous to a Martian sample. 47th Lunar Planet Sci Conf 47:2568
Fukushi K, Sverjensky DA (2007) A predictive model (ETLM) for arsenate adsorption and surface speciation on oxides consistent with spectroscopic and theoretical molecular evidence. Geochim Cosmochim Acta 71:3717–3745
Galli D, Palla F (1998) The chemistry of the early universe. Astron Astrophys 335:403–420
Gallori E, Branciamore S (2012) Origin and evolution of self-replicating polymers on mineral habitats. In: Seckbach J (ed) Genesis – in the beginning. Cellular origin, life in extreme habitats and astrobiology. Springer, Dordrecht, pp 55–66

Gambino GL, Grassi A, Marletta G (2006) Molecular modeling of interactions between l-Lysine and functionalized quartz surfaces. J Phys Chem B 110:4836–4845
Garry JRC, Ten Kate IL, Martins Z et al (2006) Analysis and survival of amino acids in Martian regolith analogs. Meteorit Planet Sci 405:391–405
Glavin DP, Dworkin JP (2009) Enrichment of the amino acid l-isovaline by aqueous alteration on CI and CM meteorite parent bodies. Proc Natl Acad Sci USA 106:5487–5492
Glavin DP, Freissinet C, Miller KE et al (2013) Evidence for perchlorates and the origin of chlorinated hydrocarbons detected by SAM at the Rocknest aeolian deposit in Gale Crater. J Geophys Res E Planets 118:1955–1973
Goetz W, Brinckerhoff WB, Arevalo RJ et al (2016) MOMA: the challenge to search for organics and biosignatures on Mars. Int J Astrobiol 15:239–250
Gomes R, Levison HF, Tsiganis K et al (2005) Origin of the cataclysmic Late Heavy Bombardment period of the terrestrial planets. Nature 435:466–469
Grotzinger JP, Sumner DY, Kah LC et al (2014) A habitable fluvio-lacustrine environment at Yellowknife Bay, Gale Crater, Mars. Science 343:1242777
Hassler DM, Zeitlin C, Wimmer-Schweingruber RF et al (2014) Mars' surface radiation environment measured with the Mars science laboratory's curiosity rover. Science 343:1244797
Hazen RM (2005) Genesis: the scientific quest for life's origin. Joseph Henry Press, Washington, DC
Hazen RM (2006) Mineral surfaces and the prebiotic selection and organization of biomolecules. Am Mineral 91:1715–1729
Hazen RM, Sverjensky DA (2010) Mineral surfaces, geochemical complexities, and the origins of life. Cold Spring Harb Perspect Biol 2:815–824
Hazen RM, Filley TR, Goodfriend GA (2001) Selective adsorption of l- and d-amino acids on calcite: implications for biochemical homochirality. Proc Natl Acad Sci USA 98:5487–5490
Herbst E, Van Dishoeck EF (2009) Complex organic interstellar molecules. Annu Rev Astron Astrophys 47:427–480
Horneck G, Mileikowsky C, Melosh HJ et al (2002) Viable transfer of microorganisms in the solar system and beyond. In: Horneck G, Baumstark-Khan C (eds) Astrobiology – the quest for the conditions of life. Springer, Berlin, pp 57–76
Jiménez-Serra E, Chiatti F, Corno M et al (2012) Glycine adsorption at nonstoichiometric (010) hydroxyapatite surfaces: a B3LYP study. J Phys Chem C 116:14561–14567
Jonsson CM, Jonsson CL, Sverjensky DA et al (2009) Attachment of l-glutamate to rutile (α-TiO2): a potentiometric, adsorption, and surface complexation study. Langmuir 25:12127–12135
Kong H, Sun Q, Wang L et al (2014) Atomic-scale investigation on the facilitation and inhibition of guanine tautomerization at Au(111) Surface. ACS Nano 8:1804–1808
Lee N, Hummer DR, Sverjensky DA et al (2012) Speciation of l-DOPA on nanorutile as a function of pH and surface coverage using Surface-Enhanced Raman Spectroscopy (SERS). Langmuir 28:17322–17330
Martin W, Baross J, Kelley D et al (2008) Hydrothermal vents and the origin of life. Nat Rev Microbiol 6:805–814
Martins Z, Watson JS, Sephton MA et al (2006) Free dicarboxylic and aromatic acids in the carbonaceous chondrites Murchison and Orgueil. Meteorit Planet Sci 41:1073–1080
McAdam AC, Franz HB, Sutter B et al (2014) Sulphur-bearing phases detected by evolved gas analysis of the Rocknest aeolian deposit, Gale Crater, Mars. J Geophys Res Planets 119:373–393
Meinert C, Cassam-Chenaï P, Jones NC et al (2015) Anisotropy-guided enantiomeric enhancement in alanineusing Far-UV circularly polarized light. Orig Life Evol B 45:149–161
Mignon P, Sodupe M (2013) Structural behaviors of cytosine into the hydrated interlayer of Na-+-montmorillonite clay. An ab initio molecular dynamics study. J Phys Chem C 117:26179–26189
Mileikowsky C, Cucinotta FA, Wilson JW et al (2000) Natural transfer of viable microbes in space. Icarus 145:391–427
Miller SL (1953) A production of amino acids under possible primitive earth conditions. Source Sci New Ser 117:528–529
Miller SL, Urey HC (1959) Origin of life. Science 130:1622–1624

Milliken RE, Swayze GA, Arvidson RE et al (2008) Opaline silica in young deposits on Mars. Geology 36:847–850
Monti S, van Duin ACT, Kim SY et al (2012) Exploration of the conformational and reactive dynamics of glycine and diglycine on TiO2: computational investigations in the gas phase and in solution. J Phys Chem C 116:5141–5150
Navarro-González R, Vargas E, de la Rosa J et al (2010) Reanalysis of the Viking results suggests perchlorate and organics at midlatitudes on Mars. J Geophys Res 115:E12010
Newman SP, Di Cristina T, Coveney PV et al (2002) Molecular dynamics simulation of cationic and anionic clays containing amino acids. Langmuir 18:2933–2939
Ody A, Poulet F, Langevin Y et al (2012) Global maps of anhydrous minerals at the surface of Mars from OMEGA/MEx. J Geophys Res E Planets 117:1–14
Oró J (1961) Mechanism of synthesis of adenine from hydrogen cyanide under possible primitive earth conditions. Nature 191:1193–1194
Panigrahi S, Bhattacharya A, Banerjee S et al (2012) Interaction of nucleobases with wrinkled graphene surface: dispersion corrected DFT and AFM studies. J Phys Chem C 116:4374–4379
Parikh SJ, Kubicki JD, Jonsson CM et al (2011) Evaluating glutamate and aspartate binding mechanisms to Rutile (α-TiO2) via ATR-FTIR spectroscopy and quantum chemical calculations. Langmuir 27:1778–1787
Patel MR, Bérces A, Kolb C et al (2003) Seasonal and diurnal variations in Martian surface ultraviolet irradiation: biological and chemical implications for the Martian regolith. Int J Astrobiol 2:21–34
Pizzarello S, Cronin JR (2000) Non-racemic amino acids in the Murray and Murchison meteorites. Geochim Cosmochim Acta 64:329–338
Pizzarello S, Huang Y, Alexandre MR (2008) Molecular asymmetry in extraterrestrial chemistry: insights from a pristine meteorite. Proc Natl Acad Sci USA 105:3700–3704
Poch O, Jaber M, Stalport F et al (2015) Effect of nontronite smectite clay on the chemical evolution of several organic molecules under simulated Martian surface ultraviolet radiation conditions. Astrobiology 15:221–237
Pollet R, Boehme C, Marx D (2006) Ab initio simulations of desorption and reactivity of glycine at a water-pyrite interface at "iron-sulphur world" prebiotic conditions. Orig Life Evol Biosph 36:363–379
Rimola A, Sodupe M, Ugliengo P (2012) Computational simulations of prebiotic processes. In: Seckbach J (ed) Genesis – in the beginning. Cellular origin, life in extreme habitats and astrobiology. Springer, Dordrecht, pp 345–362
Ruff SW, Farmer JD (2016) Silica deposits on Mars with features resembling hot spring biosignatures at El Tatio in Chile. Nat Commun 7:13554
Ruiz-Mirazo K, Peretó J, Moreno A (2004) A universal definition of life: autonomy and open-ended evolution. Orig Life Evol Biosph 34:323–346
Rull F, Maurice S, Hutchinson I et al (2017) The Raman laser spectrometer for the ExoMars Rover Mission to Mars. Astrobiology 17:627–654
Saladino R, Carota E, Botta G et al (2015) Meteorite-catalyzed syntheses of nucleosides and of other prebiotic compounds from formamide under proton irradiation. Proc Natl Acad Sci USA 112:E2746–E2755
Sandford SA, Aléon J, Alexander CMD et al (2006) Organics captured from comet 81P/Wild 2 by the Stardust spacecraft. Science 314:1720–1724
Sato I, Urabe H, Ishiguro S et al (2003) Amplification of Chirality from extremely low to greater than 99.5% ee by asymmetric autocatalysis. Angew Chem 115:329–331
Scappini F, Casadei F, Zamboni R et al (2004) Protective effect of clay minerals on adsorbed nucleic acid against UV radiation: possible role in the origin of life. Int J Astrobiol 3:17–19
Schoonen M, Smirnov A, Cohn C (2004) A perspective on the role of minerals in prebiotic synthesis. AMBIO A J Hum Environ 33:539–551
Smith ML, Claire MW, Catling DC et al (2014) The formation of sulphate, nitrate and perchlorate salts in the Martian atmosphere. Icarus 231:51–64

Sowerby S, Heckl W (1998) The role of self-assembled monolayers of the purine and pyrimidine bases in the emergence of life. Orig Life Evol Biosph 28:283–310

Sowerby SJ, Edelwirth M, Heckl WM (1998) Self-assembly at the prebiotic solid–liquid interface: structures of self-assembled monolayers of adenine and guanine bases formed on inorganic surfaces. J Phys Chem B 102:5914–5922

Squyres SW (2004) In situ evidence for an ancient aqueous environment at Meridiani Planum, Mars. Science 306:1709–1714

Tarasevych AV, Sorochinsky AE, Kukhar VP et al (2015) High temperature sublimation of α-amino acids: a realistic prebiotic process leading to large enantiomeric excess. Chem Commun 51:7054–7057

ten Kate IL, Garry JRC, Peeters Z et al (2005) Amino acid photostability on the Martian surface. Meteorit Planet Sci 40:1185–1193

ten Kate IL, Garry JRC, Peeters Z et al (2006) The effects of Martian near surface conditions on the photochemistry of amino acids. Planet Space Sci 54:296–302

Tolstoy VP, Chernyshova IV, Skryshevsky VA (2003) Handbook of infrared spectroscopy of ultrathin films. Wiley, Hoboken, NJ

Valocchi AJ (1985) Validity of the local equilibrium assumption for modeling sorbing solute transport through homogeneous soils. Water Resour Res 21:808–820

Viviano-Beck CE, Seelos FP, Murchie SL et al (2014) Revised CRISM spectral parameters and summary products based on the currently detected mineral diversity on Mars. J Geophys Res Planets 119:1403–1431

Wächtershäuser G (1990) Evolution of the first metabolic cycles (chemoautotrophy/reductive citric acid cycle/origin of life/pyrite). Evolution 87:200–204

Westall JC (1987) Reactions at the oxide-solution interface: chemical and electrostatic models. In: Davis JA, Hayes KF (eds) Geochemical processes at mineral surfaces. American Chemical Society, Washington, DC, pp 54–78

Westall F, Loizeau D, Foucher F et al (2013) Habitability on Mars from a microbial point of view. Astrobiology 13:887–897

Yun Y, Gellman AJ (2015) Adsorption-induced auto-amplification of enantiomeric excess on an achiral surface. Nat Chem 7:520–525

Zhao YL, Köppen S, Frauenheim T (2011) An SCC-DFTB/MD study of the adsorption of Zwitterionic glycine on a Geminal hydroxylated silica surface in an explicit water environment. J Phys Chem C 115:9615–9621

Chapter 10
Photochemistry and Photoreactions of Organic Molecules in Space

Avinash Vicholous Dass, Hervé Cottin, and André Brack

Abstract The primary aim of exobiology research is to recognize the routes leading to the initiation of life on Earth and its plausibility elsewhere in the universe. How would we recognize life if we encounter it or its remnants on an extraterrestrial body? This is the critical question of biosignature research to which astrochemical studies can contribute. Our understanding of preserved fossils and contemporary terrestrial life serves as a guide in the search for biosignatures in the universe. Of the various life-detection techniques available, carbon chemistry is particularly pertinent and perhaps the most significant biosignature (Summons et al., Astrobiology 11 (2):157–181; 2011). 'Life' as we know it is based on C, H, N, O, P, S chemistry and the organic matter derived from its remains is ubiquitous on Earth, constituting an extensive chemical and isotopic record of past life that surpasses by a huge margin what is recorded by visible (and microscopic) fossils. Biosignatures are highly subjective to the geological conditions in which they form and the subsequent diagenetic and metamorphic events that reprocess them (Sleep, Cold Spring Harb Perspect Biol. 2(6): a002527; 2010) and thus need careful assessing before coming to concrete conclusions concerning biogenicity. However, chemistry alone is inadequate to detect life and collaborative efforts from all of the relevant investigations, combined with considerations of geological and environmental factors, will likely provide the best evidence for the presence or absence of life, in localities of interest.

A. V. Dass (✉) · A. Brack
Centre de Biophysique Moléculaire, CNRS, Orléans, France
e-mail: avinash.dass@cnrs-orleans.fr; brack@cnrs-orleans.fr

H. Cottin
Laboratoire Interuniversitaire des Systèmes Atmosphériques (LISA), UMR CNRS 7583, Université Paris Est Créteil et Université Paris Diderot, Institut Pierre Simon Laplace, Paris, France
e-mail: herve.cottin@lisa.u-pec.fr

B. Cavalazzi, F. Westall (eds.), *Biosignatures for Astrobiology*, Advances in Astrobiology and Biogeophysics, https://doi.org/10.1007/978-3-319-96175-0_10

10.1 Why Are We Interested in Radiation Chemistry for Biosignatures?

Radiation is an important factor that influences organic molecules (Swallow 1960) and the antagonistic nature of radiation over a long period of time makes direct detection of biosignatures difficult, as the search for organic signatures on Mars has amply documented (e.g. Freissinet et al. 2015).

Spectroscopic studies have revealed an abundance of small molecules, such as CH_3OH, CH_3CH_2OH, HCHO, HCOOH and HCN, in the interstellar medium and the role these molecules play in the formation of organic polymers or complex organic material on exposure to UV radiation and/or thermal activation (Tielens and Charnley 1997). Moreover, the major source of organic molecules on the early Earth is credited to extraterrestrial matter, impact shocks and UV or electrical discharges (Chyba and Sagan 1992). Radiation can lead to molecular synthesis, as well as destruction and thus, the understanding of radiation-molecule interactions becomes pertinent to biosignature investigations.

10.2 Modes of Production and Delivery of Organic Material

10.2.1 Endogenous Delivery

The present understanding is that atmospheric conditions on our primitive Earth were very different to the conditions simulated in the Miller–Urey experiment (Miller and Urey 1959), the atmosphere being rather neutral to mildly reducing (e.g. Kasting et al. 1993). The Miller–Urey experiment demonstrated the ease of production of several amino acids and other prebiotically relevant organic compounds from small molecules, like water (H_2O), methane (CH_4), ammonia (NH_3) and hydrogen (H_2). Recent studies also reveal that more than 40 amino acids and other amines were synthesized with similar experiments by conditions that simulated volcanic eruptions and lighting on the primitive Earth environment (Bada 2013).

Organic molecules in the subsurface of present Earth are mainly found in the zones influenced by hydrothermal activity. Central to origins of life hypotheses focused on hydrothermal vents are their associated reactions and products, such as minerals in vent walls and small molecules including CO_2, CH_4 and H_2 (Shock and Schulte 1998), formed by Fischer Tropsch-type (FTT) synthesis. FTT synthesis involves the interaction of hot hydrothermal fluids with (ultra) mafic rocks, i.e., rocks rich in Fe and Mg minerals (olivines, pyroxenes), a process termed serpentinisation. Fischer Tropsch-type synthesis on hydrothermally formed alloys has been demonstrated to catalyse the formation of alkanes (C1–C3) in alkaline hydrothermal conditions (Sherwood Lollar et al. 1993; Horita and Berndt 1999). The formation of C2 and C3 alkanes apparently depends on the presence of metallic catalysts, such as Cr, an element common in peridotite rocks associated with oceanic crusts (Foustoukos and Seyfried 2004).

Common serpentinisation reactions include:

$$3Fe_2SiO_4(\text{fayalite}) + 2H_2O \rightarrow 2Fe_3O_4(\text{magnetite}) + 3SiO_2 + H_2$$
$$3Mg_2SiO_4(\text{forsterite}) + SiO_2 + 4H_2O \rightarrow 2Mg_3Si_2O_5(OH)_4(\text{serpentine})$$
$$4H_2 + CO_2 \rightarrow CH_4(\text{methane}) + 2H_2O(\text{magnetite-catalysed reaction}).$$

Although the fact that these kinds of molecules could be produced by and in hydrothermal processes initially suggested that the subsurface could be a potentially substantial source of organics for prebiotic reactions, McCollom and Seewald (2001, 2003) estimated that the potential for abiogenic methanogenesis, for instance, on the early Earth may have been minimal. Furthermore, a recent observation suggests that methane (CH_4) and higher hydrocarbons are, in fact, derived from fluid inclusions trapped in plutonic rocks, which are liberated upon cooling at the seawater-hydrothermal fluid interface, thus generating hydrocarbon-rich fluids (McDermott et al. 2015), rather than being formed by subsurface Fischer Tropsch synthesis. In each of these scenarios, however, the subsurface is an endogenous source of organic molecules.

10.2.2 Exogenous Delivery

10.2.2.1 Extraterrestrial Delivery

There is abundant evidence for the presence of organic molecules in comets, asteroids, interplanetary dust particles, meteorites (specifically carbonaceous chondrites, which contain up to 4% C and micrometeorites. The wide variety of organic compounds, including macromolecular carbon, amino acids and monocarboxylic acids among many others, in these extraterrestrial bodies is well-documented and beyond the scope of this chapter, although we direct readers to comprehensive reviews by Sephton (2002, 2005), and references therein. During its life time, and especially in the early formative ~600 million years, the Earth has been bombarded by a large variety of impactors ranging from the huge Mars-sized impactor (Theia) that created the Moon to cosmic dust less than 1 μm in size (Peucker-Ehrenbrink and Schmitz 2012). This impact phenomenon continues even to this day but the flux of impactors is no longer as high as it was during the Hadean and Early Archaean eras. Some of the carbon in meteorites is so primitive that it is speculated to have a pre-solar origin (Martins et al. 2015).

Estimates of the flux of extraterrestrial material to the present-day Earth are highly variable, ranging from 30,000–60,000 tonnes per year (±15,000 tonnes); (Bland et al. 1996), the error being related to poor understanding of the complete size range of incoming materials, for instance the mass of incoming micrometeorites (Maurette 2006).

The value of ~0.1 kg/km^2/year (Dass et al. 2016) for the modern Earth impactor flux is modelled to have been 1000 times greater during the hypothesized Late Heavy Bombardment (LHB), ~3.9–3.8 Ga (Frey 1980; Ryder et al. 2000; Ryder 2002), therefore roughly 100 kg/km^2/year. This flux represents the absolute maximum for extraterrestrial input to Earth at any point during its history, perhaps equaled solely in the initial stages of planetary accretion (Koeberl 2006). A rough estimation for the amount of extraterrestrial flux during the Hadean between 4.5 and 4.0 Ga indicates that it averaged $<$11.76 kg/km^2/year over that time period (from the lunar cratering record, cf. Koeberl 2006). Note, however, that the LHB has since been re-assessed and appears to have been an artefact of the original lunar sampling programme, rather than an actual event (Zellner 2017).

Laboratory studies to simulate the formation of organic molecules under interstellar conditions were conducted extensively (Briggs et al. 1992; Kobayashi et al. 1998) and the experimental results revealed the formation of amino acids when ice mixtures containing CO, CH_4, CH_3OH, NH_3, H_2O were irradiated (see also Bernstein et al. 2002). Moreover, recently, synthesis of ribose and other sugar was achieved after photolysis of an icy mixture of H_2O, CH_3OH and NH_3 (Meinert et al. 2016). In order to test these results in realistic space conditions and to understand the cumulative effects of vacuum, high-energy particles, near-weightlessness and extreme low temperatures on organic molecules, space as an astrobiology tool became pertinent (Hashimoto et al. 2002).

10.3 Radiation Studies on Earth *versus* Space

Limited access to space (Martins et al. 2017) poses one of the major challenges in studying the realistic space radiation effects on molecules. Photochemical studies in laboratories are useful but the challenge arises in mimicking the complete range of solar radiation using a typical lamp source. Table 10.1, provides an overview of the types of lamp sources that have been used in laboratories to attain a specific range of UV wavelength (see Cottin et al. (2017) for a detailed review of artificial light sources).

Table 10.1 Sources of UV lamps used to generate specific ranges of UV wavelengths

UV source	λ in nm	References
H_2/He	122	Cottin et al. (2000, 2008), Es-sebbar et al. (2015)
Kr	145	Kuzicheva and Gontareva (2003)
Xe	147	Cottin et al. (2000, 2008)
CH_4/He	193	Cottin et al. (2000, 2008)
H_2	122–160	Ten Kate et al. (2005)
D_2 or Xe	190–400	Ten Kate et al. (2005), Stalport et al. (2009)
Synchrotron	25–300, varied wavelength	Schwell et al. (2006, 2008), Nahon et al. (2012)

However, even this comes at a price and the synchrotron facilities have their own shortcomings as pointed out by Cottin et al. (2017). For example; the integrated solar UV flux for a range of 100 and 300 nm is of the order of 2×10^{13} ph cm^{-2} s^{-1} (Thuillier et al. 2004) while, in the same range of wavelengths, the flux generated by the DESIRS synchrotron source can reach 10^{15} ph cm^{-2} s^{-1}, i.e. about 50 times more intense than from the Sun.

10.4 Space Experiments Relevant to Exposure of Organics

In this chapter, we will restrict our discussions to the space expose experiments that were carried out with organic molecules and, if the reader is interested to explore various other space-expose experiments (e.g. Biological experiments), we refer you to (Horneck 2007; Horneck et al. 2010; Cottin et al. 2017) and the references therein.

During the 1970s, onboard the Salut-6 orbital mission, for the first time, the formation of adenosine, deoxyadenosine, and thymidine nucleosides was demonstrated (Kuzicheva and Gontareva 2003). This, in turn, stirred a series of follow up investigations resulting in a cascade of explorations in space-related chemistry experiments. Two follow-up experiments on board the Salut-7 space craft were carried out for the durations of 13 and 16 months respectively (Kuzicheva and Gontareva 2003). A specially designed cassette-like device called "Menduza" fitted with quartz glasses with a UV cut-off at 220 nm (see Fig. 10.1) was used for the two missions. The in-flight temperature on this device varied between −50 °C and 65 °C.

Similar flight missions called Bion-9 (Cosmos-2044/Kosmos-2044) and Bion-11 with flight durations of 14 days were carried out using devices similar to the Menduza, called "outside container" (OC). The OC resembles a circular suitcase with an overhead lid and a base plate. The base plate could be fitted with quartz windows and the overhead lid locks onto the base plate with automatic unlocking on deployment in orbit. The OC of Bion-9 experienced a range of temperatures from −13 °C to +67 °C and the Bion-11 OC experienced temperatures between −30 °C and 100 °C. In order to substantiate the results obtained during the Salyut-7 missions,

Fig. 10.1 Cassette-like device called 'Menduza' used onboard Salut-7. Source: Modified from Kuzicheva and Gontareva (2003)

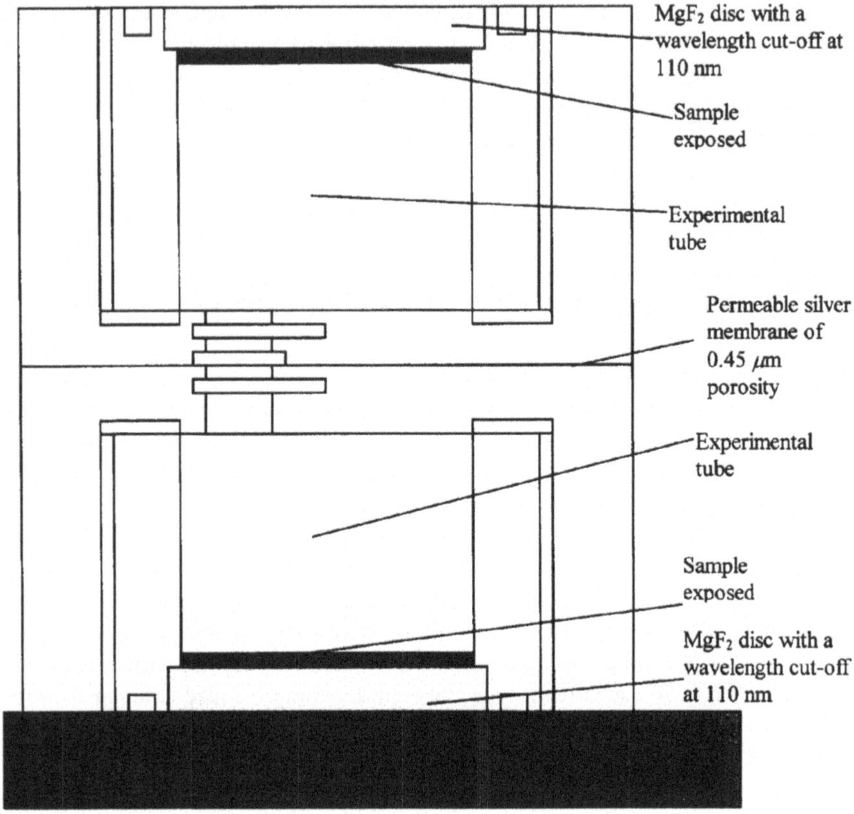

Fig. 10.2 A general scheme of the cell holder during the "Pereus Exobiology mission", where the top stage was exposed to radiation, and the lower stage served as the in-flight control. Source: Reprinted with permission from Kuzicheva and Gontareva (2001)

a new experiment called "Perseus exobiology' onboard the Mir space station was carried out. The Perseus experiment was also used for simultaneous studies on amino acids and peptides (Kuzicheva and Gontareva 2001, 2003; Boillot et al. 2002). For this flight, a specially designed, two-stage monoblock device was designed in France.

During the experiment, the top stage was exposed to the UV radiation of space, whereas the lower stage was used as flight-control. The monoblock consisted of 66 cell holders. Each holder (see Fig. 10.2) had two stages, with a temperature sensor each. The samples were deposited on MgF_2 windows with a UV cut-off at 120 nm, which was an improvement from the Salyut-7 missions. During the course of the mission, the monoblock experienced a high temperature of 41 °C.

For each of the missions following the Salyut-6 mission, the aim was to photosynthesize nucleotides (e.g. adenosine nucleotide, Fig. 10.3) by reaction of nucleosides (sugar like ribose + bases like adenine) with orthophosphates (like KH_2PO_4). Adenosine, deoxyadenosine, cytidine, uridine, thymidine and deoxythymidine

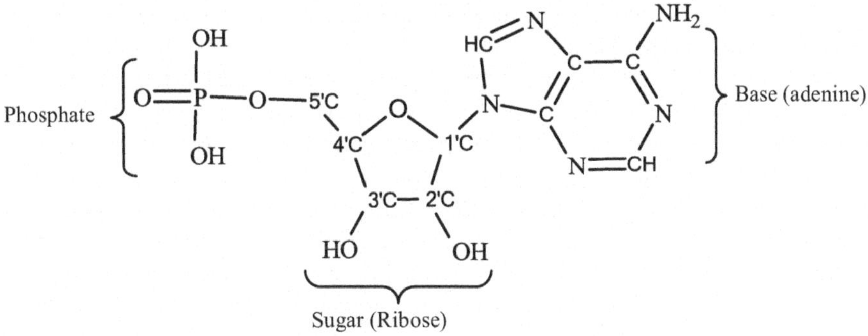

Fig. 10.3 Adenosine nucleotide with 5′ phosphate (i.e. phosphate on carbon 5′), the other possible sites for phosphorylation are 2′ and 3′ carbons. Source: A V Dass

nucleosides were homogenized in solution with orthophosphates and evaporated as thin films on MgF_2 (Perseus) or quartz windows (Salyut 7, Bion-9, 11).

The results of the missions demonstrated that, due to a favourable spatial orientation, 5′ nucleotide was the predominant product in comparison to 2′, 3′ nucleotides. They also observed that purine nucleotides (with bicyclic bases) were less prone to degradation, possibly due to an extended conjugation (double bond) system, in comparison to pyrimidine nucleotides. Additionally, it was observed that, in all the experiments, the rate of degradation of adenosine nucleotides was lowest in comparison to uridine and cytidine nucleotides (this could also be due to bicyclic system influence). Interestingly, 5′ ATP was also detected at trace levels (Kuzicheva and Gontareva 2003) after the Perseus mission. This was attributed to the ambient temperatures recorded on the Perseus monoblock device (~41 °C), thus reducing the extent of hydrolysis of 5′ATP (highly energetic molecule).

Parallel experiments carried out on ground, aided better understanding of the conditions of photoreactions and it was concluded during that, Vacuum UV and elevated temperatures (~160 °C) were more effective in the synthesis of nucleotides in comparison to γ-radiations (highly energetic). Although, this could be a result of simultaneous destruction of the formed nucleotides in the presence of high-energy γ-radiations. This conclusion could be derived from the reported results of effective synthesis 5′AMP with reduced flux of γ-radiation (~3 times lower flux to VUV). This reinforces the importance of the thermodynamic stability of 5′AMP to 5′ATP in surviving hydrolysis. The researchers also observed that shorter flight times (like Bion-9, 11) resulted in greater molecular survival quantitatively. This was accredited to the fact that, during prolonged exposure times, the rate of formation of the products was far exceeded by the rate of degradation. The other conclusion of the study was that infrared energy was the most effective in the synthesis of nucleotides (canonical), followed by VUV of 145 nm wavelength in comparison to 254 nm radiation. It was pointed out that lower yields of nucleotides are due to the fact that 254 nm radiation is mainly absorbed by nucleobases, whereas at 145 nm, the sugar and the phosphates are photochemically excited, thus resulting in nucleotide synthesis and hence higher yields.

As mentioned above, during the Perseus mission, amino acids and peptides were also exposed and studied. The experiment was carried out for 97 days in protected (with clay, basalt and meteoritic powder) and unprotected mode (pure molecules). Two amino acids L-leucine, α-methyl L-leucine, one cyclic dipeptide, L-leucine diketopiperazine (DKP) and one activated tripeptide, and tri-L-leucine thioethylester were used for the experiment (Boillot et al. 2002). The samples were mainly composed of leucine and its derivatives since this molecule occurs in the Murchison meteorite. Methyl leucine was chosen in order to compare the relative stabilities of substituted (alkyl group on the α-carbon) and unsubstituted amino acid (only hydrogen on the α-carbon) to radiation. The dipeptides were used to study the stability of peptide bonds and thioethylester of the tripeptide was used due to its activated state in order to study oligomerisation and the prebiotic relevance of thioesters (Wächtershäuser 2000, 2006). After the conclusion of the experiment, the results revealed that the cyclic peptide (DKP) was relatively inert towards space radiation as a whole. It is important to note that cyclic peptides are not composed of a carboxyl terminal and thus are not prone to the decarboxylation; the most common photoreaction is observed in solid-state samples under the effect of UV radiation. During sample analysis, it was observed that there was no diversity of chemical photoproducts for peptides and yet loss of starting material, which led to the interpretation that peptides were more susceptible to sublimation under high vacuum space-conditions (Boillot et al. 2002). In the case of the thioethylester tripeptide, the loss of starting material was attributed to the additive effects of sublimation and degradation at mildly elevated temperature (40 °C) over the period of the Perseus experiment. Moreover, the peptide bond decomposition process is considered to hinder the oligomerization of thioethylester peptide. The authors concluded that, in the case of rest of the samples, decarboxylation and decarbonylation were the most common photoreactions. This conclusion was supported by the laboratory studies involving solid-state samples irradiated with UV radiation. For the samples using mineral surfaces to mimic a natural protection mechanism, it was determined that the threshold for efficient protection against radiation was about 5 μm. This information was used to buttress the idea that a 50–100 μm micrometeorite fraction could deliver organic compounds to Earth like bodies.

In 1982, ESA's Directorate of Manned Spaceflight and Microgravity offered an interesting opportunity to study the response of microorganisms to outer space with the exobiology radiation assembly (ERA) on board the European Retrievable Carrier (EURECA). Hereafter, ESA developed the BIOPAN facility in 1994, a multipurpose space exposure system devoted to research in space biology. BIOPAN was installed on the external surface of the FOTON descent capsule protruding from the thermal blanket that envelops the satellite. It was a large circular container with a deployable lid, which is opened in orbit upon command and remained closed and sealed during launch and re-entry (Fig. 10.4). BIOPAN-1 was fixed outside the Russian FOTON-9 satellite (for a review, refer to Horneck 2007). Three chemistry experiments were carried out on the BIOPAN facility.

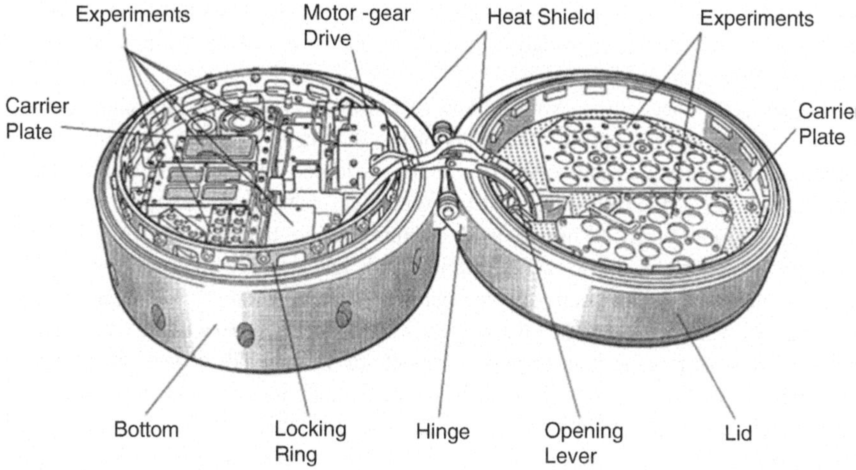

Fig. 10.4 A scheme of the BIOPAN facility. Source: Reprinted with permission from Demets et al. (2005)

The Biopan-1 and Biopan-2 experiments were carried out as templates to test molecules and hardware for future long-duration experiments outside space stations. Each chemistry-related experiment was named with respect to the platform used during a specific mission. For example, Dust-1 was the name given to the chemistry segment carried out on the Biopan facility during the Biopan-1 mission.

For Biopan-1 (Dust-1), L-forms of Gly, Ala, Leu, Val, Asp, Glu, all of which were reported in Murchison meteorite, were used to study the potential racemisation and stability of amino acids in space conditions (Barbier et al. 1998). In order to study possible photochemical oligomerisation and due to its photosensitivity, L-tyrosine was also chosen, although this molecule was not detected in Murchison samples. A dipeptide L-alanyl-L-alanine was also selected to test the stability of the peptide bond. After the conclusion of the experiment, there was no analytical evidence that pointed to inversion of configuration (i.e. no racemisation). Post-flight analysis of the amino acids showed that L-Asp and L-Glu had undergone partial decomposition (10–15%). Not so surprisingly, when shielded in a 5–7 μm thick layer of mineral matrix (montmorillonite and kaolinite), the same amino acids survived the duration of the space exposure experiment. The other amino acids samples remained unaffected by the flight, especially L-Gly, an achiral molecule, which remained virtually unaffected, and L-Ala, L-Leu, L-Val, the aliphatic amino acids, which also showed excellent stability. L-tyrosine displayed no oligomerisation and the dipeptide L-Alanyl-L-Alanine showed greater stability (Barbier et al. 1998). Surprisingly, there were no reports of loss of the peptide samples to space vacuum, indicating the difference between linear peptides and cyclic peptides as reported in the Perseus mission (Boillot et al. 2002). This might be due to the ability of linear chains to remain in zwitterionic form and therefore more stable in comparison to their cyclic counterparts. Alternatively, the authors suggested that the overall stability of the samples during the mission was due to shorter exposure times, non-synchronous orbitals and absence of sum-pointing device and thus milder degradation and therefore recommended longer exposure times (Barbier et al. 1998).

Biopan-2 (Dust-2) with an exposure time of 10 days and exposing Gly, Glu, Asp and Tyr was carried out to confirm the results of the Biopan-1. Additionally a new set of compounds, including various esters of glutamic acid and leucine, such as nitrobenzyl (–ONb), benzyl (–OBzl), thioethyl (–SEt) and methyl (–OMe), were exposed in order to examine the possibility of amino acid oligomerization under solar UV radiation. Thioethyl esters are known to activate amino acids and thus were chosen to study the nature of oligomerisation in activated compared to inactivated esters. A tripeptide $(Glu)_3$ was selected to study the peptide linkage stability and $(Leu)_3$–ONb was selected to study the possibility of peptide condensation. Tripeptides were selected over dipeptides to prevent the formation of diketopiperazines, as cyclic dipeptides are generally considered as relatively inert in oligomerisation processes. The thickness of the films was maintained at 0.5 microns to be able to irradiate most of the exposed molecules. The molecules were also exposed in clay, shielded environment, with a 50 microns thick film, representing a micrometeorite-like protection.

The analytical results confirmed the absence of racemization of the exposed molecules and also the high sensitivity of acidic amino acids towards UV radiation already observed in the DUST 1experiment. Surprisingly, the degradation of glycine was observed to be enhanced when protected in montmorillonite. This behaviour was attributed to photoionization of residual water molecules in the clay sheets, although this effect was observed only in the case of glycine. Laboratory studies confirmed that loss of glycine was not due to sublimation and thus the cause remained unexplained. Oligomerisation was observed with the esters of amino acids, although it was noted that, in the case of the benzyl ester of Leucine, the unexposed samples showed the presence of cyclic peptides indicating that space vacuum had a major role in the formation of such products. Additionally, it was observed that esters of the amino acids (since they lack the zwitterionic state) were lost due to sublimation in space vacuum. The results with the tripeptides also confirmed that photolysis reactions were the main cause of the loss of CO_2 and CO and not peptide bond hydrolysis, even when embedded in a mineral matrix. Overall, peptides showed greater stability in comparison to amino acids.

The ORGANICS experiment was flown on Biopan-5 in 2005 (Ehrenfreund et al. 2007). Polycyclic aromatic hydrocarbons (PAHs) and fullerenes (C_{60}) (see Fig. 10.5) were exposed to a total fluence of 602.45 kJ m^{-2} (for photons in the range 170–280 nm). The experiment was a precursor hardware test-flight for the long-term exposure experiment (Survival of organics in space) on the EXPOSE facility on the International Space Station (ISS). Thin films of the selected organic molecules were used for expose studies lasting 16 days in the LEO.

The space exposure experiments were compared to the laboratory studies carried out using a hydrogen discharge lamp with photon wavelengths ranging between 150–195 nm. In space, the most widespread form of carbon is found in the form of PAH's and fullerenes. They form the bulk of the organic matter found in meteorites and micrometeorites (Maurette 2006). Post-flight analysis on the samples revealed

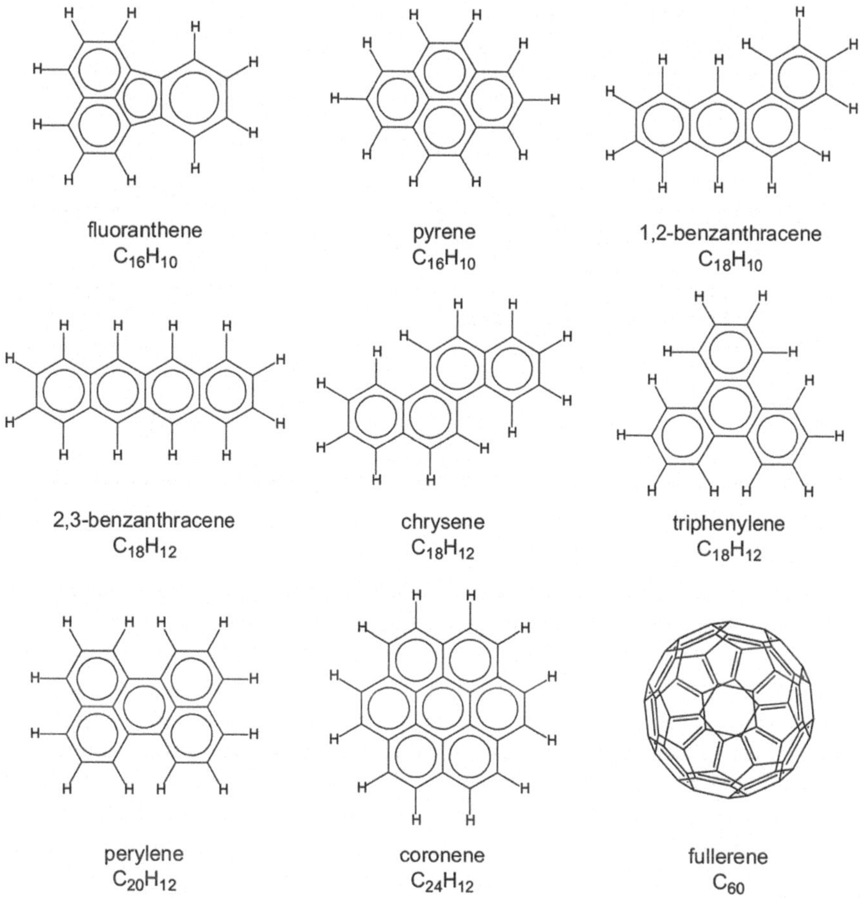

Fig. 10.5 Showing the polycyclic aromatic hydrocarbons and Fullerene (C_{60}) selected for flight on Biopan 5. Source: Reprinted with permission from Ehrenfreund et al. (2007)

no dehydrogenation reaction, even though the selected molecules had a gas-phase ionizing potential of up to 7 eV. The results revealed that there was a difference of only 1% loss of starting material between the flight-exposed and flight control samples, which falls within the measure of an analytical error bar.

The results show that the type of photon fluence (order of 10^{17}) observed during Biopan-5 was insufficient to cause effective photolysis of the samples over the 16 days period. It was predicted that a photon fluence of the order of $>10^{20}$ with an effective energy greater than 10 eV would be necessary to initiate a photochemical response with this set of molecules. The laboratory studies confirmed these observations and were in agreement with the results of Biopan-5.

At this point, the reader is familiar with the diverse categories of molecules studied under space conditions (sugars, amino acids and PAH's) and the following experiments involved a detailed study of specific molecules within these categories

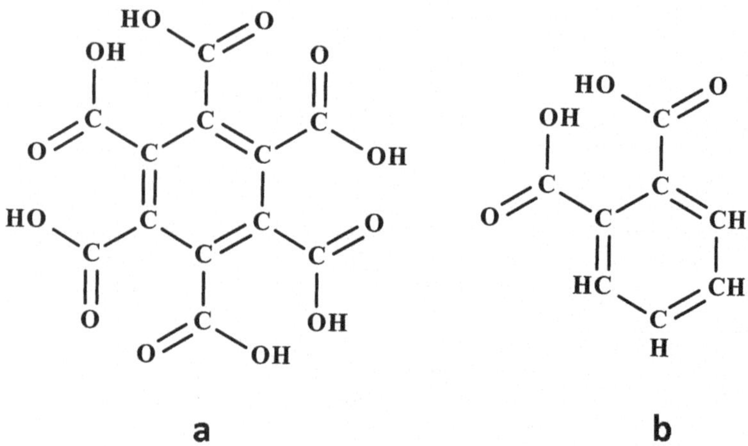

Fig. 10.6 Some molecules selected for UVloution experiment (**a**) Mellitic acid and (**b**) Phthalic acid. Source: A V Dass

in order to understand individual types of molecules. Therefore, in the following, we will concentrate only on the important discussions and conclusions.

In 2007, the UVolution experiment was carried on the Biopan-6 mission over an orbital period of (Stalport et al. 2010). The main aim of this experiment was to study the behaviour of molecules in Mars-like conditions, since Mars is amongst the prime locations for astrobiology exploration. The surface layers of Mars are intermixed with oxidizing agents (e.g. peroxides) and, thus, any organic molecules could be potentially destroyed even if present. It was hypothesised that metastable molecules (Fig. 10.6) could potentially be part of the organic inventory on the Martian surface or in its subsurface. Most of the acids used had a carboxylic acid substitution at different positions on the benzene ring. α-aminoisobutyric acid (AIB) was also selected as it is also present in both meteoritic and micrometeoritic material.

Stalport et al. (2010) concluded that the exposed molecules were less photostable as they were partially destroyed in a short period of time. In particular, the behaviour of mellitic acid showed a greater difference in comparison to laboratory simulation experiments (Stalport et al. 2009, 2010) as it was almost completely decomposed with no byproducts reported. Gauging from the previous chemistry expose experiments, and given the abundance of the carboxylic acid functionality of the selected molecules, decarboxylation could be the predominant reaction. However, it is important to study the sublimation pattern of these molecules more rigorously before arriving at concrete conclusions. The greater photolysis of the selected molecules in the presence of analogous martian soil led to the authors to conclude that a search for such molecules on the martial surface could be potentially futile. Nevertheless, it would be interesting to know the byproducts of some of these molecules, since the residual products might be of interest for biosignature studies on Mars.

The next generation of long-term expose experiments was carried out on the EXPOSE facility of the International Space Station (ISS). The EXPOSE facility is located on an external platform secured to the outer hull of the Zvezda Service Module of the ISS in the Russian segment. Three experiments have been carried out successfully on this facility, including the most recent one called the PSS (photochemistry on the space station). Depending on the type of molecules to be exposed, cells of different configuration are used. Traditional open cells and semi-hermetic cells are used routinely, but for gaseous samples, special cells named "CNES closed cells" have been employed. For comprehensive information on the hardware, please refer to (Cottin et al. 2017).

The first experiment was emplaced as a part of the EXPOSE payload in 2008 and lasted almost 1.5 years. The experiment was carried out to study the chemical behaviour of amino acids and dileucine, exposed unprotected and embedded in meteorite powder, to irradiation. It was concluded that both types of samples (with or without meteorite powder) were affected chemically to varying degrees when exposed to solar radiation. This study also included the quantification of organic degradation (Bertrand et al. 2012). The results revealed that resistance to irradiation depended on (1) the chemical nature (structure and composition) of the exposed molecules, (2) the emission spectrum of the UV source (laboratory lamp source versus solar radiation), and (3) the shielding effect of the meteorite powder. Moreover, it was observed that amino acids with two-acid group functionality (e.g. aspartic acid) was more sensitive to UV radiation than amino acids with alkyl side-chains (e.g. valine). The dipeptide with an amide bond was almost completely degraded when unprotected by a mineral coating. Additionally, amino acids with a substituted chain, such as valine, were more stable than those with a linear chain, as in the case of aminobutyric acid. As part of the PROCESS experiment mission on board the space station, a study was performed on the degradation of organics under filtered electromagnetic radiation that mimics Mars-like surface UV radiation. The mission served as a test-bed for further understanding of the nature of certain PAH's and to confirm the results of UVolution experiment on a long duration flight. Glycine, serine, phthalic acid, mellitic acid were amongst the molecules selected for the experiment and the researchers concluded that most of these molecules underwent extensive photodegradation (Noblet et al. 2012). The presence of a martian soil did not prevent the degradation of these molecules. The results do not follow the usual trend of PAH's which display greater levels of photostability under extreme UV conditions. There appear to be inconsistencies with exposure regime (as glycine was recovered well preserved post-flight) and interference of the cell material during this mission that might have contributed to non-uniform observations.

The ORGANIC and AMINO experiments were exposed on the EXPOSE facility in 2009. The objectives of the ORGANIC experiment on EXPOSE-R were (1) to study the photostability of selected PAH's and fullerene-type molecules in an interplanetary environment to allow a comparison with space data; and (2) to allow a quantitative estimation of dissociation regimes for organic molecules that could be extrapolated to different space environments (interstellar medium, interplanetary, Earth atmosphere, etc.) (Bryson et al. 2011). The ORGANIC experiment on EXPOSE-R spent 682 days outside the ISS and the fourteen samples (11 polycyclic aromatic hydrocarbons (PAHs) and three fullerenes) received an irradiation dose of the order of 14,000 MJ m^{-2} over 2900 h of unshadowed solar illumination (Bryson et al. 2015). Analyses on the returned samples and ground control measurements showed limited spectral changes in most cases, pointing to the stability of PAHs and fullerenes under space exposure conditions. However, some molecules were strongly depleted and the experiments confirm the known trend in the stability of PAH species according to molecular structure: compact PAHs are more stable than non-compact PAHs and the least stable are PAHs containing heteroatoms (Bryson et al. 2015).

AMINO was the second experiment carried out on the EXPOSE platform in 2009. For the experiment, the amino acids glycine, D-alanine, D-valine, D-aspartic acid, amino isobutyric acid, 2-amino butyric acid and the dipeptide dileucine were exposed to space conditions in the free form (i.e. non protected by meteorite powder) and embedded in meteorite powder (Bertrand et al. 2015; Cottin et al. 2017). This experiment was mainly carried out to confirm the previous results obtained from the PROCESS experiment, showing indeed that amino acids with two-acid functional groups (e.g. aspartic acid) are more sensitive to UV radiation than amino acids with a hydrocarbon chain (e.g. aspartic acid). It was also confirmed that the dileucine amide bond had a high susceptibility to degradation, only surviving when protected by a mineral matrix of ground meteorite. High resolution mass spectrometry analyses demonstrated, for the first time, that some degradation is due to decarbonylation and decarboxylation caused by exposure to UV radiation. The new compounds formed by losing their carboxylic group were then more resistant to UV radiation (Bertrand et al. 2015). The conclusion of the experiment was that it would be necessary to investigate additional samples in the laboratory and in low Earth orbit (LEO) in order to obtain better understanding of the effects of photochemistry in UV radiation and to establish the link between organic matter synthesized in space and the first living organisms on Earth. Thus, it was decided that a new set of organic molecules with different structural configurations and chemical compositions would be used for the next mission.

The most recent experiment to be carried out on the EXPOSE platform was the PSS (Photochemistry on the Space Station). The updated platform can accommodate 75 samples thus allowing for gathering numerically higher analytical data with more diverse molecules. At the time of compilation of this chapter, the exposed samples are being analysed. The PSS experiment was a long-term experiment lasting almost 531 days. For this mission, several types of molecules including peptides and nucleobases have been exposed.

10.5 Conclusions and Perspectives

This chapter has provided a brief overview of experiments concerning the photoreactivity of organic molecules of general interest to the origins of life—and this from the point of view of the exogenous delivery of organic molecules to Earth during the Hadean. We have seen that, while it is impossible to recreate the full range of Vacuum UV conditions of space in a laboratory setting, ground-based simulations are generally useful for providing a preliminary indication of what could happen in space (and in the environmental conditions of another planet, such as Mars), as well as for testing experimental procedures and protocols. The space environment, however, is invaluable for photochemical experiments and the long series of experiments undertaken on the various types of support—the Photon capsules, the MIR and ISS stations—have provided a wealth of information on what affects the photoreactivity and stability of various molecules found in carbonaceous meteorites and micrometeorites: these include, for example, the structural conformation and types of bonding exhibited by the original molecules, for example peptides being more stable than amino acids. The processes of alteration of the molecules could also be determined, with decarboxylation and decarbonylation reactions as the major contributors to the degradation of molecules. Finally, a number of expose experiments have clearly demonstrated the protective power of even a thin coating of mineral matrix, thus confirming that extraterrestrial organic molecules can be delivered (not considering atmospheric entry), some of them unscathed, to Earth and that they would have been an important contribution to the organic inventory that was available for prebiotic reactions leading to the emergence of life on Earth in the Hadean Era.

The experiments described have shown that the space environment is an excellent vehicle for simulating conditions at the surface of Mars, where the lack of a significant atmosphere and, especially, the lack of an ozone layer, allows UV radiation to reach the surface of the planet with subsequent deleterious effects on the organic molecules. These kinds of experiments are essential for preparing for what kinds of organic molecules could potentially be preserved in the martian surface materials, and this in view of the on-going Mars Science Laboratory mission to Gale Crater and also the future missions, the European/Russian ExoMars 2020 and the international Mars Sample Return missions.

In terms of perspectives, there are a number of aspects of space experimentation that could be improved and new scientific paths to be followed. As the reader might have noticed, all of the aforementioned missions have only two points of realistic analytical data; one before flight and another after flight. A wide range of chemical dynamics data is unaccounted for during the actual flight, thus a need for real time data gathering on chemical transformations of exposed molecules arises. In order to do so, a more futuristic and much needed piece of technology has been conceived which resembles a nanosatellite, capable of in-situ analysis on a concentric conveyer belt-like system fitted with a step motor. The cells used in this instrument can contain their own microenvironment, independent of that of the neighbouring cells. UV-Vis spectroscopy is the mode of analysis on the miniaturized system. For hardware design and technical information of the subject, the reader is directed to (Ehrenfreund et al. 2014; Cottin et al. 2017). A NASA cubesat Organism/Organic Exposure to Orbital Stresses (O/OREOS) was launched in November 19, 2010 on a 6-month mission as a cost-effective technology demonstration flight. This technology will be utilized in a more recent project called OREOcube (Elsaesser et al. 2014).

References

Bada JL (2013) New insights into prebiotic chemistry from Stanley Miller's spark discharge experiments. Chem Soc Rev 42:2186

Barbier B, Chabin A, Chaput D et al (1998) Photochemical processing of amino acids in Earth orbit. Planet Space Sci 46:391–398

Bernstein MP, Dworkin JP, Sandford SA et al (2002) Racemic amino acids from the ultraviolet photolysis of interstellar ice analogues. Nature 416:401–403

Bertrand M, Chabin A, Brack A et al (2012) The PROCESS experiment: exposure of amino acids in the EXPOSE-E experiment on the international space station and in laboratory simulations. Astrobiology 12:426–435

Bertrand M, Chabin A, Colas C et al (2015) The AMINO experiment: exposure of amino acids in the EXPOSE-R experiment on the International Space Station and in laboratory. Int J Astrobiol 14:89–97

Bland PA, Berry FJ, Smith TB et al (1996) The flux of meteorites to the Earth and weathering in hot desert ordinary chondrite finds. Geochim Cosmochim Acta 60:2053–2059

Boillot F, Chabin A, Buré C et al (2002) The perseus exobiology mission on MIR behaviour of amino acids and peptides in Earth Orbit. Orig Life Evol Biosph 32:359–385

Briggs R, Ertem G, Ferris JP et al (1992) Comet Halley as an aggregate of interstellar dust and further evidence for the photochemical formation of organics in the interstellar medium. Orig Life Evol Biosph 22:287–307

Bryson KL, Peeters Z, Salama F et al (2011) The ORGANIC experiment on EXPOSE-R on the ISS: flight sample preparation and ground control spectroscopy. Adv Space Res 48:1980–1996

Bryson KL, Salama F, Elsaesser A et al (2015) First results of the ORGANIC experiment on EXPOSE-R on the ISS. Int J Astrobiol 14:55–66

Chyba C, Sagan C (1992) Endogenous production, exogenous delivery and impact-shock synthesis of organic molecules: an inventory for the origins of life. Nature 355:125–132

Cottin H, Gazeau M-C, Doussin J-F et al (2000) An experimental study of the photodegradation of polyoxymethylene at 122, 147 and 193 nm. J Photochem Photobiol Chem 135:53–64

Cottin H, Coll P, Coscia D et al (2008) Heterogeneous solid/gas chemistry of organic compounds related to comets, meteorites, Titan, and Mars: laboratory and in lower Earth orbit experiments. Adv Space Res 42:2019–2035

Cottin H, Kotler JM, Billi D et al (2017) Space as a tool for astrobiology: review and recommendations for experimentations in earth orbit and beyond. Space Sci Rev 209:83–181

Dass AV, Hickman-Lewis K, Brack A et al (2016) Stochastic prebiotic chemistry within realistic geological systems. Chemistry Select 1:4906–4926

Demets R, Schulte W, Baglioni P (2005) The past, present and future of biopan. Adv Space Res 36 (2):311–316

Ehrenfreund P, Ruiterkamp R, Peeters Z et al (2007) The ORGANICS experiment on BIOPAN V: UV and space exposure of aromatic compounds. Planet Space Sci 55:383–400

Ehrenfreund P, Ricco AJ, Squires D et al (2014) The O/OREOS mission—astrobiology in low Earth orbit. Acta Astronaut 93:501–508

Elsaesser A, Quinn RC, Ehrenfreund P et al (2014) Organics exposure in orbit (OREOcube): a next-generation space exposure platform. Langmuir 30:13217–13227

Es-sebbar E, Bénilan Y, Fray N et al (2015) Optimization of a solar simulator for planetary-photochemical studies. Astrophys J Suppl Ser 218:19
Foustoukos DI, Seyfried WE (2004) Hydrocarbons in hydrothermal vent fluids: the role of chromium-bearing catalysts. Science 304:1002–1005
Freissinet C, Glavin DP, Mahaffy PR et al (2015) Organic molecules in the Sheepbed Mudstone, Gale Crater, Mars. J Geophys Res Planets 120:495–514
Frey H (1980) Crustal evolution of the early earth: the role of major impacts. Precambrian Res 10:195–216
Hashimoto H, Ushio K, Kaneko T et al (2002) Formation of prebiotic organics in space: its simulation on ground and conceptual design of space experiment in earth orbit. Adv Space Res 30:1495–1500
Horita J, Berndt ME (1999) Abiogenic methane formation and isotopic fractionation under hydrothermal conditions. Science 285:1055–1057
Horneck G (ed) (2007) Complete course in astrobiology. Weinheim, Wiley-VCH
Horneck G, Klaus DM, Mancinelli RL (2010) Space microbiology. Microbiol Mol Biol Rev 74:121–156
Kasting JF, Whitmire DP, Reynolds RT (1993) Habitable zones around main sequence stars. Icarus 101:108–128
Kobayashi K, Kaneko T, Saito T et al (1998) Amino acid formation in gas mixtures by high energy particle irradiation. Orig Life Evol Biosph 28:155–165
Koeberl C (2006) The record of impact processes on the early Earth: a review of the first 2.5 billion years. In: Reimold WU, Gibson RL (eds) Processes on the Early Earth. GSA Special Papers 405, pp 1–22
Kuzicheva EA, Gontareva NB (2001) Study of the peptide prebiotic synthesis in context of exobiological investigations on earth orbit. Adv Space Res 28:713–718
Kuzicheva EA, Gontareva NB (2003) Exobiological investigations on Russian spacecrafts. Astrobiology 3:253–261
Martins Z, Modica P, Zanda B et al (2015) The amino acid and hydrocarbon contents of the Paris meteorite: insights into the most primitive CM chondrite. Meteorit Planet Sci 50:926–943
Martins Z, Cottin H, Kotler JM et al (2017) Earth as a tool for astrobiology—a European perspective. Space Sci Rev 209:43–81
Maurette M (2006) Cometary micrometeorites in planetology, exobiology, and early climatology. In: Thomas PJ, Hicks RD, Chyba CF et al (eds) Comets and the origin and evolution of life. Advances in astrobiology and biogeophysics. Springer, Berlin, pp 69–111
McCollom TM, Seewald JS (2001) A reassessment of the potential for reduction of dissolved CO_2 to hydrocarbons during serpentinization of olivine. Geochim Cosmochim Acta 65:3769–3778
McCollom TM, Seewald JS (2003) Experimental constraints on the hydrothermal reactivity of organic acids and acid anions: I. Formic acid and formate. Geochim Cosmochim Acta 67:3625–3644
McDermott JM, Seewald JS, German CR et al (2015) Pathways for abiotic organic synthesis at submarine hydrothermal fields. Proc Natl Acad Sci USA 112:7668–7672
Meinert C, Myrgorodska I, de Marcellus P et al (2016) Ribose and related sugars from ultraviolet irradiation of interstellar ice analogs. Science 352:208–212
Miller SL, Urey HC (1959) Organic compound synthesis on the primitive earth. Science 130:245–251
Nahon L, de Oliveira N, Garcia GA et al (2012) DESIRS: a state-of-the-art VUV beamline featuring high resolution and variable polarization for spectroscopy and dichroism at SOLEIL. J Synchrotron Radiat 19:508–520
Noblet A, Stalport F, Guan YY et al (2012) The PROCESS experiment: amino and carboxylic acids under Mars-like surface UV radiation conditions in low-earth orbit. Astrobiology 12:436–444
Peucker-Ehrenbrink B, Schmitz B (2012) Accretion of extraterrestrial matter throughout Earth's history. Springer Science & Business Media
Ryder G (2002) Mass flux in the ancient Earth-Moon system and benign implications for the origin of life on Earth. J Geophys Res Planets 107:6–1

Ryder G, Koeberl C, Mojzsis SJ (2000) Heavy bombardment of the Earth at ~3.85 Ga: The search for petrographic and geochemical evidence. Orig Earth Moon 475
Schwell M, Jochims H-W, Baumgärtel H et al (2006) VUV photochemistry of small biomolecules. Planet Space Sci 54:1073–1085
Schwell M, Jochims H-W, Baumgärtel H et al (2008) VUV photophysics and dissociative photoionization of pyrimidine, purine, imidazole and benzimidazole in the 7–18eV photon energy range. Chem Phys 353:145–162
Sephton MA (2002) Organic compounds in carbonaceous meteorites. Nat Prod Rep 19:292–311
Sephton MA (2005) Organic matter in carbonaceous meteorites: past, present and future research. Philos Trans R Soc Math Phys Eng Sci 363:2729–2742
Sherwood Lollar B, Frape SK, Weise SM et al (1993) Abiogenic methanogenesis in crystalline rocks. Geochim Cosmochim Acta 57:5087–5097
Shock EL, Schulte MD (1998) Organic synthesis during fluid mixing in hydrothermal systems. J Geophys Res Planets 103:28513–28527
Stalport F, Coll P, Szopa C et al (2009) Investigating the photostability of carboxylic acids exposed to Mars surface ultraviolet radiation conditions. Astrobiology 9:543–549
Stalport F, Guan YY, Coll P et al (2010) UVolution, a photochemistry experiment in low Earth orbit: investigation of the photostability of carboxylic acids exposed to Mars surface UV radiation conditions. Astrobiology 10:449–461
Swallow AJ (1960) Radiation chemistry of organic compounds, vol 2. Pergamon Press, London
Ten Kate IL, Garry JRC, Peeters Z et al (2005) Amino acid photostability on the Martian surface. Meteorit Planet Sci 40:1185–1193
Thuillier G, Floyd L, Woods TN et al (2004) Solar irradiance reference spectra. In: Pap JM, Fox P, Fröhlich C et al (eds) Solar variability and its effect on climate. AGU/Geophys Monogr 141:171–194
Tielens AGGM, Charnley SB (1997) Circumstellar and interstellar synthesis of organic molecules. Orig Life Evol Biosph 27:23–51
Wächtershäuser G (2000) Life as we don't know it. Science 289:1307–1308
Wächtershäuser G (2006) From volcanic origins of chemoautotrophic life to Bacteria, Archaea and Eukarya. Philos Trans R Soc Lond Ser B Biol Sci 361:1787–1808
Zellner NEB (2017) Cataclysm no more: new views on the timing and delivery of lunar impactors. Orig Life Evol Biosph 47:261–280

Chapter 11
Exoplanetary Biosignatures for Astrobiology

John Lee Grenfell

Abstract Since life evolved on our planet there have been subtle interplays between biology and Earth System Components (atmosphere-lithosphere-ocean-interior). Life, for example, can impact weathering rates which, in turn, influence climate stabilizing feedback cycles on Earth. Photosynthesis is ultimately responsible for our oxygen-rich atmosphere, which favours the formation of the protective ozone layer. The recent rise of exoplanetary science has led to a re-examination of such feedbacks and their main drivers under different planetary conditions. In this work we present a brief overview of potential biosignatures (indicators of life) and review knowledge of the main processes, which influence them in an exoplanetary context. Biosignature methods can be broadly split into two areas, namely "in-situ" and "remote". Criteria employed to detect biosignatures are diverse and include fossil morphology, isotope ratios, patterns in the chemical constituents of cells, degree of chirality, shifts from thermal or redox equilibrium, and changes in the abundance of atmospheric species. For the purposes of this review, our main focus lies upon gas-phase species present in Earth-like atmospheres, which could be detected remotely by spectroscopy. We summarize current knowledge based on the modern (and early) Earth and the Solar System then review atmospheric model studies for Earth-like planets, which predict climate, photochemistry and potential spectral signals of biosignature species.

11.1 Introduction

The search for life beyond the Earth has been a fascination, which can be traced back to the dawn of human reasoning. The atomist school of Democritus in ancient Greece, for example, postulated the existence of many worlds—both like and unlike the Earth (see discussion in Dick 1984). In modern times, the dawning of the

J. L. Grenfell (✉)
Department of Extrasolar Planets and Atmospheres (EPA), Institute for Planetary Research (PF), German Aerospace Centre (DLR), Berlin, Germany
e-mail: lee.grenfell@dlr.de

B. Cavalazzi, F. Westall (eds.), *Biosignatures for Astrobiology*, Advances in Astrobiology and Biogeophysics, https://doi.org/10.1007/978-3-319-96175-0_11

exploration of the Solar System and, more recently, the onset of the burgeoning exoplanet era have marked exciting times in which the age-old question "Are we alone?" is now moving into the realm of modern science.

This chapter reviews principles and techniques discussed in the exoplanetary literature for the remote detection of potential atmospheric biosignatures. We do not focus here on in-situ biosignatures involving microbial morphology, molecular biology etc., although these are fascinating and expanding fields (see Horneck et al. 2016 and references therein). Although exoplanetary science has made great headway since its inception over 20 years ago, relatively little is known about the atmospheric properties (such as mass and composition) of potentially habitable, rocky exoplanets. Therefore, a common approach in the study of potential exoplanetary biosignatures is to apply numerical atmospheric models developed from Earth's present and past and then to vary boundary conditions, such as the spectral class of the central star, the planet-star distance etc., in order to investigate the response of planetary climate and potential biosignatures. These models are mostly constrained by knowledge gained from the modern (and Early) Earth and the Solar System. Therefore, we also briefly review knowledge of Solar System atmospheres and how this expertise can be applied in an exoplanet context. In this chapter we do not focus on the Search for Extraterrestrial Intelligence (SETI) Program nor discuss in detail the concept of "technosignatures" (signs of advanced life) but instead refer the reader to some of the main studies, such as Lin et al. (2014); Stevens et al. (2016); Griffith et al. (2015) and Korpela et al. (2015).

A comprehensive introduction to the literature is provided by five detailed review papers in remote biosignature science which recently appeared as part of the NASA Nexus for Exoplanet System Science (NExSS) and Astrobiology Program. First, Schwieterman et al. (2018) review remotely detectable potential signs of life. Second, Meadows et al. (2018) discuss atmospheric molecular oxygen in the context of its environment. Third, Catling et al. (2018) propose a framework for assessing potential biosignature signals based upon Bayesian statistics. Fourth, Walker et al. (2018) discuss a flexible framework to guide future biosignature search programs. Finally, Fujii et al. (2018) summarize future observational prospects. The present work compliments these reviews by focusing mainly on the photochemical and climate responses of atmospheric biosignatures. Note that Grenfell (2017) also summarizes the topic of exoplanetary biosignatures but with less focus on biological responses and more emphasis on physical aspects.

This chapter is organized as follows. Section 11.2 discusses briefly the challenge of defining life and its needs. Section 11.3 describes the process of preparing suitable planetary targets for biosignature assessment and the importance of understanding habitability. Section 11.4 describes currently proposed exoplanetary biosignature techniques. Section 11.5 discusses spectroscopic exoplanetary biosignatures, including climate, photochemical and spectral responses of gas-phase biosignature species in the context of modern and Early Earth, the Solar System and Earth-like exoplanets. Section 11.6 describes the "dead Earth" as a benchmark to compare against when investigating biosignature candidates. Section 11.7 presents a brief discussion and conclusions.

11.2 Challenges of Defining Life and Its Needs

Commonly quoted is the NASA-based definition (Joyce et al. 1994), which proposes that life is a "self-sustaining chemical system capable of Darwinian evolution." Benner (2010) discuss the challenges of defining a system undergoing Darwinian evolution and its general applicability beyond the Earth. Smith (2016) provides a recent summary of the philosophical challenges of defining life. A common issue is to construct a universal definition—for example the animal kingdom features hybrid crosses which are clearly alive but which cannot reproduce. Walker et al. (2018) summarize theories for defining life including the idea of biosignatures based on network theory of chemical reactions in planetary atmospheres. Life-as-we-know it has basic requirements in order to establish itself and thrive. These are, firstly, an energy source; secondly, a liquid water solvent and thirdly, a suitable supply of the chemical elements CHNOPS, as reviewed in Cockell (2016).

11.3 Preparing a Planetary Target List for Biosignature Assessment

11.3.1 Steps to Forming an Earth-Like Planet

We define two basic steps, which are necessary to form an Earth-like planet, which can host life. The first step is for a rocky planet to form with suitable water, organic and nutrient inventory. Earlier numerical simulations of volatile delivery during planetary formation (see for example Raymond et al. 2007) suggest that Earth-like planets with initially large water inventories could be rather common in nature, although this issue is still debated. The second step is for the planet to possess an atmosphere with surface climate conditions suitable to maintain liquid water over a sufficiently long period such that life can develop and thrive. The extent to which Earth-like planets can maintain mantle convection and outgas significant atmospheres (see for example Stamenkovic et al. 2012; Tackley et al. 2013; Tosi et al. 2017; Noack et al. 2017) is currently debated. Also, whether such atmospheres can be maintained against loss processes depends upon the incoming stellar wind and the planetary mass (see for example Kislyakova et al. 2014). The "eta-Earth" (ζ_{earth}) parameter (see for example Gaidos 2013) designates the fraction of stars, which possess an Earth-like planet lying in the Habitable Zone (see Sect. 11.3.2). Haghighipor (2015) define eta-Earth as an occurrence rate i.e. the mean number of rocky planets with 1–2 Earth radii per star. Better constraining eta-Earth is one of the major challenges in exoplanetary science. Traub (2015) discuss steps towards estimating ζ_{earth} using Kepler data. Estimates of ζ_{earth} have been performed for Earth-size planets orbiting cooler (M-dwarf) stars using Kepler data. Results are in the range (0.2–0.8) (see Morton and Swift 2014; Dressing and Charbonneau 2015) depending on the extent of the HZ assumed. Kane et al. (2016) discuss the challenges involved.

11.3.2 Habitability and the Habitable Zone (HZ)

Habitability refers to the potential of an environment to sustain life (see review by Cockell 2016). Note that it does not refer to the potential for life to emerge or to whether life is actually present.

11.3.2.1 Classical HZ

The classical HZ (Huang 1959; Kasting et al. 1993) is defined to be the annulus region around a star where liquid water can exist on a rocky planet's surface. Numerous works have estimated the HZ boundaries by applying radiative-convective atmospheric models (Kaltenegger and Sasselov 2011; Yang et al. 2013; Kopparapu et al. 2014; Godolt et al. 2016; Kitzmann 2016) as well as the climate and photochemical responses of planetary atmospheres across the HZ (Segura et al. 2003; Grenfell et al. 2007; Kaltenegger and Sasselov 2011). The inner and outer boundaries depend on factors, such as the stellar type, the planetary atmosphere, and on climate feedbacks involving, for example, cloud formation. Recent estimations (see e.g. Kane et al. 2016, their Fig. 1) suggest that, for a Sun-like star, the HZ boundaries range from about $1.8F_{Earth}$ (inner HZ) out to about $0.3F_{Earth}$ (outer HZ) (where "F_{Earth}" denotes the mean solar flux received by the modern Earth) and possibly even further out (Pierrehumbert and Gaidos 2011).

11.3.2.2 Non-Classical HZ

Considering solvents other than water (e.g. water-ammonia, water-sulphuric acid mixtures) in the HZ definition can clearly extend the HZ boundaries (see for example Ludwig et al. 2016). Also, considering possible sub-surface oceans on moons, which lie beyond the snow-line (Vogel 1999) can extend the classical HZ outwards (although the chances of detecting potential sub-surface life remotely on such worlds seem rather remote). Such oceans have been suggested to be associated with heat sources arising either from long-term radioactive elements or/and from tidal forces (see for example Tyler 2008) and maintained by thermal blanketing provided by water-ice sheets. These issues are of particular interest in the context of the forth-coming JUICE (JUpiter ICy moons Explorer) mission (see for example Grasset et al. 2012). Heller and Barnes (2013) suggested potential habitability due to tidal heating of hypothetical moons orbiting known giant exoplanets at ~10 planetary radii. Lammer et al. (2009) proposed four classes of planetary habitability—ranging from Earth-like planets lying in the classical HZ out to icy moons possessing sub-surface oceans. The four classes are based on stellar, orbital and geophysical criteria. Also, atmospheres rich in molecular hydrogen can lead to additional green-house warming hence to an extension of the HZ outwards (see for example Pierrehumbert and Gaidos 2011; Ramirez et al. 2014).

11.3.3 HZ Around M-Dwarf Stars

M-dwarf stars are a central focus in exoplanetary science, firstly because they are numerous in the Solar Neighborhood, secondly because their relatively weak luminosity leads to favourable (planet/star) contrast ratios and, thirdly because the close-in HZ means a more favourable geometrical transit probability and suggests a fast planetary orbital period hence faster data collection. The close proximity of the HZ to the star however, suggests that planets lying in this region could be (a) tidally-locked i.e. having a constant day and night-side and (b) strongly bombarded by cosmic rays. The habitability of planets in the HZ of M-dwarf stars was reviewed by Scalo et al. (2007) and more recently by Shields et al. (2016). The effect of cosmic rays upon biosignatures was investigated by Grenfell et al. (2012) and Tabataba-Vakili et al. (2016).

When preparing a target list for biosignature assessment an important criterion is that the planet should lie in the HZ. The value of the $\zeta\eta_{earth}$ parameter will therefore depend on our knowledge of the extent of the HZ.

11.3.4 Exoplanet Missions Providing Planetary Targets

Forthcoming missions such as the Transiting Exoplanet Space Survey (TESS) (Ricker et al. 2014) and the CHaracterising ExOPlanets Satellite (CHEOPS) (Fortier et al. 2014) will expand our knowledge of hot Jupiters, mini gas planets and hot super-Earths. The PLATO 2.0 (Rauer et al. 2014) mission will provide accurate age and mean planetary density for a considerable sample of terrestrial planets lying in the HZ. The James Webb Space Telescope (JWST) (Lightsey et al. 2012) will carry out atmospheric spectroscopy on mini gas planets and super-Earths at a hitherto unprecedented level of resolution and accuracy—although finding atmospheric biosignatures on Earth-like planets will present a major challenge even for this mission. Planned ground-based telescopes, such as the European Extremely Large Telescope (see for example Snellen 2014) will perform high resolution spectroscopy of Earth-like exoplanets, although detecting atmospheric biosignatures is anticipated to be challenging also for this mission. Further afield, the Large Aperture UV-Optical-Infrared (LUVOIR) NASA spaced-based telescope (Bolcar et al. 2016) plans to survey atmospheric biosignatures on Earth-like planets. The Habitable Planet Imaging (HabEx) mission (Mennesson et al. 2016) is currently being considered and plans to hunt spectroscopically for signs of life on nearby cool rocky worlds and has a suggested launch date in the 2030s. The mission will operate with a >3.5 m optical mirror for direct imaging and spectroscopy in order to search for exoplanetary atmospheric biosignatures. In general, knowledge of the planetary environment and its evolution, for example by measuring the stellar spectrum etc., is desirable in order to better assess candidate biosignature detections. It has been proposed that the detection of ocean "glint" could also indicate whether such targeted worlds feature liquid oceans (see for example Williams and Gaidos 2008) although achieving such detection is very challenging (Cowan et al. 2012).

In summary, the pathway to compiling a suitable target list of planets suitable for biosignature assessment involves the detection of rocky Earth-like planets lying in the HZ of their central star. Horneck et al. (2016) discuss an astrobiological European roadmap in which a key goal is to define the pathway towards remote detection of life outside the Solar System.

11.4 Remote Biosignature Techniques

Transit spectrophotometry was discussed in an Earth-like exoplanetary context by Schneider (1994) and by numerous theoretical works since (see for example Rauer et al. 2011). Spectropolarimetry (see for example Stam 2008; Sterzik et al. 2012) is also commonly discussed in the exoplanetary literature as a possible technique to search for remote exoplanetary biosignatures. Retrieving planetary conditions, such as atmospheric composition and climate, from the spectroscopic data can be associated with degeneracies (see for example Benneke and Seager 2012) and other processes (e.g. band overlap, false positives etc. as we will discuss), which have to be considered and discounted. Spectroscopic-based topics in the context of remote biosignatures include the 'vegetation red edge', 'isotopic signals' and 'atmospheric gas-phase species' which we will now discuss.

11.5 Spectroscopic Exoplanetary Biosignatures

11.5.1 Reflectivity Increase with Wavelength

The "vegetation red edge" refers to the abrupt increase in reflectance of electromagnetic radiation in the wavelength range ~700–750 nm arising due to vegetation. This phenomenon is related to the marked difference between, on the one hand, strong absorption of chlorophyll in the red wavelength region and, on the other hand, high internal scattering which leads to strong reflectivity (R) by leaves in the near infrared (see for example Horler et al. 1983; Seager et al. 2005). The Horler et al. (1983) study discussed the dependence of the spectral form of the red edge, as determined by remote-sensing, on the modern Earth upon, for example species-type, developmental stage, chlorophyll content and vertical leaf-stacking. They noted in particular the dependence of λ_{max} (i.e. the wavelength where $dR/d\lambda$ attains its maximum). They also discussed some of the main challenges of performing and interpreting Earth-based remote sensing of the red-edge, such as the relatively high spectral resolution required and possible pitfalls when interpreting the spectra due to, for example, different forms of chlorophyll.

For exoplanets, Seager et al. (2005) summarized current theories regarding the potential for Earth-like planets to develop O_2-producing photosynthesis. Several studies, such as Seager et al. (2005), Pallè et al. (2009), Kaltenegger et al. (2012), have discussed the vegetation red edge in an exoplanet context, including the challenge of detectability and potential false positives due to phenomena such as surface mineral reflectance. Numerous works (such as Woolf et al. 2002; Arnold et al. 2002; Sterzik et al. 2012) discussed the red edge spectral feature in modern Earthshine spectra and the challenges of applying this technique in an exoplanetary framework.

Analogous to the 'red edge' but shifted to weaker energies is the spectral feature related to purple bacteria (Sanroma et al. 2014). These could have been present during the Archaean Eon on the Early Earth. They are non-oxygenic phototrophs that gain energy by cleaving H_2 or H_2S (instead of H_2O for the oxygenic case), hence their photosynthetically-active wavelength region extends out into the infra-red range as far as ~1025 nm. Applying a similar principle as the red edge, Schwieterman et al. (2015) considered non-photosynthetic pigments in general as remote biosignatures.

11.5.2 Isotopic Ratios

Biochemical processes can be highly specific and tend to favor the lighter isotope (cf. Chap. 3). Biological isotopic fractionation takes place for chemical elements such as C, N, S, Fe, Cu and Zn. It is hypothesized that passive diffusion across the pores of semi-permeable membranes during photosynthesis occurs more quickly for the lighter isotope. Biofractionation (which is relevant for all metabolisms, not just photosynthesis) in geochemical and biological environments has been studied in detail in works by Schidlowski (1988) (who discusses the isotopic carbon record in sedimentary rock); Boschker and Middleburg (2002) (who study stable isotope biomarker signals in the context of microbial ecology) and Fujii et al. (2014) (who investigate biomarkers related to transition metal isotope signals).

In an exoplanet context, however we will focus here only on the lighter elements from the above list because these could, in theory, accumulate in planetary atmospheres and hence be detected remotely via spectroscopy. Oxygen, whose atmospheric isotopic ratios are influenced by a rather wide range of both abiotic and biotic processes, is nevertheless a particular focus in an exoplanetary context since its isotopic distribution in Earth's (oxygen-rich) atmosphere has been measured remotely by high resolution spectroscopy of Earthshine (Yan et al. 2015). There is no clear consensus about the extent to which life influences the atmospheric oxygen isotope ratio. Addressing this question requires consistent Earth system models featuring abiotic processes (such as ocean-atmosphere exchange, temperature, photochemistry etc.) and biotic processes (e.g. oxygen exchange within chloroplasts) in the carbon-oxygen-nitrogen cycles. Farquhar and Lloyd (1993), for example, discuss the influence of vegetation effects upon the isotopic composition of oxygen in atmospheric CO_2. Abiotic effects have been established for some time, for example Kroopnick and Craig (1972) discuss the temperature-dependence of oxygen isotopic fractionation in seawater.

Hedelt et al. (2011) showed in a proof-of-concept that isotopes in Venus' CO_2 atmosphere could be distinguished by IR spectroscopy performed from the ground on Earth. The challenges inherent in detecting gas-phase species and their isotopes via infrared spectroscopy in exoplanetary atmospheres (mostly hot Jupiters) are discussed in works, such as Encrenaz (2014). In an exoplanet context however, it is very challenging to detect isotope fractionations which arise due to potential biotic activity (Snellen 2014; Holmen 1992) using spectrophotometry. A discussion of possible atmospheric isotopic ratios for a "dead Earth" in an exoplanet context is provided in Sect. 11.6.

11.5.3 Gas-Phase Species as Biosignatures

Here we present the main focus of this chapter—a review of photochemical and climate processes affecting key atmospheric species, such as O_2, O_3, and N_2O, in an exoplanetary context. These molecules can be interpreted to be remote biosignatures if they occur in an exoplanetary atmosphere at an abundance (or feature a time-dependence), which is inexplicable in the context of the exoplanetary environment and its evolution without invoking biology. There are two principal of criteria that a gas-phase species must fulfil to qualify as an atmospheric biosignature: it should be uniquely attributable to life (i.e. lacking abiotic sources) and it should feature a strong and unique (i.e. without band overlap) spectral signal which is straightforward to retrieve (see e.g. Meadows et al. 2018).

Examples of potential biosignatures include, for example, the Earth's rich-oxygen-atmosphere, or its strong ozone shield. In each case, potential abiotic sources (so-called false positives) should be discounted taking into account the planet's environmental evolution, as we will discuss. A review of atmospheric biosignatures was provided by Seager et al. (2013, 2016). Those works grouped atmospheric biosignatures into those produced by metabolism and those produced by stress or signalling. A challenge of the approach (noted in those works) is that biomass on our planet emits a wide range of relatively simple molecules which have both biotic and non-biotic sources.

Due to the lack of knowledge (in for example mass, composition) of potentially habitable Earth-like atmospheres in exoplanetary science, a common approach is therefore to investigate responses of proposed biosignatures based on modern Earth, Early Earth, Solar System and then to extrapolate the knowledge gained (using, for example, atmospheric model studies) to Earth-like exoplanets. The following sections therefore discuss potential gas-phase biosignatures species on modern Earth, on Early Earth, in the Solar System and on Earth-like exoplanets.

11.5.3.1 Modern Earth

Earth's atmosphere influences the biosphere by modifying climate, protecting the surface from harmful radiation and cosmic rays and enables the existence of surface liquid water. Conversely, life has modified Earth's atmospheric composition over time, for example via photosynthetic and biomass emissions. Over geological time, life could have significantly influenced weathering rates and hence mantle hydration and plate tectonics (Höning et al. 2014). Life may have strongly influenced the Earth's nitrogen cycle (e.g. Krissansen-Totton et al. 2016).

Molecular Oxygen (O_2) Earth's rich mantle of atmospheric O_2(g) (volume mixing ratio (vmr), $\xi_{o2} = 0.21$; total atmospheric mass ~1.2×10^{21} g) is mainly of biotic origin. Atmospheric abundances are mostly constant in latitude and longitude and in altitude up to ~80 km above which molecular oxygen starts to photolyse into its constituent atoms. Regeneration of the resulting O_2(g) is associated with the oxygen airglow (a phenomenon which is also seen in the atmospheres of Venus and Mars) (Slanger and Copeland 2003).

The main source of molecular oxygen is via photosynthesis regulated by geological burial and amounts to ~320 Tg/year (see e.g. Lasaga and Ohmoto 2002) with a smaller source which arises due to photolysis of water vapor followed by atmospheric escape of the resulting hydrogen atoms—this process constitutes therefore a net source for atmospheric oxygen. Non-biological sources of O_2(g) in modern Earth's atmosphere are, by comparison, negligible. The main sinks are reaction with reducing gases (such as CH_4 and H_2S) in the atmosphere (Catling and Claire 2005) and surface weathering (Holland 2002).

Ozone (O_3) Earth's ozone layer forms a protective shield against harmful incoming solar radiation and high-energy particles. Ozone is formed in Earth's atmosphere starting with photolysis of molecular oxygen (see below), therefore it can be interpreted as a potential biosignature for Earth-like planets, if abiotic sources on the planet are demonstrated to be weak.

Stratosphere On Earth the O_3(g) layer peaks in the mid-stratosphere reaching ~10 parts per million (ppm) at ~30 km and with a total atmospheric mass ~3×10^{15} g O_3(g). Ozone is formed by the so-called Chapman mechanism (Chapman 1930):

$$\begin{aligned}
&O_2 + h\nu_{(\lambda<242.4\,\text{nm})} \rightarrow O + O \\
&O + O_2 + M \rightarrow O_3 + M \\
&O_3 + O \rightarrow 2O_2 \\
&O_3 + h\nu_{(\lambda<\sim320\,\text{nm})} \rightarrow O_2 + O
\end{aligned}$$

The main sinks for O_3(g) are via gas-phase catalytic cycles split according to so-called chemical "families", for example via hydrogen-oxides (HOx) (in the mid to upper atmosphere, Bates and Nicolet 1950), ClOx (mostly occurring in the upper stratosphere; Stolarski and Cicerone 1974), and NOx (mostly in the lower stratosphere; Crutzen 1970). An overview of these cycles and how they impact ozone on the modern Earth is provided by Wayne (1993).

Ozone is affected mainly by transport in the lower stratosphere and by photochemistry in the upper stratosphere where its lifetime is shorter (World Meteorological Organization (WMO) Report 1995). Its main formation region is in the Earth's tropics from where it is transported polewards via the Brewer-Dobson circulation in the middle atmosphere. The weakened photolytic sink of ozone in winter leads to a characteristic annual variation with an amplitude in the ozone column of (15–20%). In addition to protecting the surface from UV, ozone heating leads to a temperature inversion from ~20 to 30 km, which is the origin of the stability of air parcels against mixing processes in the stratosphere. A wide range of studies exists which have investigated the interactions of ozone with climate change (see e.g. the Intergovernmental Panel on Climate Change (IPCC) Fourth Assessment Report 2007).

Troposphere In this region (from the surface up to the region between 10 and 20 km) ozone is formed mainly via the so-called "smog-mechanism" (Haagen-Smit 1952), which proceeds via the NOx catalysed oxidation of volatile organic compounds in the presence of UV. Production via this mechanism accounts for ~10% of the ozone on the modern Earth.

Mesosphere In this region (from ~50 to 80 km) ozone is removed in the atmosphere mainly via photolysis and by destructive catalytic cycles involving principally HOx, whose rates are influenced by the abundance of water vapour. Hence, there is a link between ozone, water and dynamical processes. Ozone features a secondary maximum in the upper mesosphere (see for example Smith and Marsh 2005), possibly associated with low temperatures there which slow its photochemical sink(s).

Nitrous Oxide (N_2O) N_2O(g) ("laughing gas") is produced mainly by (de)nitrifying bacteria as part of the nitrogen cycle. It is an important greenhouse gas (in Earth's modern atmosphere it has an abundance, $\xi_{N2O} = 3.3 \times 10^{-7}$ vmr (IPCC Fourth Assessment Report 2007) with a total mass in Earth's atmosphere of ~2.6×10^{15} g) and is a dynamical tracer of air motions in the lower atmosphere due to its long lifetime of several hundred years against chemical destruction. It is removed in-situ in Earth's atmosphere mainly by the following gas-phase reactions (see for example McElroy and McConnell 1971):

$$\begin{aligned} &N_2O + h\nu \rightarrow N_2 + O^* \quad (\sim 90\%) \\ &N_2O + O^* \rightarrow 2NO \quad (\sim 6\%) \\ &N_2O + O^* \rightarrow N_2 + O_2 \quad (\sim 4\%) \end{aligned}$$

where O^* denotes electronically-excited oxygen atoms. Global production and loss are reviewed in Syakila and Kroeze (2011). Non-biological sources are essentially negligible on the modern Earth. Suggested abiotic sources include production by lightning (Levine et al. 1979; Levine and Shaw 1983) and reaction of dissolved nitrite on iron-containing minerals (Samarkin et al. 2010).

Methane (CH_4) About 90% of modern Earth's CH_4(g) (ξ_{CH4} ~1.8×10^{-6} vmr with a total mass of ~5.1×10^{15} g) source (amounting to about 500 Tg/year) is produced biologically via methanogenic bacteria. The remaining, abiotic contributions (Schoell 1988 provides an overview) arise from volcanic emissions or other geological processes. Atmospheric methane is destroyed mainly in the troposphere by gas-phase reaction with the hydroxyl (OH) radical:

$$CH_4 + OH \rightarrow CH_3 + H_2O$$

This is the first step in a rather complex sequence of reactions which ultimately lead to methane being oxidized into water and carbon dioxide. Atmospheric OH is generated mainly via the reaction:

$$H_2O + O^* \rightarrow 2OH$$

where O* mainly arises via photolysis of ozone. As discussed for N_2O(g) above, CH_4(g) is also a tracer of dynamical motions and has a lifetime against chemical loss of ~8 years (IPCC Fourth Assessment Report 2007).

11.5.3.2 Early Earth

During its 4.56 billion years (Ga) history, Earth's atmosphere has undergone considerable changes—evolving from the hot, neutral to slightly reducing conditions of the Hadean up to the present-day oxidizing atmosphere. The evolution of Earth's atmosphere over geological periods is commonly applied in exoplanetary science as a proxy for investigating the development of Earth-like atmospheres over time.

Oxygen A wide range of geological and atmospheric proxy data (Holland 2006) suggest that Earth's atmosphere was marked by two oxidation events in its history, the first ("Great Oxidation Event", GOE, Kopp et al. 2005; Catling and Claire 2005) occurred at ~2.5 Ga ago and atmospheric oxygen levels rose to at least ~1% of the modern day abundance. During the "Second Oxidation Event" (SOE, see for example Campbell and Squire 2010) at ~0.6 Ga ago oxygen levels rose to ~10–100% of the modern day abundance. Kump et al. (2011) suggested the GOE may have been related to an increase in the burial rate of organic carbon at the ocean floor possibly associated with changes in continental distribution. Gaillard et al. (2011) suggested it could have been related to a change in the oxidation state of volcanically-emitted sulphur species. Sverjensky and Lee (2010) proposed that oxidation of Earth's atmosphere could have led to a rapid increase in mineral diversification on the surface. The study by Anbar et al. (2007) suggested small, possibly localized excursions in atmospheric oxygen could have occurred in some regions on the Early Earth even before the GOE. Changing oxygen in the atmosphere can affect the abundance of greenhouse gases, such as methane, and, hence, impact climate. The 'Faint Young Sun" (FYS) problem (or paradox) refers to the inconsistency between proxy (paleosol) data (which suggests habitable conditions on the Early Earth surface) and model simulations (which suggest a globally frozen surface or "Snowball Earth"). Feulner (2012) provide a review of the FYS. The study by Gebauer et al. (2017) applied a diagnostic package ("pathway analysis program") to the output of a coupled climate-biogeochemical column model of the Early Earth atmosphere, whereas Gebauer et al (2018) performed a similar study but for the atmospheric evolution of an Earth-like planet orbiting in the HZ of an M-dwarf star. Those works identified and quantified the main chemical pathways influencing O_2(g). Results suggest that oxygen is mainly destroyed by rather complex oxidation pathways in the lower atmosphere, whereas in the mid to upper stratosphere, O_2(g) can be produced via photolysis of CO_2.

Ozone Model estimates suggest that the protective shield of ozone was mostly in place in its present form after the Great Oxidation Event (see for example Gebauer et al. 2017). Ozone abundances are relatively stable against changes in atmospheric oxygen increases. Why is this? An initial increase in oxygen would first favour an increase in ozone via the Chapman mechanism as shown above. The increased ozone however blocks UV at lower atmospheric levels where less oxygen is photolysed and so less ozone is formed. As this opposes the original change in ozone, a negative feedback is said to be operating.

Nitrous Oxide Although a relatively minor greenhouse in modern Earth's atmosphere, on the Early Earth nitrous oxide could have played a major role in affecting climate as suggested by Buick (2007) (see also the atmospheric model studies of Grenfell et al. 2011; Roberson et al. 2011). Recent work (Airapetian et al. 2016) has suggested that strong stellar activity could lead to non-biological formation of nitrous oxide in Earth-like atmospheres.

Methane Methane has been proposed as a greenhouse gas resulting in a relatively warm the Early Earth (e.g. Pavlov et al. 2000) with methane abundance up to several orders of magnitude more than the present day atmosphere. High methane abundances were likely favoured due to neutral to slightly reducing conditions on Early Earth where methanogenic bacteria could thrive. Also, volcanic emissions of methane were likely enhanced on the Early Earth (Kasting and Catling 2003). Haqq-Misra et al. (2009) noted, however, that methane's warming potential could be limited by haze formation, which would lead to an opposing, cooling effect.

11.5.3.3 Solar System

Oxygen CO_2-dominated atmospheres, such as those found on Mars and Venus, lead to the formation of low abundances of O_2(g) ($\xi_{o2} = 1.4 \times 10^{-3}$ vmr on Venus and $\xi_{o2} = 3 \times 10^{-7}$ vmr on Mars (see Yung and DeMore 1999). O_2(g) in these environments is formed by photolysis of CO_2(g) into CO+O followed by the combination of two O atoms (a so-called "third body" is also required to be present e.g. N_2 or O_2 in order to remove excess vibrational energy of the reactants). The amount of abiotic O_2 formed depends, therefore, on the rate of CO_2(g) photolysis and on the fate of the resulting O-atoms. Potentially complex, catalytic cycles involving HOx (formed e.g. from water vapor) and NOx (formed e.g. from lightning and cosmic rays in the presence of N_2(g)-O_2(g)) control how efficiently CO and O are recycled back into CO_2(g). If such cycles operate quickly then the formation of abiotic O_2(g) is slowed. Although abiotic formation of O_2(g) on Mars and Venus is weak compared with the biological sources on Earth, it has been suggested that this process can proceed much more quickly in exoplanet conditions if the UV-environment is favourable (see below). Small amounts of O_2(g) can also be formed abiotically by the action of high energy particles on rocky or icy surfaces. This mechanism has been suggested to operate on Mercury (Hunten 1988) and the moons Europa (Hall et al. 1995), Ganymede (Hall et al. 1998) and Rhea (Teolis et al. 2010).

Ozone As discussed above, the atmospheres of Mars and Venus feature small amounts of O_2(g) and O(g). These species can react together in a three-body reaction to form O_3(g), albeit in low amounts. On Venus the ozone column (Montmessin et al. 2011) is estimated to be ~150 times weaker than on Earth, whereas on Mars it is (100–1000) times weaker (Perrier et al. 2006). Weak ozone signals have also been detected in the outer solar system on some of the Galilean and Saturnian moons (see for example Noll et al. 1997). Its formation is proposed to occur via a mechanism involving high energy particles impinging upon icy surfaces.

Nitrous Oxide Abiotic sources of N_2O(g) could have operated on the Early Earth and early Venus and Mars. The mechanism involves high energy particles breaking up N_2(g) to form reactive nitrogen radicals in the presence of oxygen-containing species (see papers such as Naa Mvondo et al. 2001).

Methane Measurements by Formisano et al. (2004) triggered a debate as to the existence of methane in the Martian atmosphere (see e.g. overview presented in Zahnle et al. 2011). Recent studies which suggested up to a couple of ppbv CH_4(g) as detected by the Rover Curiosity (Webster et al. 2015), have added confidence to the argument that CH_4(g) is indeed present in the Martian atmosphere. The origin (Atreya et al. 2007) and the lifetime against chemical loss (Lefèvre and Forget 2009) of Martian methane, however, are not well understood (see Chap. 12).

11.5.3.4 Earth-Like Exoplanets

To estimate potential gas-phase biosignatures (and their false positives) on Earth-like planets, a common approach (given the paucity of data) is to extrapolate knowledge gained from the (Early) Earth and Solar System to an Earth-like exoplanet context, by performing studies with coupled convection-climate-photochemistry models. Assuming the Earth's development and biomass, one typically varies key parameters, such as the stellar spectrum, the orbital parameters etc., and calculates the resulting atmospheric and climate responses, hence the potential effect upon atmospheric spectral bands (see for example Grenfell et al. 2007; Segura et al. 2003; Rauer et al. 2011). In an alternative approach, which is somewhat less Earth-based, models simulating planetary formation and water delivery are applied (as studied in

Raymond et al. 2007) in order to learn about the occurrence and nature of rocky planets in the Habitable Zone. Other potentially important factors affecting habitability include whether the planet has a large moon (Laskar et al. 1993) or the presence of gas giants (Horner and Jones 2008). We now discuss some of the main results for proposed atmospheric biosignatures in an Earth-like exoplanetary context.

Oxygen The case of O_2(g) has illustrated the importance of elucidating the planetary environment (for example, stellar insolation etc.), primarily in order to disregard false positives when assessing potential biosignatures. A potentially important false positive for O_2(g) involves H_2O(g) photolysis followed by escape of atomic hydrogen. This process operates efficiently for low mass exoplanets subject to strong EUV input from their central star (which can be especially active in the pre-main sequence phase), as discussed in e.g. Luger and Barnes (2015). Wordsworth and Pierrehumbert (2014) noted that the rate of O_2(g) abiotic formation for rocky planets in the HZ depends on the ability of the atmosphere's "cold trap" (the temperature minimum at the tropopause) to freeze-out and retain water in the troposphere and, hence, prevent it photolysing. This ability is, in turn, related to the atmosphere's inventory of non-condensable gases (such as N_2(g)). Another potentially significant false positive for O_2(g) involves CO_2(g) photolysis followed by self-reaction of the resulting oxygen atoms, as investigated by Tian et al. (2014) and Domagal-Goldman et al. (2014). The resulting abiotic O_2(g) source on Earth-like planets is sensitive to the incoming NUV from the star and could, under certain conditions, even exceed the abundance of O_2(g) on the Earth. Since CO_2(g) photolysis leads to CO (g) formation, the presence of this species has been suggested to be an "anti-biosignature", i.e. an indication of abiotic oxygen formation (Wang et al. 2016; see also Schwieterman et al. 2016). The propensity to form abiotic O_2(g) could be related to the elemental composition of the planet, which is mainly determined during formation (Hu and Seager 2014).

Regarding potential biological sources of O_2(g) on Earth-like planets, possible photosynthesis analogues for rocky worlds orbiting in the HZ of M-dwarf stars were discussed in Kiang et al. (2007). Their work implied up to half of Earth's photosynthetic activity occurring in visible wavelengths but possibly stronger than Earth's activity in the IR range. Note that the release of biological O_2 into Earth's atmosphere depends upon geological processes, such as burial of organic material, which is efficient near continental shelves. Therefore, in an exoplanet context, the abundance of atmospheric O_2(g) could be influenced by continental distribution and subduction—parameters which are poorly constrained for Earth-like planets. Regarding sinks of O_2(g) on Earth-like planets, reaction with reducing gases in the planet's atmosphere are estimated to be potentially important, for example some studies (Segura et al. 2005; Grenfell et al. 2012) have suggested elevated abundances of atmospheric methane (up to 1000 times those on Earth) for rocky planets orbiting in the HZ of M-dwarf stars. This is due to weaker UV output of the star, which leads to reduced production of OH, an important sink for methane. With respect to changes in the O_2(g) biosignature over geological lifetimes, studying the Early Earth suggests that Earth-like atmospheres could feature strong variations in

O_2(g) as the planetary atmosphere evolves. Feedbacks between atmospheric evolution, climate and the development of life were discussed in Grenfell et al. (2010). Theoretical spectra for Earth-like planets calculated at different geological eons were presented and discussed in Schindler and Kasting (2000) and Kaltenegger et al. (2007).

O_2 Atmospheric Spectra Often discussed in an exoplanetary context is the rather thin O_2(g) absorption band which occurs at 0.76 microns in the visible spectrum. Detecting such spectral features in the atmospheres of Earth-like planets is expected to be extremely challenging—even for bright targets in the Solar neighborhood using next generation instruments (see e.g. Snellen 2014). Des Marais et al. (2002) presented a seminal study, which investigated theoretical spectral features of Earth-like atmospheres. Rodler and López-Morales (2014) suggested that detecting the 0.76 micron band would require several tens of transits assuming an Earth-twin orbiting a bright M-star with medium (M3–M6) spectral class. The study by Kawahara et al. (2012) however suggested instead to focus on the 1.27 microns band due to more straightforward adaptive optics from the ground at these wavelengths. Misra et al. (2014) suggested to look for O_2-O_2 dimer band features in the IR as a proxy for detecting O_2(g).

Ozone On Earth-like planets the ozone abundance will be determined firstly by the incoming UV from the central star, which regulates the Chapman cycle forming ozone and, secondly, by trace species such as hydrogen oxides (formed for example from photolysis of water vapor) and nitrogen oxides (formed from processes such as lightning and cosmic rays) which regulate the catalytic ozone loss cycles (see above). Since ozone strongly absorbs UV it also indirectly controls the abundance of other atmospheric biosignatures, such as nitrous oxide, which are efficiently removed by photolysis in the UV. Earth-like planets orbiting M-dwarf stars are important targets in exoplanetary science as already mentioned above. A key issue is whether the close-in habitable zone, together with the planet's possibly slowed rotation rate (hence, potentially weaker magnetosphere) due to tidal-locking, could lead to strong perturbations of the planet's atmosphere by cosmic rays and stellar flares. The effect of these upon atmospheric biosignatures has been demonstrated by model studies to be potentially important (see Grenfell et al. 2012; Tabataba-Vakili et al. 2016; Segura et al. 2010). Regarding the evolution of atmospheric ozone in time, several studies (for example Kaltenegger et al. 2007 and Des Marais et al. 2002; Rugheimer et al. 2015) have investigated this effect by calculating theoretical atmospheric spectra.

Ozone Atmospheric Spectra Ozone features spectral bands in the IR at 9.6 microns as well as in the visible and UV in the Chappuis and Hartley regions. The strength of certain spectral features depends not only on the ozone abundance in the atmosphere but also on the difference in temperature between the troposphere and stratosphere. Furthermore, since ozone can be produced abiotically in steam atmospheres and in carbon dioxide-dominated atmospheres, Selsis et al. (2002) put forward the concept of the "triple signature". This proposes to search for the simultaneous presence of ozone, water and carbon dioxide (and not only ozone) as a more reliable potential biosignature. The reasoning is as follows: finding ozone in the presence of water (which can photolyse into HOx, see above, which efficiently destroys ozone) is a clue that such an ozone signal must be strongly produced in order to overcome the strong HOx sink. Atmospheres with large carbon dioxide amounts produce ozone abiotically and would tend to mask the ozone band. Retrieving atmospheric information from spectral observations is a challenging process. The study by von Paris et al. (2013) highlighted the difficulties of extracting temperature and biosignature information for nearby Earth-like planets. For ozone, a potential false positive which must be considered is band overlap with CO_2 (von Paris et al. 2011) and interfering effects due to the presence of clouds (Kitzmann et al. 2011). Rauer et al. (2011) and Hedelt et al. (2013) discussed the detection signals possible for next generation missions for atmospheric CO_2 and O_3 for such worlds considering photon noise. Barstow and Irwin (2016) assumed an Earth-like ozone layer for the exoplanets TRAPPIST-1c and 1d and suggested that this species could be detected by averaging several tens of transits with the James Web Space Telescope.

Nitrous Oxide An important factor influencing the biosignature nitrous oxide is the UV output from the central star. This is expected to represent the main atmospheric sink for this species, whereas the main source on Earth is through biological activity. Theoretical studies applying atmospheric models with coupled climate-photochemistry (see for example Segura et al. 2005; Grenfell et al. 2014) suggest that Earth-like planets in weak UV environments (i.e. orbiting cooler stars) could feature a significant build-up in nitrous oxide in the planetary atmosphere and, therefore, an enhanced detectability in its spectral signature. On Earth, nitrous oxide is a tracer of dynamical motions due to its long lifetime against chemical removal processes in the atmosphere. Responses in the global atmospheric circulation on Earth-like exoplanets has been investigated by several 3D model studies, such as Merlis and Schneider (2010); Godolt et al. (2015). Regarding the evolution of atmospheric nitrous oxide on Earth-like planets with time, studies from the Earth's Proterozoic (such as Buick 2007, see above) suggest that this species could have been elevated due to enhanced biological activity. For planets that experience enhanced input of high energy particles, the study by Airapetian et al. (2016) suggested a significant abiotic source of nitrous oxide in nitrogen-oxygen atmospheres. This suggests a trade-off between UV from super-flares (destroying nitrous oxide directly via photolysis) and high-energy particles associated with flares producing nitrous oxide abiotically, which requires further investigation. Regarding spectral features, nitrous oxide absorbs rather weakly in the IR at 7.8, 4.5 and 3.7 microns (Muller 2013). In addition to its spectral features being sensitive to UV, the study by Grenfell et al. (2014) also noted that reducing methane biomass emissions led to a cooling in the middle atmosphere due to less shortwave absorption, which produced an enhancement in the spectral features of nitrous oxide.

Methane Sources and sinks for CH_4 on Earth-like exoplanets are not known. One can, however perform studies varying those processes which are known to be important on Earth, such as surface biomass emissions and abiotic sources from geological processes. In the atmosphere, methane can be produced by degradation of higher hydrocarbons and destroyed mainly (on Earth) by reaction with hydroxyl, whose concentration is sensitive to the incoming UV and water vapour abundance. For Earth-like planets orbiting in the habitable zone of M-dwarf stars, an important effect suggested by model studies (e.g. Segura et al. 2005; Rauer et al. 2011; Grenfell et al. 2014) is that weak incoming UV can lead to reduced OH in the atmosphere, which enables methane to build-up to values of up to $\times 1000$ the methane abundance on modern Earth. Guzmán-Marmolejo et al. (2013) suggested that abiotic methane sources in rocky exoplanets from geological processes and volcanism could produce up to 10 ppmv i.e. about $\times 5$ more than the abundance on the modern Earth. Elevated methane (and other hydrocarbons) can form hazes the properties of which could be interpreted as a biosignature, as proposed by the Arney et al. (2016) study. Theoretical studies (such as that by Schindler and Kasting 2000) suggested that atmospheric methane with abundance in excess of about $\times 50$ that of modern Earth could be detected on nearby Earth-like exoplanets by future space missions. Regarding its spectral properties, CH_4(g) features rotational-vibrational absorption bands in the IR at ~3.4 μm and ~7.7 μm (see for example Rauer et al. 2011; Werner et al. 2016). At wavelength resolutions ($R = \lambda/\delta\lambda$) coarser than $R = 20$, the methane bands could become indistinguishable from water absorption bands (des Marais et al. 2002).

11.5.4 Atmospheric Redox Disequilibrium as a Biosignature

Up to now this section has focused on the role of individual gas-phase species as atmospheric biosignatures. Alternatively, one can search for combinations of species, which are present in redox disequilibrium, i.e. the simultaneous detection of an oxidizing and a reducing chemical species whose relative abundance cannot be accounted for by abiotic processes alone. This "redox disequilibrium (or imbalance)" approach offers a new avenue to explore. It is based on the principle that strongly oxidizing and reducing species in atmospheres are not compatible and would tend to chemically react unless they are re-supplied as metabolic by-products from life. The essential idea is based on earlier works by Lovelock (1965) and Lederberg (1965). They suggested that the abundances of methane (a reducer) and oxygen (an oxidizer) in Earth's atmosphere could suggest the presence of life. Without methanogenic bacteria supplying methane and cyanobacteria supplying oxygen, these two species would react within a few thousand years to reach abundances several orders of magnitude lower than their present day atmospheric values. The idea was discussed further in the context of the Galileo spacecraft mission in a well-known paper by Sagan et al. (1993). Simoncini et al. (2013) calculated that the CH_4-O_2 redox imbalance on modern Earth needs an input of ~0.7 TW from life on our planet.

The above ideas were extended further by calculations in Krissansen-Totton et al. (2016). Their work calculated redox disequilibrium for the Earth and solar system planets. Their results suggested that (N_2-O_2) in the presence of liquid water is an important indication of chemical disequilibrium in Earth's atmosphere since, without life, atmospheric nitrogen would be transformed by lightning into nitrogen oxides, which would eventually be washed out and exist in the chemical stable form of nitrate in the ocean (although some molecular nitrogen could be returned to the atmosphere via mid ocean ridges). When assessing biosignature candidates using redox equilibrium, several false positives have been suggested, for example from ablating micrometeorites (Court and Sephton 2012) and the presence of a moon (Rein et al. 2014).

11.5.5 Additional Atmospheric Biosignatures

11.5.5.1 Species Containing Sulphur

Species such as dimethyl sulphide (CH_3SCH_3) have been suggested to build up as atmospheric biosignatures (Domagal-Goldman et al. 2011; Pilcher 2004) especially for Earth-like planets orbiting M-dwarf stars, which are weak emitters of UV. Global estimates of the atmospheric sources and sinks of sulphur compounds based on the modern Earth have been estimated (see for example Seinfeld and Pandis 2016).

11.5.5.2 Chloromethane

This species has been investigated as a biosignature (Segura et al. 2005; Grenfell et al. 2014). In the atmosphere this rather reactive species has an abundance of ~0.6 ppbv and is removed mainly by reaction with OH and by photolysis with a half-life of 1–2 years (IPCC Fourth Assessment Report 2007). Recently, Agúndez (2017) suggested an abiotic source of CH_3Cl via comet delivery. Regarding spectral absorption bands, chloromethane absorbs rather weakly in the IR, for example at ~13.7 microns.

11.6 Earth-Like Planets Without Life ('Dead Earths')

When investigating potential biosignature candidates it is clearly useful as a benchmark for comparison to calculate atmospheric spectra for planets where life did not evolve but which otherwise have similar properties (such as mass, radius, central star, orbit etc.) to the Earth—such worlds are sometimes referred to as "dead Earths". There are different approaches (Margulis and Lovelock 1974). First, starting with the modern Earth, one can switch off the biomass fluxes and calculate the response in atmospheric composition and climate. Second, one can estimate the evolution of an "Earth" atmosphere where life never evolved. The second approach is more challenging than the first but has the advantage that it includes the feedback that life had upon Earth's evolution.

Margulis and Lovelock (1974) estimated the change in Earth's atmospheric composition by removing life on Earth. Atmospheric oxygen is removed mainly via weathering of surface minerals whereas atmospheric nitrogen is removed as follows. First it undergoes molecular dissociation via lightning or cosmic rays to form atomic nitrogen which quickly forms NOx in the atmosphere. NOx is then oxidized in-situ mainly into nitric acid, which is removed for example by rainout to form nitrate in the ocean. Molecular nitrogen can be returned into the atmosphere via outgassing at mid ocean ridges. The atmospheric $CO_2(g)$ abundance in the atmosphere of a Dead Earth is challenging to estimate. The study by Margulis and Lovelock (1974) suggested 0.3–1000 mb at the surface. More recent studies however (see for example Morrison and Owen 2003) suggested that all the Earth's CO_2 (their study suggested ~69 bars which on the modern Earth resides mainly in the lithosphere) would, instead, reside in the atmosphere in the case of a dead Earth. In order to constrain these large uncertainties, improved information regarding the flux responses in Earth's carbon cycle are required. In a somewhat different approach, the study by O'Malley-James et al. (2014) investigated the future Earth where the biosphere dies at ~2.8 Gr as the luminosity of the sun increases. Their work suggested an intensification in the hydrological cycle with photosynthesis stopping when carbon dioxide levels in the atmosphere decrease below 10 ppmv, i.e. about $\times 40$ times lower than the present day values. It is desirable in the biosignature literature to re-investigate such issues with the modern suite of 1D and 3D coupled atmospheric models.

The massive extinction event at the end of the Permian 251 million years ago led to the loss of ~95% of Earth's species. Geochemical data (see for example Benton and Twitchet 2003) suggest a decrease of ~6 parts per thousand in $\delta^{18}O$ as well as a negative excursion in $\delta^{13}C$ from +2 to +4 down to −2 parts per thousand in the ocean (with possibly a similar change in the lower atmosphere) (see also Kump 1991). It is not clear, however to what extent these signals could have arisen due to the loss of life (via a reduction in the global organic carbon burial fluxes) or, due to other abiotic processes, for example, possibly large releases of methane from gas hydrates at the time. In an exoplanet context, however, such small isotopic signals are not likely to be measured remote spectroscopy within the next few decades. An additional major challenge is to determine which isotopic background to compare against when searching for signs of life.

11.7 Discussion and Conclusions

The search for signs of life beyond the Earth is one of the greatest challenges of our times. Scientific developments in exoplanetary science have been particularly rapid over the last few decades. Significant progress has already been made in observing the occurrence and nature of the hotter, gassy planets. For the cooler, rocky planets the determination of their occurrence and characterization is still in a relatively early stage. Regarding theoretical studies of the potential distribution of life beyond the Earth there is still much headway to be made. Some of the key issues are to investigate the factors which affect how life is initiated and better constrain conditions under which it can thrive, survive and possibly even spread beyond its planet of origin. In a more concrete sense, these issues are related to the distribution of habitable environments in the Universe, which are likely linked with the occurrence of rocky planets and the width of the habitable zone. Such issues are active areas of research in exoplanetary science.

Techniques to search for potential exoplanetary life include hunting for spectral signals of vegetation's strong reflection in visible wavelengths (the "red edge") and the luminescence of, for example, biopigments. Another technique involves searching for atmospheric species associated with life in planetary atmospheres via spectrophotometry and spectropolarimetry. Regarding atmospheres there are two main possibilities—either one can search for individual gas-phase species (such as ozone, nitrous oxide etc.) or one can search for combinations of reducing-oxidizing gas-phase species with abundances which are incompatible without life ("redox disequilibrium" method).

Our work has focused mainly on gas-phase atmospheric species and their potential to act as biosignatures. Ozone is a key biosignature species because it protects the lower atmosphere from UV hence control to a large extent whether other atmospheric biosignatures can survive. Oxygen has received considerable attention lately, whereby various works discuss possible abiotic sources and which emphasize the role of the environment when assessing atmospheric biosignatures. Nitrous oxide is usually acknowledged to arise almost exclusively from biology, although caveats to this idea have recently appeared (see for example Airapetian et al. 2016) which suggest abiotic production due to cosmic rays associated with stellar flares.

Hedelt et al. (2013) calculated theoretical transit depth signals estimated for nearby Earth-like planets as well as atmospheric absorption features for e.g. climate relevant species such as the broad CO_2 feature (at 15 microns) and potential biosignatures such as ozone (at 9.6 microns).

Regarding future exoplanetary missions, TESS and CHEOPS will improve understanding of gas planets and hot Super-Earths. PLATO 2.0 will provide a target list of habitable, rocky worlds in the HZ for follow-up spectroscopy. It is possible that the JWST and E-ELT could deliver the first spectroscopic observations of atmospheric biosignatures if a list of suitable targets (for example rocky planets in the HZ of nearby, M-dwarf stars) is available. Missions planned for farther into the future such as LUVOIR and HabEx aim to observe atmospheric biosignatures for a wider sample of cool, rocky planets.

As a final note we should keep in mind that the first exoplanetary biosignatures to be detected will likely be those with clear spectral signals on planets where strong biomass emissions have globally modified the atmospheric composition. A lack of detection of such signals, however, does not necessarily mean that the planet is not inhabited.

We live in exciting times—where age-old debates move into the realm of modern science.

References

Agúndez M (2017) Organohalogens in space. Nat Astron 1:655–656

Airapetian VS, Glocer A, Gronoff G et al (2016) Prebiotic chemistry and atmospheric warming of early Earth by an active young sun. Nat Geosci 9:452–455

Anbar AD, Duan Y, Lyons TW et al (2007) A whiff of oxygen before the great oxidizing event. Nature 317:1903–1906

Arney G, Domagal-Goldman S, Meadows VS et al (2016) The pale orange dot: the spectrum and habitability of Hazy Archean Earth. Astrobiology 16:873–899

Arnold L, Gillet S, Lardière et al (2002) A test for the search for life on extrasolar planets. Looking for the terrestrial vegetation signature in Earthshine spectrum. Astron Astrophys 92:231–237

Atreya SK, Mahaffy PR, Wong AS (2007) Methane and related trace species on Mars: origin, loss, implications for life and habitability. Planet Space Sci 55:358–369

Barstow J, Irwin PGJ (2016) Habitable worlds with JWST: transit spectroscopy of the TRAPPIST-1 system? MNRAS 461:L92–L96

Bates DR, Nicolet M (1950) The photochemistry of atmospheric water vapor. J Geophys Res 55:301–327

Benneke B, Seager S (2012) Atmospheric retrieval for Super-Earths. Astrophys J 753:2

Benner SA (2010) Defining life. Astrobiology 10:1021–2030

Benton MJ, Twitchet RJ (2003) How to kill (almost) all life: the end-Permian extinction event. Trends Ecol Evol 18:358–365

Bolcar MR, Balasubramanian K, Crooke J et al (2016) Technology gap assessment for a future large-aperture ultraviolet-optical infrared space telescope. Astron Telesc Instrum Syst 2:041209

Boschker HTS, Middleburg JJ (2002) Stable isotopes and biomarkers in microbial ecology. FEMS Microbiol 40:85–95

Buick R (2007) Did the Proterozoic 'Canfield ocean' cause a laughing gas greenhouse? Geobiology 5:97–100

Campbell H, Squire RJ (2010) The mountains that triggered the Late Neoproterozoic increase in oxygen: the second great oxidation event. Geochim Cosmochim Acta 74:4187–4206

Catling DC, Claire MW (2005) How Earth's atmosphere evolved to an oxic state: a status report. Earth Planet Sci Lett 237:1–20

Catling DC, Krissansen-Totton J, Kiang NY, Crisp D, Robinson TD et al (2018) Exoplanet biosignatures: a framework for their assessment. Astrobiology 18:709–738

Chapman S (1930) On ozone and atomic oxygen in the upper atmosphere. The Lond Edin Dub Philps Mag J Sci 10:369–383

Cockell C (2016) Habitability: a review. Astrobiology 16:89–117

Court RW, Sephton MA (2012) Extrasolar planets and false atmospheric biosignatures: the role of micrometeoroids. Planet Space Sci 73:233–242

Cowan NB, Abbot DS, Voigt A (2012) A false positive for ocean glint on exoplanets: the latitude-albedo effect. Astrophys J 752:L3

Crutzen PJ (1970) The influence of nitrogen oxides upon the atmospheric ozone content. Q J R Met S 96:320–325

Des Marais DJ, Harwit MO, Jucks KW (2002) Remote sensing of planetary properties and biosignatures on extrasolar terrestrial planets. Astrobiology 2:153–181

Dick SJ (1984) The plurality of worlds: the extra-terrestrial life debate from Democritus to Kant. Cambridge University Press, Cambridge, UK

Domagal-Goldman S, Meadows VS, Claire MW (2011) Using biogenic sulfur gases as remotely detectable biosignatures on anoxic planets. Astrobiology 11:419–441

Domagal-Goldman S, Segura A, Claire MW (2014) Abiotic ozone and oxygen in atmospheres similar to prebiotic Earth. Astrphys J Lett 787:2

Dressing C, Charbonneau D (2015) The occurrence of potentially habitable planets orbiting M dwarfs estimated from the full Kepler dataset and an empirical measurement of the detection sensitivity. Astrophys J 807:45

Encrenaz T (2014) Infrared spectroscopy of exoplanets: observational constraints. Philos Trans R Soc A 372:20130083
Farquhar GD, Lloyd J (1993) Carbon and oxygen isotope effects in the exchange of carbon dioxide between terrestrial planets and the atmosphere. In: Ehleringer JR, Hall AE, Farquhar GD (eds) Stable isotopes and plant carbon-water relations. Elsevier, New York, pp 47–70
Feulner G (2012) The faint young Sun problem. Rev Geophys 50:1–29
Formisano V, Atreya S, Encrenaz T (2004) Detection of methane in the atmosphere of Mars. Nature 306:1758–1761
Fortier A, Beck T, Benz W (2014) CHEOPS: a space telescope for ultra-high precision photometry of exoplanet transits. J Astron Telesc Instrum Syst 9143
Fujii T, Moynier F, Blichert-Toft J et al (2014) Density functional theory estimation of isotope fractionation of Fe, Ni, Cu, and Zn among species relevant to geochemical and biological environments. Geochim Cosmochim Acta 140:553–576
Fujii Y, Angerhausen D, Deitrick R et al (2018) Exoplanet biosignatures: observational prospects. Astrobiology 18(6). doi: https://doi.org/10.1089/ast.2017.1733
Gaidos E (2013) Candidate planets in the habitable zones of Kepler stars. Astophys J 770:2
Gaillard F, Scaillet B, Arndt NT (2011) Atmospheric oxygenation caused by a change in volcanic degassing pressure. Nature 478:229–232
Gebauer S, Grenfell JL, Stock JW et al (2017) Evolution of Earth-like extrasolar planetary atmospheres. Astrobiology 17:27–54
Gebauer S, Grenfell JL, Lehmann R, Rauer H (2018) Evolution of Earth-like planetary atmospheres around M Dwarf Stars: assessing the atmospheres and biospheres with a coupled atmosphere biogeochemical model. Astrobiology 18:856–872
Godolt M, Grenfell JL, Hamann-Reinus A et al (2015) 3D climate modeling of Earth-like extrasolar planets orbiting different types of host stars. Planet Space Sci 111:62–76
Godolt M, Grenfell JL, Kitzmann D et al (2016) Assessing the habitability of planets with Earth-like atmospheres with 1D and 3D climate modeling. Astron Astrophys 592:A36
Grasset O, Dougherty MK, Coustenis A et al (2012) JUpiter ICy moons Explorer (JUICE): an ESA mission to orbit Ganymede and to characterise the Jupiter system. Planet Space Sci 78:1–21
Grenfell JL (2017) A review of exoplanetary biosignatures. Phys. Rep. 713:1–17
Grenfell JL, Stracke B, von Paris P et al (2007) The response of atmospheric chemistry on earthlike planets around F, G and K stars to small variations in orbital distance. Planet Space Sci 55:661–671
Grenfell JL, Rauer H, Selsis F et al (2010) Co-evolution of atmospheres, life and climate. Astrobiology 10:77–88
Grenfell JL, Gebuaer S, von Paris P et al (2011) Sensitivity of biomarkers to changes in chemical emissions in Earth's Proterozoic atmosphere. Icarus 211:81–88
Grenfell JL, Griessmeier J-M, von Paris P et al (2012) Response of atmospheric biomarkers to NO_x-induced photochemistry generated by stellar cosmic rays for Earth-like planets in the habitable zone of M dwarf stars. Astrobiology 12:1109–1122
Grenfell JL, Gebauer S, von Paris et al (2014) Sensitivity of biosignatures on Earth-like planets orbiting in the habitable zone of cool M-dwarf stars to varying stellar UV radiation and surface biomass emissions. Planet Space Sci 98:66–76
Griffith RL, Wright JT, Maldonado J et al (2015) The G infrared search for extraterrestrial civilizations with large energy supplies. Astrophys J 217(2)
Guzmán-Marmolejo A, Segura A, Escobar-Briones E (2013) Abiotic production of methane in terrestrial planets. Astrobiology 13:550–559
Haagen-Smit AJ (1952) Chemistry and physiology of Los Angeles smog. Ind Eng Chem 44:1342–1346
Haghighipor N (2015) Eta-Earth. Encyclopedia of astrobiology. Springer, Heidelberg
Hall DT, Strobel DF, Feldman PD et al (1995) Detection of an oxygen atmosphere on Jupiter's moon Europa. Nature 373:677–681

Hall DT, Feldman PD, McGrath MIA et al (1998) The far-ultraviolet oxygen airglow of Europa and Ganymede. Astrophys J 449:475–481
Haqq-Misra JD, Domagal-Goldman SD, Kasting PJ et al (2009) A revised, hazy methane greenhouse for the Archean Earth. Astrobiology 8:1127–1137
Hedelt P, Alonso R, Brown T et al (2011) Venus transit 2004: illustrating the capacity of exoplanet transmission spectroscopy. Astron Astrophys 533:A136
Hedelt P, von Paris P, Godolt M et al (2013) Spectral features of Earth-like planets and their detectability at different orbital distances around F, G, and K-type stars. Astron Astrophys 553: A9
Heller R, Barnes R (2013) Exomoon habitability constrained by illumination and tidal heating. Astrobiology 13:18–46
Holland HD (2002) Volcanic gases, black smokers and the great oxidation event. Geochim Cosmochim Acta 66:3811–3826
Holland HD (2006) The oxygenation of the atmosphere and oceans. Philos Trans R Soc Lond Biol Sci 361:903–915
Holmen K (1992) The global carbon cycle. London Academic Press, London, pp 237–262
Höning D, Hansen-Goos H, Airo A (2014) Biotic vs. abiotic Earth: a model for mantle hydration and continental coverage. Planet Space Sci 98:5–13
Horler DNH, Dockray M, Barber J (1983) The red edge of plant leaf reflectance. Int J Remote Sens 4:273–288
Horneck G, Walter N, Westall F et al (2016) AstRoMap European Astrobiology Roadmap. Astrobiology 16:201–243
Horner J, Jones BW (2008) Jupiter – friend or foe? I: the asteroids. Int J Astrobiol 7:251–261
Hu R, Seager S (2014) Photochemistry in terrestrial planet atmospheres III. ApJ 784:1
Huang S (1959) Occurrence of life in the universe. Am Sci 47:397–402
Hunten DM (1988) Mercury. University of Arizona Press, Tucson, AZ
International Panel on Climate Change (IPCC) Climate Change (2007) In: Solomon S et al (eds) The physical basis. IPCC, Geneva
Joyce G, Deamer DW, Fleischaker GR (1994) In: Deamer DW, Fleichacker GR (eds) Origins of life: the central concepts. Jones and Bartlett, Boston, pp xi–xii
Kaltenegger L, Sasselov D (2011) Exploring the habitable zone for Kepler planetary candidates. ApJ 736:2
Kaltenegger L, Traub WA, Jucks KW et al (2007) Spectral evolution of an Earth-like planet. ApJ 658:1
Kaltenegger L, Miguel Y, Rugheimer S (2012) Rocky exoplanet characterization and atmospheres. Int J Astrobiol 11:297–307
Kane SR, Hill ML, Kasting JF et al (2016) A catalogue of Kepler habitable zone exoplanet candidates. ApJ 830(1)
Kasting JF, Catling DC (2003) Evolution of a habitable planet. Annu Rev Astron Astrophys 41:429–463
Kasting JF, Whitmire DP, Reynolds RT (1993) Habitable zones around main sequence stars. Icarus 101:108–128
Kawahara H, Matsuo T, Takami M et al (2012) Can ground-based telescopes detect the 1.27 micron absorption feature as a biomarker in exoplanets? ApJ 758:1
Kiang NY, Segura A, Tinetti G et al (2007) Spectral signatures of photosynthesis. II. Coevolution with other stars and the atmosphere on extrasolar worlds. Astrobiology. 7:252–274
Kislyakova KG, Johnstone CP, Odert P et al (2014) Stellar wind interaction and pick-up ion escape of the Kepler-11 "super-Earths". Astron Astrophys 562:A116
Kitzmann D (2016) Revisiting the scattering greenhouse effect of CO_2 ice clouds. ApJL 817:2
Kitzmann D, Patzer ABC, von Paris P et al (2011) Clouds in the atmospheres of extrasolar planets. Astron Astrophys 531:A62

Kopp RE, Kirschvink JL, Hilburn IA et al (2005) The Paleoproterozoic snowball Earth: a climate disaster triggered by the evolution of oxygenic photosynthesis. Proc Natl Acad Sci USA 102:11131–11136

Kopparapu RK, Ramses M, Schttelkotte J et al (2014) Habitable zones around main sequence stars: dependence upon planetary mass. ApJL 787:2

Korpela EJ, Sallmen SM, Greene DL (2015) Modeling indications of technology in planetary transit light curves – dark-side illumination. ApJ 809:2

Krissansen-Totton J, Bergsman DS, Catling DC (2016) On detecting biosignatures from chemical thermodynamic disequilibrium in planetary atmospheres. Astrobiology 16:39–67

Kroopnick P, Craig H (1972) Atmospheric oxygen: isotopic composition and solubility fractionation. Science 175:54–55

Kump LR (1991) Interpreting carbon-isotope excursions: Strangelove oceans. Geology 19:299–302

Kump LR, Junium C, Arthur MC et al (2011) Isotopic evidence for massive oxidation of organic matter following the Great Oxidation Event. Science 334:1694–1696

Lammer H, Bredehöft JH, Coustenis A et al (2009) What makes a planet habitable? Astron Astrophys Rev 17:181–189

Lasaga AC, Ohmoto H (2002) The oxygen geochemical cycle: dynamics and stability. Geochim Cosmochim Acta 66:361–381

Laskar J, Joutel F, Roboutal P et al (1993) Stabilization of the Earth's obliquity by the Moon. Nature 361:615–617

Lederberg J (1965) Signs of life. Nature 207:9–13

Lefèvre F, Forget F (2009) Observed variations of methane on Mars unexplained by known atmospheric chemistry and physics. Nature 460:720–723

Levine JS, Shaw EF (1983) In situ aircraft measurements of enhanced levels of N_2O associated with thunderstorm lightning. Nature 303:312–314

Levine JS, Hughes RE, Chameides WL et al (1979) N_2O and CO production by electric discharge: atmospheric implications. Geophys Res Lett 6:557–559

Lightsey PA, Atkinson CB, Clampin MC et al (2012) James Webb Space Telescope: large deployable telescope in space. Opt Eng 51:1

Lin HW, Abad GG, Loeb A (2014) Detecting industrial pollution in the atmospheres of Earth-planets. Astrophys J Lett 791:1

Lovelock JE (1965) A physical basis for life detection experiments. Nature 207:568–570

Ludwig W, Eggl S, Neubauer D et al (2016) Effective stellar flux calculations for limits of life-supporting zones of exoplanets. MNRAS 458:3752–3759

Luger R, Barnes R (2015) Extreme water loss and abiotic O_2 buildup on planets throughout the habitable zone on M-dwarfs. Astrobiology 15:119–143

Margulis LM, Lovelock JE (1974) Biological modulation of the Earth's atmosphere. Icarus 21:471–489

McElroy MB, McConnell JC (1971) Nitrous oxide: a natural source of NO. Am Met Soc 28:1095–1098

Meadows VS, Reinhard CT, Arney GN et al (2018) Exoplanet biosignatures: understanding oxygen as a biosignature in the context of its environment. Astrobiology 18(6):630–662

Merlis TM, Schneider T (2010) Atmospheric dynamics of Earth-like tidally-locked aquaplanets. J Adv Mod Earth Sys 2:13

Mennesson B, Gaudi S, Seager S et al (2016) The Habitable Exoplanet (HabEx) Imaging Mission: preliminary science drivers and technical requirements. J Astron Telesc Instrum Syst 9904

Misra A, Meadows VS, Claire MW et al (2014) Using dimers to measure biosignatures and atmospheric pressure for terrestrial exoplanets. Astrobiology 14:67–86

Montmessin F, Bertaux JL, Lefèvre F et al (2011) A layer of ozone detected in the nightside upper atmosphere of Venus. Icarus 216:82–85

Morrison D, Owen T (2003) The planetary system, 3rd edn. Addison-Wesley, Reading, MA

Morton TD, Swift J (2014) The radius distribution of planets around cool stars. Astrophys J 791:10

Muller C (2013) N_2O as a biomarker: from the Earth and solar system to exoplanets. Astrophys Spa Sci Proc 35:99–106

Naa Mvondo D, Navarro-Gonzalez R, McKay CP et al (2001) The production of nitrogen oxides by lightning and coronal discharges in simulated early Earth, Venus and Mars environments. Adv Space Res 27:217–223

Noack L, Rivoldini A, Van Hoolst T (2017) Volcanism and outgassing of stagnant-lid planets: implications for the habitable zone. PEP 269:40–57

Noll KS, Roush TL, Cruikshank DP et al (1997) Detection of ozone on Saturn's satellites Rhea and Dione. Nature 388:45–47

O'Malley-James JT, Greaves JS, Raven JA et al (2014) Swansong Biospheres II: the final signs of life on terrestrial exoplanets near the end of their habitable lifetimes. Int J Astrobiol 13:229–243

Pallé E, Osorio MRZ, Barena R et al (2009) Earth's transmission spectrum from lunar eclipse measurements. Nature 459:814–816

Pavlov AA, Kasting JF, Brown LL et al (2000) Greenhouse warming by CH_4 in the atmosphere of Early Earth. J Geophys Res 105:11,981–11,990

Perrier S, Bertaux JL, Lefèvre F et al (2006) Global distribution of total ozone on Mars from SPCAM/MEX UV measurements. J Geophys Res 111:E9

Pierrehumbert R, Gaidos E (2011) Hydrogen greenhouse planets beyond the habitable zone. Astrophys J Lett 734:L13

Pilcher CB (2004) Biosignatures of Early Earths. Astrobiology 3:471–486

Ramirez RM, Kopparapu R, Zugger ME et al (2014) Warming early Mars with CO_2 and H_2. Nat Geosci 7:59–63

Rauer H, Gebauer S, von Paris P et al (2011) Potential biosignatures in super-Earth atmospheres. I. Spectral appearance of super-Earths around M dwarfs. Astron Astrophys 529:A8

Rauer H, Catala C, Aerts C et al (2014) The PLATO 2.0 Mission. Exp Astron 38:249–330

Raymond SN, Quinn T, Lunine JI (2007) High-resolution simulations of the final assembly of Earth-like planets. 2. Water delivery and planetary habitability. Astrobiology 7:66–84

Rein H, Fujii Y, Spiegel DS (2014) Some inconvenient truths about biosignatures involving two chemical species on Earth-like exoplanets. Proc Natl Acad Sci USA 111:6871–6875

Ricker GR, Winn JN, Vanderspeck R et al (2014) Transiting exoplanet survey satellite. J Astron Telesc Instrum Syst 1:014003

Roberson AL, Roadt J, Halevy I et al (2011) Greenhouse warming by nitrous oxide and methane in the Proterozoic eon. Geobiology 9:313–320

Rodler F, López-Morales M (2014) Feasibility studies for the detection of O_2 in an Earth-like exoplanet. Astrophys J 781:1

Rugheimer S, Kaltenegger L, Segura A et al (2015) Effect of UV on the spectral fingerprints of Earth-like planets orbiting M-stars. Astrobiology 809:1–16

Sagan C, Thompson WR, Carlson R et al (1993) A search for life on Earth from the Galileo spacecraft. Nature 365:375–377

Samarkin VA, Madigan MT, Bowles MW et al (2010) Abiotic nitrous oxide emission from the hypersaline Don Juan Pond in Antarctica. Nat Geophys 3:341–344

Sanroma E, Palle E, Parenteau MN et al (2014) Characterizing the purple Earth: measuring the globally-integrated spectral variability of the Archaean Earth. Astrophys J 780(1)

Scalo J, Segura A, Fridlund M et al (2007) M stars as targets for terrestrial exoplanet searches and biosignature detection. Astrobiology 7:85–166

Schidlowski M (1988) A 3800 million-year isotopic record of life from carbon in sedimentary rocks. Nature. 333:313–318

Schindler TL, Kasting JF (2000) Synthetic spectra of simulated terrestrial atmospheres containing possible biomarker gases. Icarus 145:262–271

Schneider J (1994) On the search for O_2 in extrasolar planets. Astrophys Space Sci 212:321–325

Schoell M (1988) Multiple origins of methane in the Earth. Chem Geol 71:1–10

Schwieterman E et al (2015) Non photosynthetic pigments as potential biosignatures. Astrobiology 15:341–361

Schwieterman EW, Cockell CS, Meadows VS et al (2016) Identifying planetary biosignature imposters: spectral features of CO and O_4 resulting from O_2/O_3 production. Astrophys J 819(1)
Schwieterman EW, Kiang NY, Parenteau MN, Harman CE, DasSarma S et al (2018) Exoplanet biosignatures: a review of remotely detectable signs of life. Astrobiology 18:663–708
Seager S, Turner E, Schafer LJ et al (2005) Vegetation's red edge: a possible spectroscopic biosignature of extraterrestrial plants. Astrobiology 5:372–390
Seager S, Bains W, Hu R (2013) Biosignature gases in H_2-dominated atmospheres on rocky planets. Astrophys J 777:2
Seager S, Bains W, Petkowski JJ (2016) Toward a list of molecules as potential biosignature gases for the search for life on exoplanets and applications to terrestrial biochemistry. Astrobiology 16:465–485
Segura A, Krelove K, Kasting JF et al (2003) Ozone concentrations and ultraviolet fluxes on Earth-like planets around other stars. Astrobiology 3:689–708
Segura A, Kasting JF, Meadows VS et al (2005) Biosignatures from Earth-like planets around M-stars. Astrobiology 5:706–725
Segura A, Walkowicz MVS et al (2010) The effect of a strong stellar flare on the atmospheric chemistry of an Earth-like planet orbiting an M-dwarf. Astrobiology 10:751–771
Seinfeld JH, Pandis SN (2016) From air pollution to climate change. Wiley, Hoboken, NJ
Selsis F, Despoit D, Parisot J-P et al (2002) Signature of life on exoplanets: can Darwin produce false positive detections? Astron Astrophys 388:985–1003
Shields AL, Ballard S, Johnson JA (2016) The habitability of planets orbiting M-dwarf stars. Phys Res 663:1–38
Simoncini E, Virgo N, Kleidon A et al (2013) Quantifying drivers of chemical disequilibrium: theory and application to methane in Earth's atmosphere. Earth Syst Dyn 4:317–331
Slanger TG, Copeland RA (2003) Energetic oxygen in the upper atmosphere and the laboratory. Chem Rev 103:4731–4766
Smith KC (2016) Life is hard: countering definitional pessimism concerning the definition of life. Int J Astrobiol 15:277–289
Smith AK, Marsh DR (2005) Processes that account for the ozone maximum at the mesopause. J Geophys Res 110:D23
Snellen I (2014) High-dispersion spectroscopy of extrasolar planets: from CO in hot Jupiters to O_2 in exo-Earths. Philos Trans R Soc A 372:20130075
Stam DM (2008) Spectropolarimetric signatures of Earth-like extrasolar planets. Astron Astrophys 482:989–1007
Stamenkovic V, Noack L, Breuer D et al (2012) The influence of pressure-dependent viscosity on the thermal evolution of Super-Earths. Astrophys J 748:1
Sterzik MF, Bagnul S, Palle E (2012) Biosignatures as revealed by spectropolarimetry of Earth-shine. Nature 483:64–66
Stevens A, Forgan D, James JOM (2016) Observational signatures of self-destructive civilizations. Int J Astrobiol 15:33–44
Stolarski RJ, Cicerone RS (1974) Stratospheric chlorine: a possible sink for ozone. Can J Chem 52:1610–1615
Sverjensky DA, Lee N (2010) The great oxidation event and mineral diversification. Elements 6:31–36
Syakila A, Kroeze C (2011) The global nitrous oxide budget revisited. Greenhouse Gas Meas Manag 1:17–26
Tabataba-Vakili F, Grenfell JL, Griessmeier J-M et al (2016) Atmospheric effects of stellar cosmic rays on Earth-like exoplanets orbiting M-dwarfs. Astron Astrophys 585:A96
Tackley PJ, Ammann M, Brodholt JP (2013) Mantle dynamics in super-Earths: post-perovskite rheology and self-regulation of viscosity. Icarus 225:50–61
Teolis BD, Jones GH, Miles PF (2010) Cassini finds an oxygen-carbon dioxide atmosphere at Saturn's icy moon Rhea. Science 333:6012

Tian F, France K, Linsky JL et al (2014) High stellar FUV/NUV ratio and oxygen contents in the atmospheres of potentially habitable planets. Earth Planet Sci 385:22–27
Tosi N, Godolt M, Stracke B et al (2017) The habitability of a stagnant-lid Earth. Astron Astrophys 605:A71
Traub WA (2015) Steps towards eta-Earth from Kepler data. Int J Astrobiol 14:359–363
Tyler RH (2008) Strong ocean tidal flow and heating on moons of the outer planets. Nature 456:770–772
Vogel G (1999) Expanding the habitable zone. Science 286:70–71
von Paris P, Cabrera J, Godolt M et al (2011) Spectroscopic characterization of the atmospheres of potentially habitable planets: Gl581d as a model case study. Astron Astrophys 534:A26
von Paris P, Hedelt P, Selsis F et al (2013) Characterization of potentially habitable planets: retrieval of atmospheric and planetary properties from emission spectra. Astron Astrophys 551:A120
Walker SI, Bains W, Cronin L et al (2018) Exoplanet biosignatures: future directions. Astrobiology 18(6):779–824
Wang Y, Tian F, Li T et al (2016) On the detection of carbon monoxide as an anti-biosignature in exoplanetary atmospheres. Icarus 266:15–23
Wayne RP (1993) Chemistry of atmospheres, 2nd edn. Oxford University Press, Oxford
Webster CR, Mahaffy P, Atreya SK (2015) Mars methane detection and variability at gale crater. Science 412:415
Werner MW, Swain MR, Vasisht G et al (2016) Extension of ATLAST/LUVOIR's capabilities to 5 μm or beyond. J Astron Telesc Instrum Syst 2:041205
Williams DM, Gaidos E (2008) Detecting the glint of starlight on the oceans of distant planets. Icarus 195:927–937
Woolf NJ, Smith PS, Traub WA et al (2002) The spectrum of Earthshine: a pale blue dot observed from the ground. Astrophys J 574:430–433
Wordsworth R, Pierrehumbert R (2014) Abiotic oxygen-dominated atmospheres on terrestrial habitable zone planets. Astrophys J Lett 785:1–4
World Meteorological Organization (WMO) (1995) Scientific assessment of ozone depletion: 1994. Report Number 37. WMO, Geneva
Yan F, Fosbury RAE, Petr-Gotzens MG et al (2015) High-resolution transmission spectrum of the Earth's atmosphere-seeing Earth as an exoplanet using a lunar eclipse. Int J Astrobiol 14:255–266
Yang J, Cowan NB, Abbot DS (2013) Stabilising cloud feedback dramatically expands the habitable zone of tidally-locked planets. ApJL 771:2
Yung YL, DeMore WB (1999) Photochemistry of planetary atmospheres. Oxford University Press, Oxford
Zahnle K, Freedman RS, Catling DC (2011) Is there methane on Mars? Icarus 212:493–503

Part III
Biosignatures, Instruments and Missions

Chapter 12
The Enigma of Methane on Mars

Franck Lefèvre

Abstract Between 2004 and 2012, four independent groups reported detections of low levels (10–60 ppbv) of methane on Mars. If true, these constitute the first observations of a potential biosignature on that planet and would be an important finding and addition to the inventory of minor species in its atmosphere. However, these claims for the presence of methane have been highly controversial. In 2014, the most robust search for methane on Mars was performed by the rover *Curiosity*. The latest measurements by *Curiosity* indicate a background CH_4 level of 0.2–0.7 ppbv, except during a two-month period between November 2013 and January 2014, when high mixing ratios of around 7 ppbv were observed. These observations immediately raise the question of the origin of methane on Mars, but also pose fundamental challenges to our current understanding of Martian atmospheric physics and chemistry.

12.1 Observations

After many years of unsuccessful search, the first report of methane detection on Mars was made by Krasnopolsky et al. (2004), who observed 10 ± 3 ppbv of methane in January 1999 at $L_s = 88°$ using the Fourier Transform Spectrometer (FTS) at the Canada-France Hawaii Telescope. This averaged value was obtained over a significant part of the Martian disk. The same year, Formisano et al. (2004) detected varying amounts of methane between 0 and 30 ppbv over a few orbits of the Planetary Fourier Spectrometer (PFS) infrared sounder on board the Mars Express spacecraft. Geminale et al. (2011) expanded on this using the same instrument and method over a 6-year baseline. Their results revealed substantial seasonal variations of methane, with local enhancements of up to 70 ppbv located at high northern latitudes in summer. The mean value derived from PFS is about 15 ppbv. Using the

F. Lefèvre (✉)
LATMOS, CNRS/Sorbonne Université, Paris, France
e-mail: franck.lefevre@latmos.ipsl.fr

B. Cavalazzi, F. Westall (eds.), *Biosignatures for Astrobiology*, Advances in Astrobiology and Biogeophysics, https://doi.org/10.1007/978-3-319-96175-0_12

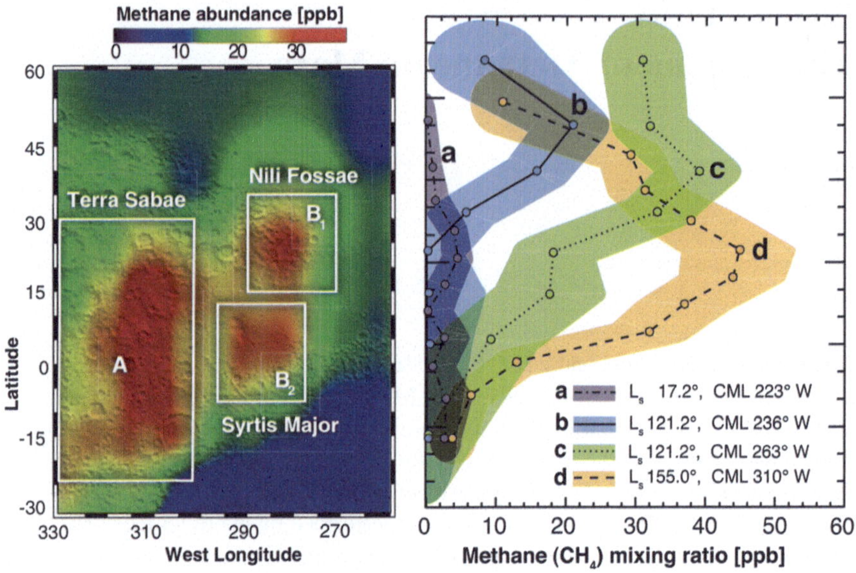

Fig. 12.1 Methane mixing ratio (ppbv) observed from Earth by Mumma et al. (2009). Left: regions where methane appeared enhanced during summer 2003 (L_s = 121°–155°). Right: latitudinal and temporal variability of methane. Profile 'a' was obtained in February 2006 at L_s = 17.2°. Profiles 'b', 'c', and 'd' were obtained in January–March 2003 between L_s = 121°–155°. Source: Reproduced with permission from Mumma et al. (2009)

CSHELL spectrometer at the IRTF (Hawaii), Mumma et al. (2009) detected, during in January–March 2003, a strong local enhancement (~50 ppbv) in the form of a "plume" of methane at low latitudes over the Syrtis Major region (Fig. 12.1). No significant amounts of methane were found in their observations performed 3 years later in January–February 2006.

Krasnopolsky (2012) reprocessed the CSHELL observations of February 2006 at L_s = 10° with refined analytical approaches and found about 10 ppbv of methane over the Valles Marineris region and ~3 ppbv outside this region. His observations for December 2009 at L_s = 20° and March 2010 at L_s = 70° showed no detection with an upper limit of 8 ppbv. Villanueva et al. (2013) also used the CSHELL spectrometer in January 2006 but could only derive an upper limit of 7.8 ppbv, in contrast to the 10 ppbv detected by Krasnopolsky (2012) one month later. Villanueva et al. (2013) did not detect methane in their later observations of November 2009 and April 2010.

All of the above studies measured CH_4 in its absorption band at 3.3 µm. Using the band at 8 µm, Fonti and Marzo (2010) performed a statistical analysis of the Thermal Emission Spectrometer (TES) spectra acquired from the Mars Global Surveyor spacecraft and extracted a weak signal attributed to methane absorption. They found seasonally variable methane at the ~10–30 ppbv level.

It is important to recognise that the claimed detections at such low levels of methane were obtained at the limits of the instrumental capabilities. PFS and TES spacecraft instruments do not have the required sensitivity and spectral resolution for the unambiguous identification of CH_4 (Zahnle et al. 2011). Instead, the detection of methane is made by summing together thousands of spectra, which does not suppress—and can even increase—instrumental effects and systematic errors. Despite their much greater spectral resolution, ground-based observations are not simpler because Martian methane must then be viewed through the atmosphere of Earth, which contains ~10^4 more methane molecules above the observer than that which is retrieved on Mars. The measurements must therefore exploit the Doppler shift of the Martian lines when Mars is approaching or receding from Earth. However, even in these conditions, the claimed detections of methane are close to the noise level. In any case, the very different geographical distributions and seasonal variations of methane obtained from the Earth (Mumma et al. 2009; Krasnopolosky 2012) and from space (Fonti and Marzo 2010; Geminale et al. 2011) are puzzling and must be considered with great care.

Because of the potential implications of the presence of methane, and in the light of the controversial observational dataset described above, the first *in situ* measurements promised by the Tunable Laser Spectrometer (TLS) aboard the *Curiosity* rover and its far superior detection capabilities were eagerly awaited. TLS uses an infrared laser at 3.27 microns to scan the same methane lines as PFS or from the Earth, but at ultrahigh spectral resolution. TLS has two measurement modes: the direct ingest mode with error bars on the order of 2 ppbv and the enriched mode achieved by scrubbing out CO_2 during a slow fill of the sample cell. This latter mode has error bars of only about 0.1 ppbv. The first atmospheric samples collected by TLS in Gale Crater (4°S, 137°E) spanned an 8-month period in spring-summer. By combining all of the individual measurements available at that time, it was concluded that methane was not detected, with an upper limit of only 1.3 ppbv (Webster et al. 2013). In a subsequent analysis, Webster et al. (2015a) presented the entire TLS dataset reprocessed over a period of almost one Martian year (605 Martian days or sols). The results indicate detection of methane at two levels of abundance (Fig. 12.2). A "background" CH_4 level of 0.7 ± 0.2 ppbv, based on high-precision methane-enriched experiments, is observed during the first 8 months and last 4 months of the dataset. Between these two periods, two episodic enhancements of methane were observed. The first was detected in June 2013 with a single measurement indicating about 6 ppbv of CH_4 followed by a drop to 2 ppbv one week later. The second was observed in November 2013–January 2014 (sols 466–526 in Fig. 12.2). During that period—corresponding to the Martian spring ($L_s = 56$–$82°$)—four sequential measurements of TLS indicated a pulse of 7 ± 2 ppbv of methane over 3 months.

The duration of this event (60 sols) followed by a sudden drop to 0.5 ppbv 47 sols later was interpreted by Webster et al. (2015a) to result from a local production of methane that, once terminated, dispersed quickly. One Martian year later (in Martian year 33, according to the calendar proposed by Clancy et al. (2000) which starts at $L_s = 0°$ on 11 April 1955), no springtime pulse of methane was observed by TLS (Fig. 12.3). On the contrary, the instrument only detected a very small background level

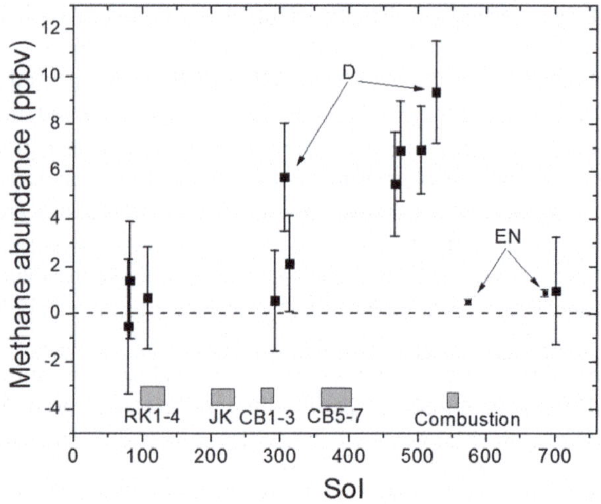

Fig. 12.2 Measurements of methane mixing ratio (ppbv) by the TLS instrument on the *Curiosity* rover, as presented by Webster et al. (2015a). 1 sol = 1 Mars day = 24 h 37 min. Martian sol 1 was on August 6, 2012. Error bars represent ±1 standard error of the mean. All measurements were made at night, except for the two marked "D" ingested during the day. The values with smaller error bars labelled EN were retrieved from the "methane enrichment" runs. Source: Reproduced with permission from Webster et al. (2015a)

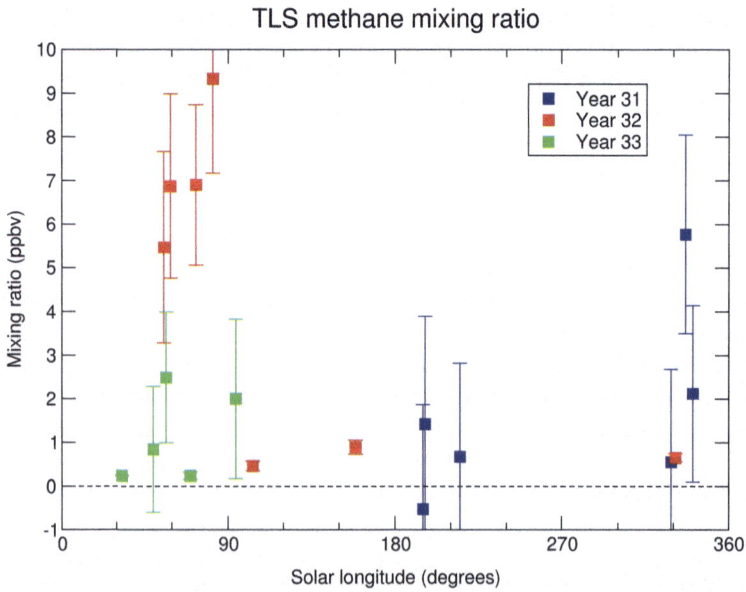

Fig. 12.3 Summary of TLS methane measurements *versus* solar longitude for the Martian years 31–33 (October 2012–January 2016). Error bars represent ±1 standard error of the mean. Source: Adapted from data by Webster et al. (2015a) and Roos-Serote et al. (2016)

of 0.2 ppbv by means of a high-precision enriched run (Webster et al. 2015b, 2018). This absence of reproducibility from one year to the next seems to rule out a seasonal effect. Thus, if the TLS findings are true, the *in situ* evidence for the presence of methane on Mars can be summarised as follows. (1) Methane is constantly present at very low levels (0.2–0.7 ppbv) in the Martian atmosphere. This equilibrium level requires a tiny source of less than 20 tonnes per year in order to balance the photochemical loss of methane over its atmospheric lifetime of 300 years (Lefèvre and Forget 2009). (2) TLS observed two pulses of methane ten times higher than the background methane level during its first Martian year on the surface. Such events were not observed in the second year. This suggests the presence of local and episodic releases of methane, which are not correlated to season. The occurrence of methane pulses at the surface has not yet been confirmed by TLS high-precision "enriched" runs.

12.2 Methane Production

The observational claims for methane on Mars naturally raise the question of sources and sinks. On Mars, methane is not produced photochemically in the atmosphere and is destroyed by ultraviolet radiation and oxidants in approximately 300 years. Therefore, the presence of methane in the atmosphere, even in very small quantities, requires on-going (or very recent) emission from the surface. On Earth, more than 90% of methane has a biological origin, in the form of living beings, organic waste or fossilized matter. It is therefore tantalising to relate the existence of methane on Mars to a possible past or extant life on the planet. For instance, Krasnopolsky (2006) calculated that methane originating only from the impacts of comets, meteorites, and interplanetary dust is insignificant, and further argued that the lack of volcanism, hot spots, and SO_2 (which is more abundant than CH_4 in terrestrial outgassing) favours the hypothesis of a biogenic origin for Martian methane. However, if a microbial source is a possibility to explain the existence of methane on Mars, it is by no means the only one. To be convinced of this, one merely has to consider the estimated mass of methane transferred to the Mars atmosphere each year—a few tens of tons are sufficient to explain the background level measured by TLS—to the abiotic CH_4 flux on Earth, which is estimated at several megatons per year (Etiope and Sherwood Lollar 2013). On Earth, methane can indeed be produced in several specific environments by chemical reactions that do not involve organic matter. These processes are found at high temperatures in volcanic or geothermal areas, and at low temperatures in aquifers *via* gas-water-rock interactions. In this latter case, Atreya et al. (2007) proposed that methane could be produced abiogenically by hydrothermal processes, such as serpentinization, i.e., the reaction between ultramafic (Mg, Fe-rich) silicates and water, producing serpentine and molecular hydrogen H_2:

$$(Mg, Fe)_2SiO_4 + H_2O \rightarrow Mg_3Si_2O_5(OH)_4 + Mg(OH)_2 + Fe_3O_4 + H_2$$

The aqueous H_2 produced above then reacts with CO_2 *via* a Fischer-Tropsch-type reaction which in turn produces methane, e.g.:

$$CO_2 + 4\,H_2 \rightarrow CH_4 + 2\,H_2O$$

At the Mid-Atlantic Ridge, a substantial production of methane by serpentinization has been detected at relatively mild temperatures (<100 °C, Kelley et al. 2005), which could also be found on Mars at depths of 2–3 km below the surface (Oze and Sharma 2005). Although the presence of such aquifers is yet to be discovered on Mars, the discovery of methane in the atmosphere increases the possibility of active hydrothermal activity and a geologically "living" planet.

Exogenous methane sources, such as comets, meteorites, or interplanetary dust particles, are also a possibility. Court and Sephton (2009) studied direct release of CH_4 by ablation and pyrolysis of carbonaceous meteorites during their atmospheric entry. They found that this process accounted for less than 10 kg of methane annually, which is a negligible fraction of the mass required to maintain the abundance of methane observed by TLS. On the other hand, Keppler et al. (2012) argued that Murchison-type carbonaceous micrometeorites might be in sufficient quantity on the surface of Mars to become a significant abiogenic source of methane when exposed to ultraviolet radiation. The extrapolation of their laboratory experiment to the global scale suggests that 10–800 tons of methane could be produced annually at the surface of Mars, which could explain a substantial fraction—if not all—of the observed abundance of methane.

However, it is important to note that none of the processes mentioned above can explain the temporal variability of Martian methane that seems to emerge from the observations. Since it has been shown that high methane episodes do not correlate with predicted meteor events (Roos-Serote et al. 2016), we may discard the hypothesis of meteor showers as a source of methane variability on Mars.

12.3 Methane Loss

According to conventional chemistry, which reproduces very well the observed methane distribution and variations on Earth, the photochemical loss of methane on Mars is accomplished either by photodissociation or by chemical reactions. The fields of action of both processes are clearly separated in altitude. In the upper atmosphere, methane is photolyzed in the Lyman-α line at 121.6 nm. The efficiency of this process peaks at about 80 km and represents 50–60% of the total loss of methane integrated over the atmospheric column. In the lower atmosphere, as on Earth, the loss of Martian methane occurs through hydrogen abstraction reactions with OH and $O(^1D)$, which initiate the methane oxidation chain. On Mars, the importance of oxidation by OH over that by $O(^1D)$ depends strongly on the season but, overall, the sum of both reactions represents the remaining 40–50% of the total loss of methane.

These processes lead to a global chemical lifetime of methane on Mars estimated by state-of-the-art models of 300–340 terrestrial years (Summers et al. 2002; Krasnopolsky et al. 2004; Lefèvre and Forget 2009). From this lifetime can be calculated the rate of globally supplied methane required to maintain a steady-state mixing ratio in the atmosphere: as mentioned above, less than 20 tons per year of methane transferred to the atmosphere are sufficient to maintain the steady-state value of about 0.5 ppbv measured by TLS. This mass may be compared with a terrestrial value on the order of 600×10^6 tons per year (Ciais et al. 2013). Can such a faint source create the variability of methane observed on Mars? Lefèvre and Forget (2009) investigated this possibility by implementing highly localized and sporadic sources with the LMD global climate model (GCM). As expected from the 300-year lifetime of methane on Mars, their results showed that methane should be homogeneously mixed by atmospheric transport across most of the planet. A striking feature of GCM simulations of methane, however, is the large enrichment in methane that results from the condensation of CO_2 gas at high latitudes in winter (Fig. 12.4). This is a well-established process, identical for all non-condensable species, as proven by observations of argon (e.g., Sprague et al. 2007) and carbon monoxide (Encrenaz et al. 2006).

Conversely, CH_4 is depleted at high-latitudes when CO_2 sublimates from the polar cap in summer and returns to the gas phase. At low latitudes, the slow seasonal modulation of methane driven by the condensation-sublimation cycle of CO_2 is greatly attenuated. This effect calculated by the LMD GCM at Gale Crater is shown in Fig. 12.5. It induces a peak-to-peak change in CH_4 of about 25% and a maximum mixing ratio obtained shortly before equinox ($L_s = 160°$). This theoretical result is here compared only to the high-precision measurements of TLS that are

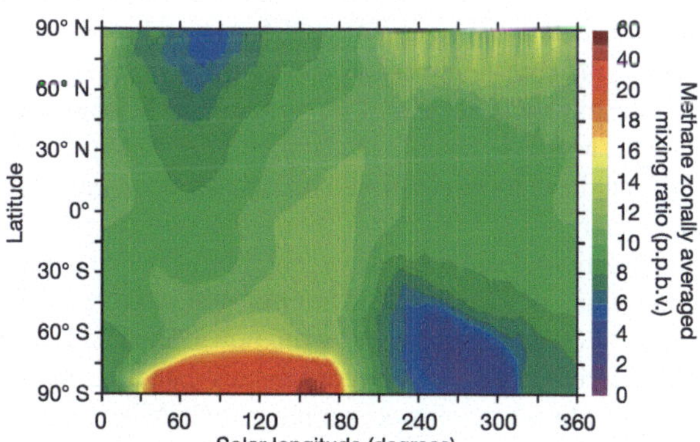

Fig. 12.4 Seasonal evolution of the zonally averaged methane mixing ratio calculated by a global climate-chemical model. The simulations include a local source at the surface near the Equator. The amount of methane released from the source balances the global photochemical loss integrated over the Martian year, assuming an equilibrium value of 10 ppbv. The lifetime of methane is 330 years as determined by conventional chemistry. Source: Adapted from Lefèvre and Forget (2009)

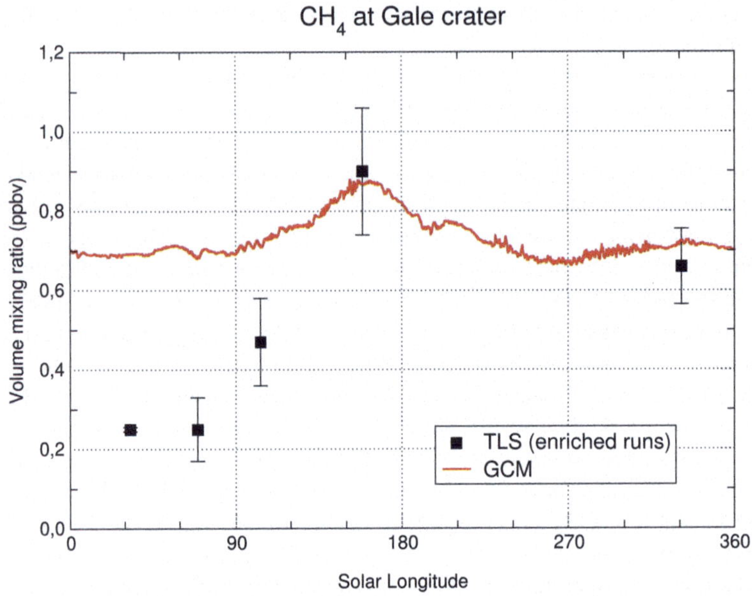

Fig. 12.5 Seasonal evolution of the background methane levels measured at Gale Crater by the TLS instrument aboard the *Curiosity* rover (only high-precision "methane enriched" runs are included) and simulated by the LMD global climate model (GCM) assuming conventional methane chemistry. The slow variation of methane visible in the model simulation is only due to the condensation-sublimation cycle of CO_2 at the Martian poles. Source: Data from Webster et al. (2015a) and Roos-Serote et al. (2016)

representative of a low background level of CH_4, which in principle should vary like those of any other non-condensable species. Figure 12.5 shows that the maximum value of background methane measured by TLS (0.9 ppbv) is also obtained near $L_s = 160°$. However, the change in background methane observed in the first half of the Martian year appears to be much stronger (by one order of magnitude) than what is expected from the GCM and from only the condensation-sublimation cycle of CO_2. This puzzling fact will have to be confirmed by new observations covering a greater part of the Martian year.

Using GCM simulations, Lefèvre and Forget (2009) showed that reproducing the Earth-based observations of methane by Mumma et al. (2009) required a methane lifetime shorter than ~200 days, and hence an unknown sink that is at least 600 times faster than the loss derived from the current kinetics data used by the atmospheric chemistry community. In what follows, we present GCM simulations of methane carried out in an attempt to reproduce the full set of TLS measurements, including the spike of 7 ppbv observed in the spring of Mars Year 32. For each model experiment, methane is continuously released from a single location chosen among nine sites located in the hemisphere centred on Gale Crater (Fig. 12.6, left).

The production of methane is assumed to be continuous, since both serpentinisation (occurring a few kilometers below the surface) and the action of

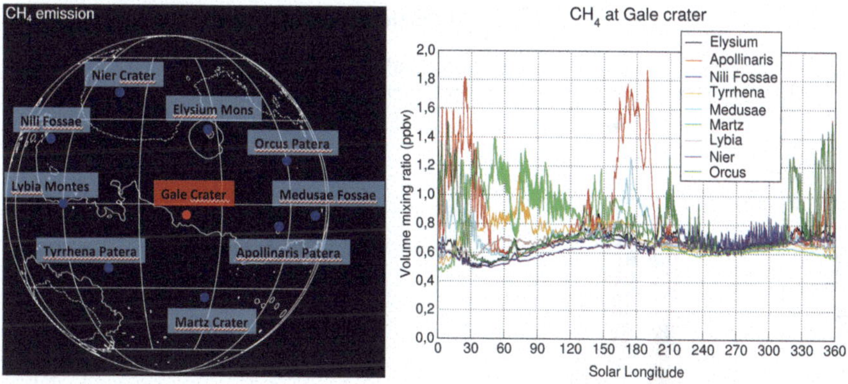

Fig. 12.6 Simulations of methane with the LMD global climate-chemical model. Left: Location of each site of emission of methane tested for the nine model experiments. In each case the emission is continuous and the methane lifetime is arbitrarily assumed to be one terrestrial year. Right: Seasonal evolution of the mixing ratio of methane obtained at Gale Crater for each of the sites of emission. Source: F Lefèvre

ultraviolet radiation on organics (widespread on the surface) are unlikely to be episodic over short timescales. In a first step, the lifetime of methane is arbitrarily assumed to be one terrestrial year. This is similar to the suggestion by Lefèvre and Forget (2009) to match the measurements of Mumma et al. (2009). In the case of the TLS measurements, a lifetime of 1 year implies an emission of 6000 tons per year in order to maintain the background level of 0.7 ppbv suggested by the instrument. Such a source would be 40 times more significant than the largest abiotic gas seep known to date on Earth (Chimaera in Turkey, Etiope and Sherwood Lollar 2013). This hypothesis is unlikely but required in order to approach the TLS measurements with model simulations. Figure 12.6 (right) shows the methane mixing ratio obtained at Gale Crater when methane is continuously released from one of the nine sites of emission. With such a short lifetime for methane (1 year), atmospheric transport can produce pulses of methane at Gale crater that are superimposed onto the slow seasonal variation due to the CO_2 condensation/sublimation cycle. The largest pulses simulated by the GCM (~1.6 ppbv) are obtained when the source is located at Apollinaris Patera, east of Gale Crater. However, they are 4 times smaller than those measured by TLS, and do not occur during the season observed by the instrument. A methane source located at Martz Crater (south of Gale Crater) provides a better result for seasonality, but requires a much larger release to match the pulse of 7 ppbv observed by TLS.

As shown in Fig. 12.7, a pulse peaking at 7 ppbv in the model simulation requires a methane lifetime of the order of 1 month in order to maintain the same background level of 0.7 ppbv. Such a short lifetime puts an enormous burden on the process that governs the destruction of methane on Mars, which must be faster than the conventional chemistry by three orders of magnitude. The fact that such a strong, unknown, chemical process would have been overlooked would be a surprise since

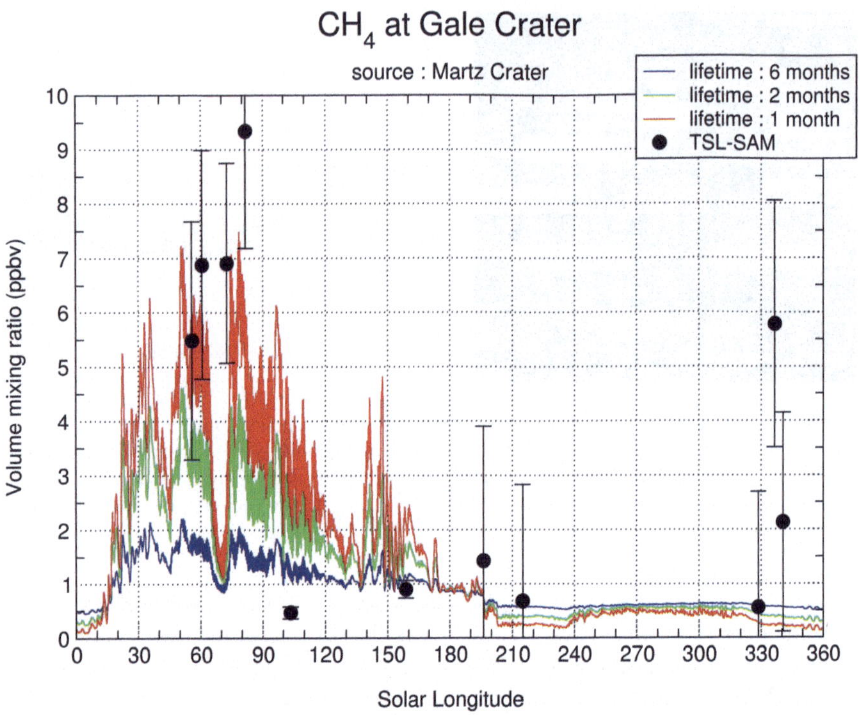

Fig. 12.7 Seasonal evolution of the mixing ratio of methane (ppbv) at Gale Crater calculated by the LMD global climate model when methane is continuously released from Martz Crater. The results are shown for arbitrary lifetimes of methane of 1, 2, and 6 months and are compared to TLS measurements by Webster et al. (2015a). Source: Adapted by Webster et al. (2015a)

conventional models do a rather good job at reproducing short-lived chemical species on Mars. For instance, species that are very sensitive to the oxidizing capacity of the atmosphere, such as hydrogen peroxide (H_2O_2) or ozone (O_3), are measured in quantities that are quite consistent with model simulations including only the standard set of chemical reactions used in Earth photochemistry (e.g., Encrenaz et al. 2015; Clancy et al. 2016).

Several studies have explored the possibility of a fast-acting sink of methane in the specific atmospheric conditions of Mars. For instance, it has been proposed that methane could be destroyed by electrochemical reactions triggered by the strong electric fields generated by local 'dust devils' or during regional-scale dust storms (Delory et al. 2006; Farrell et al. 2006). During such events, energized electrons are expected to dissociate methane directly but also to dissociate H_2O and produce vast amounts of H_2O_2. This latter effect is problematic, however, since the large-scale production of H_2O_2 by electric fields is contrary to the good agreement currently noted between observations and conventional models of H_2O_2 and O_3. Lefèvre and Forget (2009) used a GCM to demonstrate that electrochemistry should, in addition to H_2O_2, also produce large amounts of CO from the dissociation of CO_2.

Nonetheless, both of these facts are difficult to reconcile with current observations of H_2O_2 and CO, which show no significant enhancement during the dusty season. Fast destruction of methane at the global scale by electrochemical processes is therefore not supported for the moment by observations.

Atreya et al. (2006, 2007) argued that the local excesses of H_2O_2 produced in dust storms could lead to its precipitation out of the atmosphere onto the Martian surface. Methane would then be scavenged by large amounts of H_2O_2 or other super-oxides embedded in the regolith. This process has the advantage of not altering the conventional atmospheric chemistry, but it is difficult to imagine that it could affect CH_4 without processing O_3 or CO in the same fashion. Furthermore, rapid loss of methane in the regolith is, at present, not supported by laboratory work: experiments on analogues of Martian soil or perchlorates show no fast oxidation of methane in the presence of H_2O_2 (Gough et al. 2011). In addition, kinetic data on the reactions between CH_4 with metal oxides and superoxide ions are extremely slow at Martian temperatures (Krasnopolsky 2006).

Another potential explanation for the apparent variability and short lifetime of methane on Mars was presented by Knak Jensen et al. (2014), who found that wind-driven agitation of quartz crystals could produce active sites that sequester methane temporarily. Crystalline quartz is very rare on Mars but is used in their experiments as an analogue for the silicate minerals found in airborne dust or at the surface. The reverse reaction that releases CH_4 back to the atmosphere has not been studied in the laboratory and, thus, the efficiency of these idealized experiments in the real Martian atmosphere is highly uncertain. Another possibility is that methane condenses in clathrates, however, according to laboratory work (Trainer et al. 2010), the trapping of CH_4 on polar ice analogues, including clathrates, appears to be negligible under modern Martian conditions and is not a viable explanation for the apparent variability of methane on Mars.

12.4 Open Questions and Future Measurements

The detection of a very low background level (~0.5 ppbv) of methane in the atmosphere of Mars by the *Curiosity* rover raises the possibility of a planet that is biologically or geologically active now. However this detection does not constitute proof of such activity: we have seen that the production of methane required to explain the TLS observations is so minor that these results could equally be due to the slow UV-alteration of the organics delivered by inert meteoric or interplanetary dust material. In fact, for the atmospheric chemist, the variations of Martian methane pose a much greater challenge than just its presence. For robust theoretical reasons, based on knowledge well established in the context of the atmosphere of the Earth, methane should have a 300-year lifetime on Mars. Due to this long lifetime with respect to mixing by transport, methane should therefore be uniformly distributed throughout the atmosphere. Similarly to other long-lived species observed on Mars (such as argon, or carbon monoxide), methane variations should only be controlled by the slow condensation/sublimation cycle of CO_2, with peak-to-peak amplitudes of about 30% over one Martian year. Yet, the observations of Martian methane by TLS suggest variations of up to one order of magnitude in less than 2 months. This implies not only a local methane source situated by extraordinary coincidence very close to the *Curiosity* rover, but also that an unknown and extraordinarily fast mechanism of methane destruction is at work. Because this finding strongly challenges our current understanding of the atmospheric chemistry and physics of Mars, the discovery of variable amounts of methane needs to be confirmed by further detections. In particular, the question "is methane variable on Mars?" must now be addressed by mapping methane at the global scale, with a sensitivity better than 1 ppbv. Evidently the answer to this question is a prerequisite before tackling the following ones, such as "where are the emission sites of methane?", "how is methane destroyed?", and ultimately "how is methane produced?". Confirming the presence and variability of methane on Mars is one of the objectives of ESA-Roscomos Trace Gas Orbiter (TGO), which launched in March 2016 and arrived at Mars in October 2016. After a long aerobraking phase, the nominal science operations of TGO will have started in 2018. On board the platform, two different instruments, the Atmospheric Chemistry Suite (ACS) (Korablev et al. 2018) and NOMAD (Vandaele et al. 2015), will use the solar occultation technique to search for methane at all locations and seasons with a sensitivity hopefully better than 0.1 ppbv. It is therefore likely that a major advance is coming in our effort to unravel an enigma that may impact Mars science and astrobiology in a fundamental way.

References

Atreya SK, Wong AS, Renno NO et al (2006) Oxidant enhancement in Martian dust devils and storms: implications for life and habitability. Astrobiology 6:439–450

Atreya SK, Mahaffy PR, Wong AS (2007) Methane and related trace species on Mars: origin, loss, implications for life and habitability. Planet Space Sci 55:358–369

Ciais P, Sabine C, Bala G et al (2013) Carbon and other biogeochemical cycles. In: Stocker TF, Qin D, Plattner GK et al (eds) Climate change 2013: the physical science basis. Contribution of working group I to the fifth assessment report of the intergovernmental panel on climate change. Cambridge University Press, Cambridge, pp 465–570

Clancy RT, Sandor BJ, Wolff MJ et al (2000) An intercomparison of ground-based millimeter, MGS TES, and viking atmospheric temperature measurements: seasonal and interannual variability of temperatures and dust loading in the global Mars atmosphere. J Geophys Res 105:9553–9572

Clancy RT, Wolff MJ, Lefèvre F et al (2016) Daily global mapping of Mars ozone column abundances with MARCI UV band imaging. Icarus 266:112–133

Court RW, Sephton MA (2009) Investigating the contribution of methane produced by ablating micrometeorites to the atmosphere of Mars. Earth Planet Sci Lett 288:382–385

Delory GT, Farrell WM, Atreya SK et al (2006) Oxidant enhancement in Martian dust devils and storms: electric fields and electron dissociative attachment. Astrobiology 6:451–462

Encrenaz T, Fouchet T, Melchiorri R et al (2006) Seasonal variations of the martian CO over Hellas as observed by OMEGA/Mars Express. Astron Astrophys 459:265–270

Encrenaz T, Greathouse TK, Lefèvre F et al (2015) Seasonal variations of hydrogen peroxide and water vapor on Mars: further indications of heterogeneous chemistry. Astron Astrophys 578: A127
Etiope G, Sherwood Lollar B (2013) Abiotic methane on Earth. Rev Geophys 51:276–299
Farrell WM, Delory GT, Atreya SK (2006) Martian dust storms as a possible sink of atmospheric methane. Geophys Res Lett 33:L21203
Fonti S, Marzo GA (2010) Mapping the methane on Mars. Astron Astrophys 512:A51
Formisano V, Atreya SK, Encrenaz T et al (2004) Detection of methane in the atmosphere of Mars. Science 306:1758–1761
Geminale A, Formisano V, Sindoni G (2011) Mapping methane in martian atmosphere with PFS-MEX data. Planet Space Sci 59:137–148
Gough RV, Turley JJ, Ferrell GR et al (2011) Can rapid loss and high variability of Martian methane be explained by surface H_2O_2? Planet Space Sci 59:238–246
Kelley DS, Karson JA, Fruh-Green GL et al (2005) A serpentinite-hosted ecosystem: the lost city hydrothermal field. Science 307:1428–1434
Keppler F, Vigano I, McLeod A et al (2012) Ultraviolet-radiation-induced methane emissions from meteorites and the Martian atmosphere. Nature 486:93–96
Knak Jensen SJ, Skibsted J, Jakobsen HJ et al (2014) A sink for methane on Mars? The answer is blowing in the wind. Icarus 236:24–27
Korablev O, Montmessin F, Trokhimovskiy A et al (2018) The atmospheric chemistry suite (ACS) of three spectrometers for the ExoMars 2016 Trace Gas Orbiter. Space Sci Rev 214:7
Krasnopolsky VA (2006) Some problems related to the origin of methane on Mars. Icarus 180: $32#359–367
Krasnopolsky VA (2012) Search for methane and upper limits to ethane and SO_2 on Mars. Icarus 217:144–152
Krasnopolsky VA, Maillard JP, Owen TC (2004) Detection of methane in the Martian atmosphere: evidence for life? Icarus 172:537–547
Lefèvre F, Forget F (2009) Observed variations of methane on Mars unexplained by known atmospheric chemistry and physics. Nature 460:720–723
Mumma MJ, Villanueva GL, Novak RE et al (2009) Strong release of methane on Mars in northern summer 2003. Science 323:1041–1045
Oze C, Sharma M (2005) Have olivine, will gas: serpentinization and the abiogenic production of methane on Mars. Geophys Res Lett 32:L10203
Roos-Serote M, Atreya SK, Webster CR et al (2016) Cometary origin of atmospheric methane variations on Mars unlikely. J Geophys Res 121:2108–2119
Sprague AL, Boynton WV, Kerry KE et al (2007) Mars' atmospheric argon: tracer for understanding Martian atmospheric circulation and dynamics. J Geophys Res 112:E03S02
Summers ME, Lieb BJ, Chapman E et al (2002) Atmospheric biomarkers of subsurface life on Mars. Geophys Res Lett 29:24
Trainer MG, Tolbert MA, McKay CP et al (2010) Limits on the trapping of atmospheric CH_4 in Martian polar ice analogs. Icarus 208:192–197
Vandaele A, Neefs E, Drummond R et al (2015) Science objectives and performances of NOMAD, a spectrometer suite for the ExoMars TGO mission. Planet Space Sci 119:233–249
Villanueva GL, Mumma MJ, Novak RE et al (2013) A sensitive search for organics (CH_4, CH_3OH, H_2CO, C_2H_6, C_2H_2, C_2H_4), hydroperoxyl (HO_2), nitrogen compounds (N_2O, NH_3, HCN) and chlorine species (HCl, CH_3Cl) on Mars using ground-based high-resolution infrared spectroscopy. Icarus 223:11–27
Webster CR, Mahaffy PR, Atreya SK et al (2013) Low upper limit to methane abundance on Mars. Science 342:355–357
Webster CR, Mahaffy PR, Atreya SK et al (2015a) Mars methane detection and variability at Gale crater. Science 347:415–427

Webster CR, Mahaffy PR, Atreya SK, et al (2015b) Mars methane detection and variability at Gale Crater measured by the TLS instrument in SAM on the Curiosity rover. Abstract P438-2110, Fall meeting AGU
Webster CR, Mahaffy PR, Atreya SK (2018) Background levels of methane in Mars' atmosphere show strong seasonal variations. Science 360:1093–1096
Zahnle K, Freedman RS, Catling DC (2011) Is there methane on Mars? Icarus 212:493–503

Chapter 13
Detection of Biosignatures Using Raman Spectroscopy

Frédéric Foucher

Abstract Raman spectroscopy is particularly suited for the study of biosignatures: it is able to detect both organic and mineral phases, is very sensitive to carbonaceous matter and biogenic pigments, and can be used in the field and for space exploration. Thus, in a few decades it has become a key method in (micro-)palaeontology, geomicrobiology and astrobiology. In this chapter, we present an overview of the different types of biosignatures that can be detected and/or characterized using Raman spectroscopy: organic molecules, microfossils, biominerals or even living cells. A particular focus is made on the role of the excitation laser wavelength on the type of biosignatures that can be studied.

13.1 Introduction

The term biosignature refers to any direct or indirect evidence of active or past life. It includes living organisms and their fossils, as well as organic compounds of biological origin, biominerals and biogases produced by metabolic activity, or physical structures created by living organisms, such as shells or stromatolites. The detection of biosignatures is of primary importance for micropalaeontology to demonstrate the biogenicity of a fossilized structure and for astrobiology, where the objective is the search for past or present extraterrestrial life. Detection of biosignatures involves the use of a large range of techniques and instruments to make observations, determine elementary and molecular composition, date the structures, and characterise the geochemical environment.

Raman spectroscopy is a versatile technique that uses a laser to detect and identify molecules and crystals. In the laboratory, the instrument can be equipped with a scanning device permitting it to display composition over a selected area of analysis. Raman instruments generally use optical microscope objectives and are confocal in order to carry out compositional mapping in 2D and 3D, from the centimetre to

F. Foucher (✉)
Centre de Biophysique Moléculaire-CBM, CNRS, Orléans, France
e-mail: frederic.foucher@cnrs.fr

B. Cavalazzi, F. Westall (eds.), *Biosignatures for Astrobiology*, Advances in Astrobiology and Biogeophysics, https://doi.org/10.1007/978-3-319-96175-0_13

sub-micrometre scale (Deing et al. 2010; Foucher et al. 2017). Miniaturized, it can be used for field investigation and planetary exploration (Culka et al. 2011, 2012; Edwards et al. 2013). It will be a key instrument during the future missions to Mars *ExoMars 2020* (ESA-Roscosmos), with the *Raman Laser Spectrometer, RLS*, (Rull-Pérez and Martinez-Frias 2006; Lopez-Reyes et al. 2013), and *Mars 2020* (NASA), with *SHERLOC* (for *Scanning Habitable Environments with Raman & Luminescence for Organics & Chemicals*) and *SuperCam* (Beegle et al. 2014).

Raman spectroscopy is well suited for studying biosignatures but, depending on the system used (field or laboratory instrument) and on the excitation laser wavelength, the types of biosignatures that can be detected may vary. In particular, luminescence of the sample may mask the Raman signal, the laser may heat and alter the sample, or the embedding phase (e.g. the mineral matrix) may be opaque at certain wavelengths.

In this chapter, the Raman effect is explained first. In particular, the focus is placed on the advantages and disadvantages of the different laser wavelengths from deep ultra violet (UV) to infrared (IR) on instrumentation and biosignature detection. The different types of biosignatures that can be detected using Raman spectroscopy are then described, from biominerals to living organisms.

13.2 Raman Effect and Instrumentation

The Raman effect was described for the first time by Chandrasekhara Venkata Rāman in 1928. It corresponds to the inelastic scattering of photons leading to atomic bond vibrations. This phenomenon requires advanced physics to be fully described, thus, this chapter will only briefly explain the effect. However, full descriptions can be found in several books such as Poilblanc and Crasnier (2006), Deing et al. (2010) or Dubessy et al. (2012).

Photons in the UV to IR range of the electromagnetic spectrum may transfer energy to molecules or crystals as vibrations or as electronic transitions, depending on their energy (see Fig. 13.1).

The vibrational energy transitions involved in the Raman effect are in the order of magnitude of a few tenths of eV, while the energy between the excited electronic states and the ground electronic state is generally of about ~2 eV. According to the Planck-Einstein equation,[1] it is easy to convert wavelength into energy. Thus, IR electromagnetic radiation extending from 700 nm to 1 mm corresponds to photon energies ranging from 1.77 to 0.001 eV, respectively (i.e. from 10 to 15,000 cm^{-1}). Photons in the range 0.06–0.5 eV (i.e. 2.5–20 μm or 500–4000 cm^{-1}) may then be absorbed by inducing ground electronic vibrational state transitions. This phenomenon is used for IR spectroscopy. In the visible range, from 400 to 700 nm (1.77–3.10 eV respectively), a

[1]The Planck-Einstein equation is given by $E = h.c/\lambda$ with E the energy, h the Plank constant, c the speed of light in vacuum, and λ the wavelength.

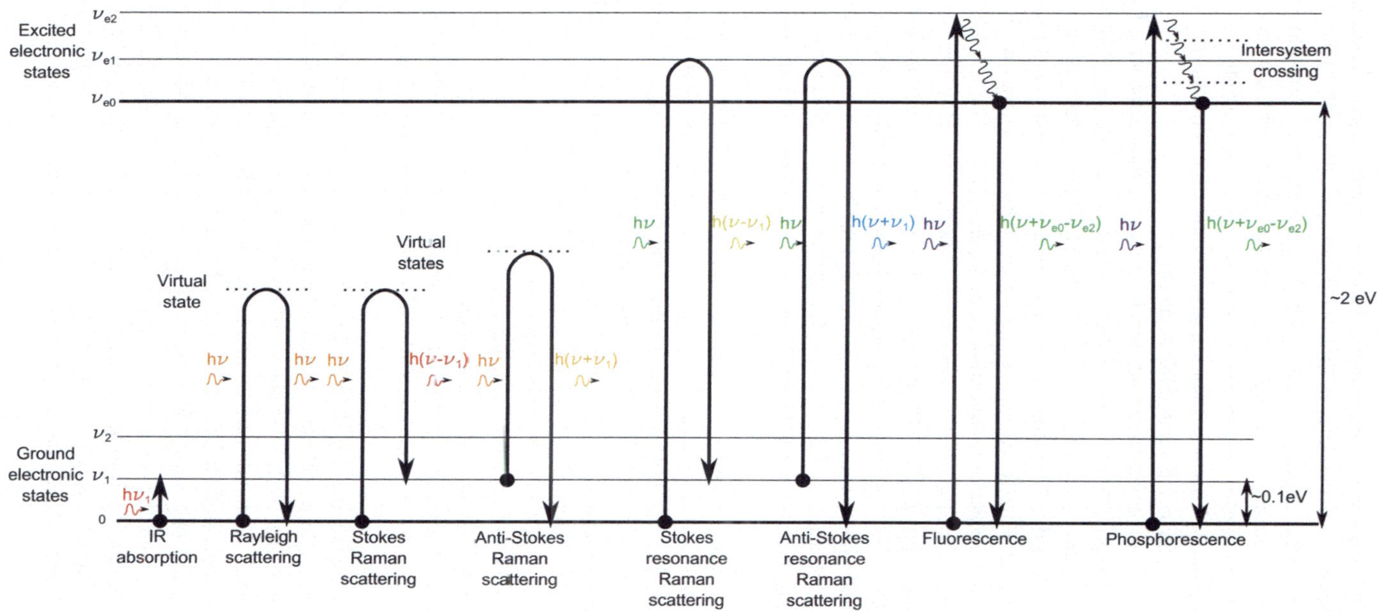

Fig. 13.1 Non-exhaustive possible photon-matter interactions in the IR to UV range of the electromagnetic spectrum. Source: F Foucher

photon could either be elastically scattered (Rayleigh effect), i.e. scattered with no loss of energy, or inelastically scattered (Raman effect), i.e. scattered with change of energy. If the photon loses energy, it undergoes Stokes Raman scattering and, on the contrary, if the photon gains energy (which is less probable), it undergoes Anti-Stokes Raman scattering. With decreasing wavelength, from about 200 to 550 nm (6.2–2.25 eV respectively), the energies of the photons are in the order of magnitude of those of excited electronic states. This may then lead to luminescence (fluorescence and phosphorescence) and resonance Raman (RR) scattering.

Fluorescence occurs when molecules reach an excited electronic state by absorbing the incident photons, de-exciting progressively by non-radiative transitions, and finally returning to the electronic ground state by emitting photons of lower energy (i.e. higher wavelength). The phosphorescent effect is similar but slower due to intersystem crossing. Finally, RR scattering occurs when the energy of the incident photons is close to an electronic transition of particular atomic bonds leading then to a resonant effect (i.e. an increase in probability of occurrence of Raman scattering). RR is characterized by strong enhancement of the intensity of certain bands of the Raman signal (by several orders of magnitude). Raman scattering is very short (~ns) in comparison with fluorescence (~μs to ms) and phosphorescence (~s to min) where the photons are absorbed and re-emitted. Raman intensity is also very weak compared to luminescence but the RR signal can be of the same order of magnitude or even higher.

The Raman effect is sensitive to atomic bonds. It permits the discrimination of polymorphic minerals, i.e. minerals having similar compositions but different crystalline structures. For example, anatase, rutile and brookite have different Raman signals despite their identical elemental composition (TiO_2).

Modern Raman spectroscopy is performed using a monochromatic light source (laser) to induce the effect. The scattered signal is diffracted using various gratings and collected using a CCD camera interfaced with the spectrometer. The collected signal consists of a spectrum displaying the number of scattered photons *versus* their wavenumber, i.e. the shift in cm^{-1} with respect to the incident beam, the excitation laser wavelength corresponding to 0 cm^{-1}. Identification of minerals or organic compounds is made by comparison with reference spectra found in the literature or in databases.

Raman spectra extend from 0 to 4000 cm^{-1} irrespective of laser wavelength. However, their range in wavelength increases with increasing excitation laser wavelength, as shown in Fig. 13.2a. The choice of laser is crucial depending on the target. For instance, organic molecules are known to be particularly fluorescent. A deep UV laser at 250 nm appears therefore to be a good solution, the Raman spectra being collected before the beginning of luminescence (Fig. 13.2a). Moreover, a short wavelength will favour the RR effect. However, in addition to being expensive and difficult to implement, UV Raman spectroscopy is limited by the depth of penetration of UV which is generally very small in solids (only a few nm), and by the possibility of burning the sample. On the other hand, the IR laser permits collection of the signal

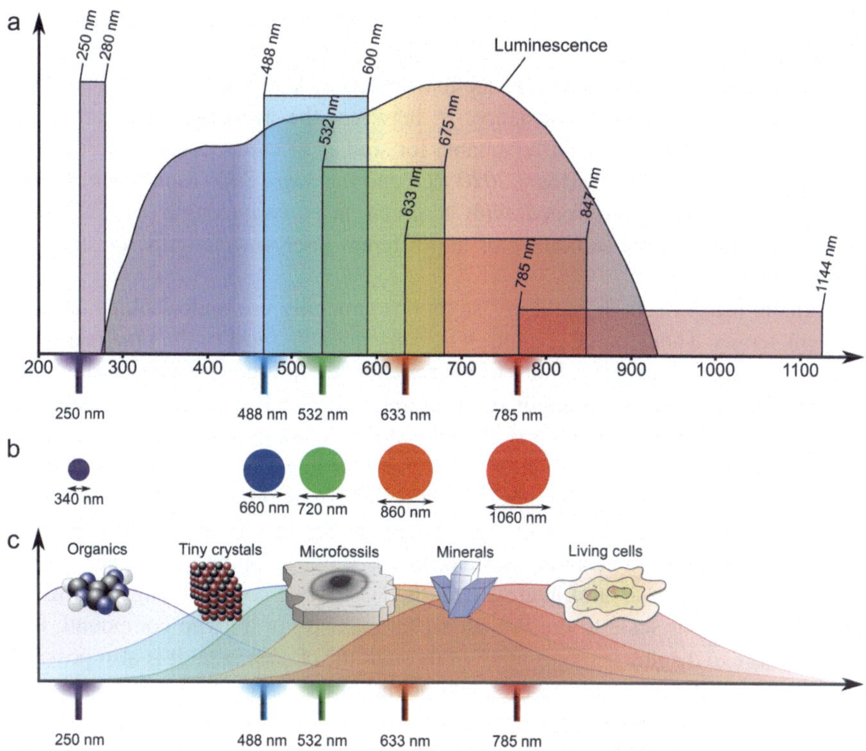

Fig. 13.2 (**a**) General luminescence envelope and Raman spectrum ranges (corresponding to 0–4000 cm^{-1}) for different excitation laser wavelengths (adapted from Beegle et al. 2014). (**b**) Laser spot size through an objective of numerical aperture 0.9. (**c**) Suitability of the different laser wavelengths to be used to study biosignatures. Source: Adapted from Beegle et al. (2014)

after the fluorescence range (Fig. 13.2a) with less risk of burning the sample. However, CCD detectors are less sensitive in the IR range and therefore the Raman signal is less intense, particularly after 900 nm (the use of 1064 nm IR excitation laser requires the use of FT-Raman systems). In order to compensate for this low signal gain it may sometimes be necessary to use high laser power, thus, increasing the risk of burning the sample. Moreover, following the Airy disk principle,[2] the laser spot diameter increases with the wavelength (Fig. 13.2b).

Finally, the depths of field and of penetration are also increased in such a way that the analysed volume may then be relatively large (several μm^3) thus requiring a larger amount of material in order to be detected.[3]

[2]The Airy disk corresponds to the best focused spot of light through an optical system. Its diameter is given by $D = 1{,}22.\lambda/NA$, with NA the numerical aperture of the objective.

[3]The depth of field of an objective is given by $\Delta z = n.\lambda/(2.NA^2)$, with n the refractive index of the material.

To summarise, a UV laser is well suited for the detection of organic molecules adsorbed on the surface and to carry out high-resolution surface analyses. An IR laser allows performing Raman at depth with less risk of burning; it is thus well suited for living tissues, for instance. In between, the available lasers are good compromises and are particularly suitable for studies of carbonaceous microfossils in a mineral matrix. The *ExoMars 2020 RLS* and the *Mars 2020 SuperCam* Raman systems will thus be equipped with a green laser (wavelength of 532 nm). Figure 13.2c shows the suitability of the different laser wavelengths for studying biosignatures.

Portable instruments for field investigations generally use optical fibres or macroscopic lenses. Their spatial resolution is, consequently, most of the time relatively low with a spot size of several tens of micrometres to millimetres (Mosier-Boss and Putnam 2013). The miniaturization of the systems also leads to a decrease in spectral resolution (several cm^{-1}) (Culka et al. 2011, 2012; Vitek et al. 2012). For practical reasons they generally use high laser wavelength, from 532 to 1064 nm (Vitek et al. 2012). By contrast, most laboratory systems are based on an optical microscope architecture, which allows significant increase in the resolution of analyses and, when interfaced with a scanning system, to carry out mapping. A Raman map is made by scanning an area of interest with the laser while accumulating spectra. By attributing a colour scale to the Raman signal intensity of a given compound, it is then possible to display its concentration through the chosen area. It is also possible to attribute a different colour to each compound in order to obtain a compositional map. More information on mapping technique can be found in Deing et al. (2010) or Foucher et al. (2017). Finally, instruments for space exploration can be seen as intermediate systems in terms of spatial and spectral resolution and in terms of capacities (quite good resolution but no mapping, for instance).

13.3 Biosignatures Raman Detection

13.3.1 Organic Molecules

Astrobiology mainly focusses the search for life on auto-replicating systems capable of Darwinian evolution and based on organic chemistry in liquid water (Lazcano 2011; Ruiz-Mirazo and Moreno 2011). All known living systems on Earth are indeed more than 95% based on the elements C, H, N, O, P and S, and this chemistry appears to be the only way to form enough various and complex molecules to enable life to appear and evolve (Brack 2001).

Organic molecules can be identified by Raman spectroscopy and various spectral databases are available. Examples of typical spectra of organic molecules are displayed in Fig. 13.3 (De Gelder et al. 2007).

A large variety of organic molecules have been found in the interstellar medium (Dickens et al. 2001), icy dust particles (Danger et al. 2013; de Marcellus et al. 2015; Meinert et al. 2016), comets (Altwegg et al. 2016) and meteorites (Schmitt-Kopplin

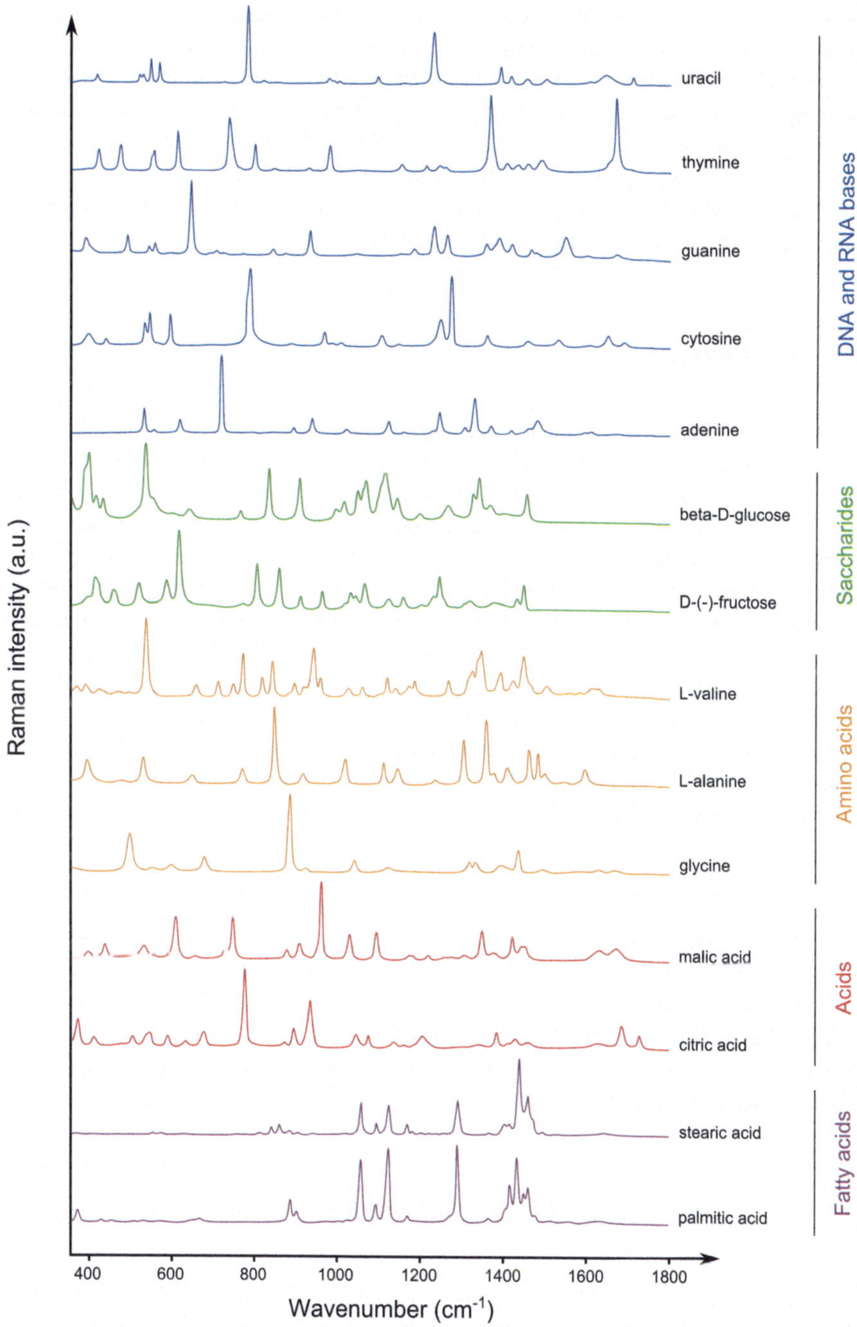

Fig. 13.3 Raman spectra of some biological molecules acquired on pure compounds using a 785 nm excitation laser wavelength. Source: Adapted De Gelder et al. (2007)

et al. 2010) demonstrating that they can be abiotically synthesised. Life uses a more restricted range of organic molecules, some of which are considered as uncontroversial evidence of life, such as DNA, RNA, and proteins. NASA thus took up the challenge to send a Raman system dedicated to the search for organics on Mars during the *Mars2020* mission: the *SHERLOC* instrument. The system will use a deep UV laser permitting it to obtain (resonant) Raman spectra below the fluorescence wavelength range (see Sect. 13.2), as well as make luminescence analysis in the visible wavelength range (Beegle et al. 2014).

13.3.2 Pigments

A particular focus will be made here on pigments. Pigments are molecules that contain at least one chromophore in their structure, i.e. a chain of alternative single and double carbon bonds. By selecting the excitation laser wavelength in order to excite the π-π* electronic transition energy range in conjugated chromophores, such as those of β-carotene, or the charge transfer transition energy in metal complex chromophores, such as in chlorophyll, it is possible to strongly enhance the Raman signal by the resonance effect (RR) of the stretching modes of the π-bonds, or of the stretching modes of the metal-ligand respectively (Merlin 1985). Among these molecules, carotenoids are those associated with the strongest RR effect when a green excitation laser is employed. The signal is still strong enough to be detected out of resonance using a 785 nm excitation laser wavelength (Merlin 1985; Jehlička et al. 2009; Vitek et al. 2009).

The RR spectrum of β-carotene is displayed in Fig. 13.4. Carotenoids are relatively common biological pigments. They act as DNA-repair agents in radiation-damaged cells, protect against UV radiation and absorb light energy used for photosynthesis (Patel et al. 2004). They are found in many organisms, such as plants, algae, bacteria and archaea. The signal of carotenoids is thus commonly observed in the field using portable instrumentation. This explains why Raman studies focussed on the detection of these molecules on Mars are numerous (e.g., Edwards et al. 2013; Baqué et al. 2016; Jehlička et al. 2016). In particular, it has been shown that β-carotene can be detected in low concentrations when mixed within a mineral matrix (Vandenabeele et al. 2012). These molecules are, thus, very interesting targets for astrobiology.

13.3.3 Carbonaceous Matter and Microfossils

Carbonaceous matter as graphite or disordered sp^2 carbon is of particular interest for Raman spectroscopy since it is always resonant, whatever the excitation laser wavelength (Ferrari 2007). The technique is thus particularly suited for the study of kerogens, i.e. insoluble organic matter of biotic origin. The typical Raman spectrum of kerogen exhibits two main bands generally labelled D, for disordered, and G, for

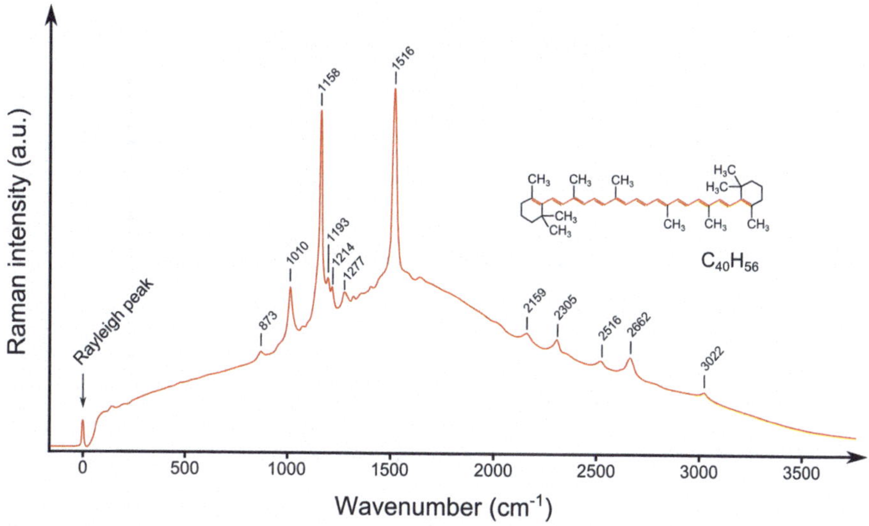

Fig. 13.4 Unprocessed Raman spectrum of pure β-carotene powder ($C_{40}H_{56}$) obtained with a 532 nm excitation laser wavelength. The RR signal is so strong that the filtered Rayleigh peak is very small in comparison and that the luminescence background is not a problem for peak detection. The chromophore is the red section in the inset molecule of β-carotene. Source: F Foucher

graphite, located respectively around ~1350 cm^{-1} and ~1600 cm^{-1} (Beyssac et al. 2002, 2003; Ferrari 2007; Foucher et al. 2015; Jehlička and Bény 1999; Jehlička et al. 2003; Lahfid et al. 2010; Quirico et al. 2009; Sforna et al. 2014). The shape of the spectrum changes with increasing metamorphism (mainly temperature) first by carbonization, then by graphitization as displayed in Fig. 13.5 (Deldicque et al. 2016; Foucher et al. 2015; Rouzaud and Oberlin 1989; Schopf et al. 2005).

Carbonization refers to the conversion of organic molecules into kerogen and is characterized by the formation of pure polyaromatic carbons forming small coherent domains. During graphitization, these coherent domains increase until, in the case of high-grade metamorphism, they form pure graphite (amphibolite facies and higher) (Bustin et al. 1995).

The Raman spectrum of kerogen was first proposed by Pflug and Jaeschke-Boyer (1979) as a tool for proving the biogenicity of carbonaceous matter in ancient sediments and then by Schopf et al. (2002a). However, spectral shapes similar to those described in these studies were observed in carbonaceous matter of abiotic origin (in meteorites, for example), thus leading to a strong debate about the origin of carbonaceous structures observed in ancient rocks (Brasier et al. 2002; Marshall et al. 2011, 2012; Marshall and Marshall 2013; Pasteris and Wopenka 2002, 2003; Schopf et al. 2002a, b). Finally, it is now acknowledged that, although the high sensitivity of Raman spectroscopy to carbonaceous matter makes it the best technique for detecting kerogens in geological samples, the shape of the Raman spectrum alone cannot be used as a proof of biogenicity.

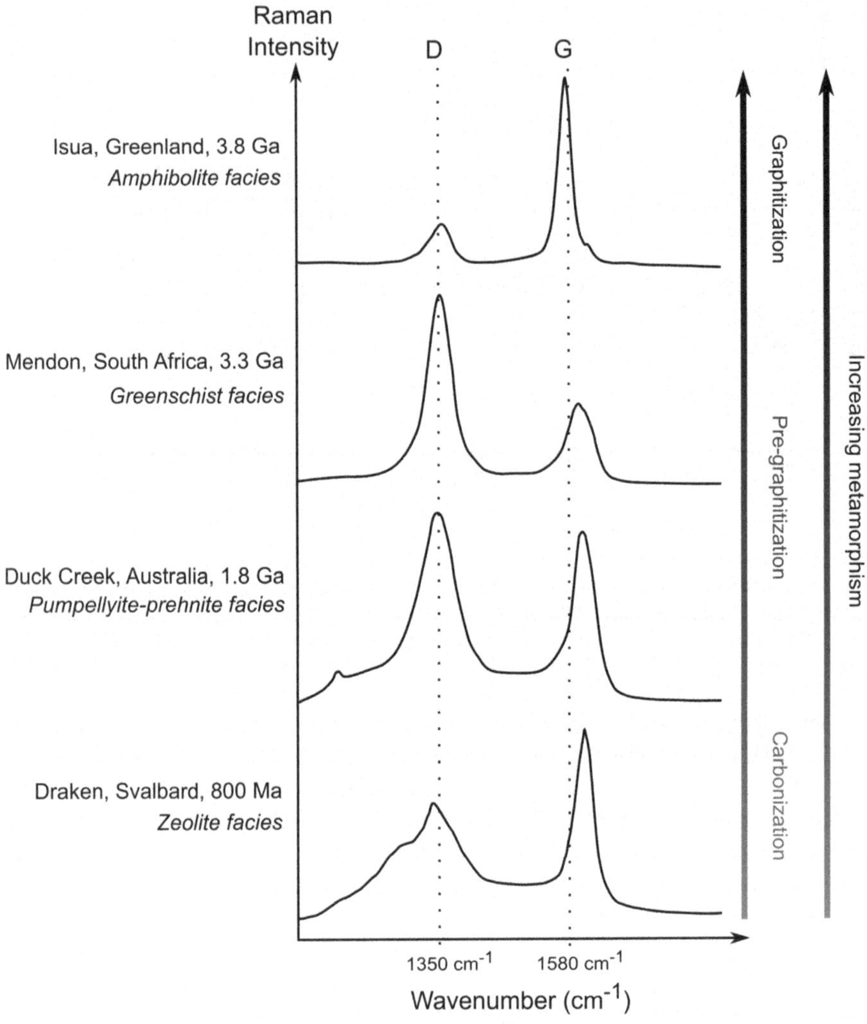

Fig. 13.5 Raman spectrum of kerogen with increasing metamorphism (temperature) obtained using a 532 nm excitation laser wavelength. Source: Adapted from Foucher et al. (2015)

On the other hand, the Raman mapping technique was recently used to demonstrate that, due to the variation of the molecular composition over a cell, and thus of the precursor of the carbonaceous matter, the spatial distribution of the changes in the D/G band intensity ratio follows the shape of the biotic structures (Foucher et al. 2015; Qu et al. 2015). By contrast, they are randomly distributed for abiotic structures, as shown in Fig. 13.6. These changes can be explained by the survival of particular functional groups in the kerogen associated with different parts of the fossil (Qu et al. 2015; Alleon et al. 2016).

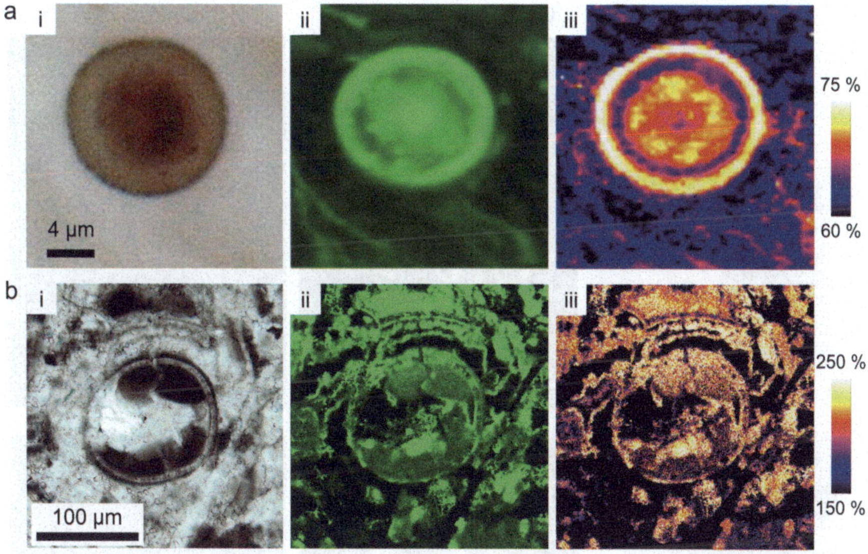

Fig. 13.6 Raman mapping of carbonaceous structures of (**a**-i, -ii, -iii) biotic and (**b**-i, -ii, -iii) abiotic origin. For each: (i) optical image, (ii) Raman map of carbonaceous matter and (iii) Raman map of the D/G peak intensity ratio obtained using a 532 nm excitation laser wavelength. Source: Adapted from Foucher et al. (2015)

13.3.4 Minerals and Microfossils

Although life is based on organic chemistry, it may also produce mineral compounds (biominerals, see Chap. 6). Large organisms biomediate the production of mineral components, such as bones, teeth or shells.

Microorganisms also produce minerals, either directly as a part of themselves, such as the frustules of diatoms, or indirectly as a result of their metabolic activities (precipitation of carbonates in the phototrophic layers forming stromatolites, for example). Contrary to organic matter that is rapidly recycled and reprocessed, mineral phases are less degraded with time making these types of biosignatures more susceptible to preservation on geological time scales and final detection in ancient sediments. On the other hand, most biominerals are known to be metastable, i.e. with metamorphism they tend to recrystallize into a more stable crystalline form. The same minerals can also form abiotically. Mineral structures must thus be placed in their mineralogical and environmental context to be considered as relevant biosignatures. Their shape (e.g. as a shell or stromatolite) and/or their association with carbonaceous matter are also very important (Rividi et al. 2010; Campbell et al. 2015).

Raman spectroscopy is particularly suited to study fossilized microorganisms since it is able to detect both carbonaceous matter and minerals at a (sub-)micrometre scale. Raman mapping is very powerful in its ability to display the relationships of mineral distributions in sediments with structures of biological origin. For instance, it has been shown that silicified carbonaceous microfossils in chert can be associated

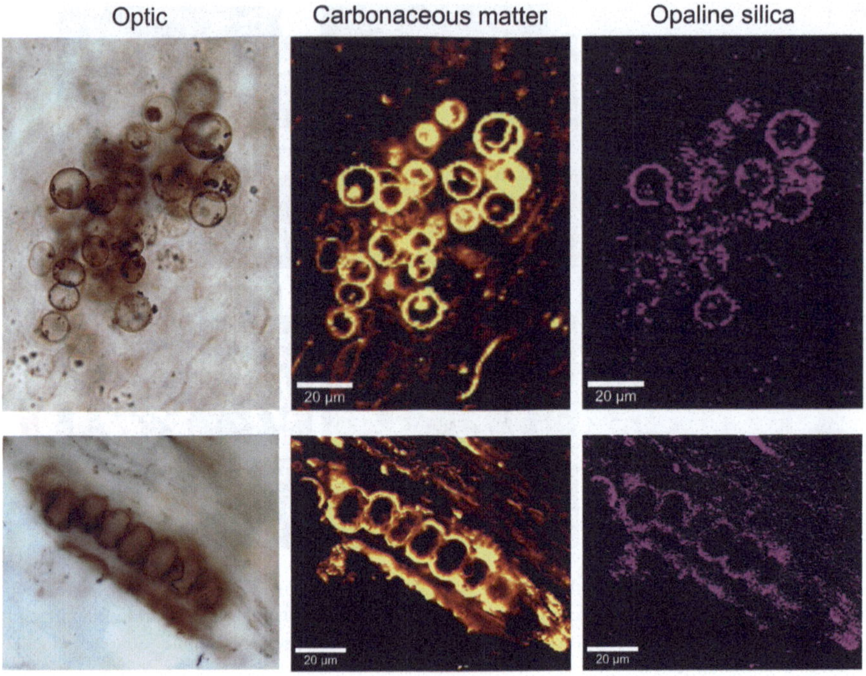

Fig. 13.7 Microfossils from the 800 Ma old Draken Formation (Svalbard) seen by optical microscopy in transmitted light, and associated Raman maps of carbonaceous matter and opaline silica obtained using a 532 nm excitation laser wavelength. Source: Adapted from Foucher and Westall (2013)

with amorphous silica despite the recrystallization of the silica matrix into quartz (Moreau and Sharp 2004) and this can be evidenced by Raman mapping, as shown in Fig. 13.7 (Foucher and Westall 2013). This interesting discovery is, however, limited to rocks of low grade metamorphism and requires the use of high resolution mapping techniques that are presently incompatible with space exploration.

13.3.5 Living Organisms

The holy grail of astrobiology remains the discovery of living extraterrestrial organisms and Raman spectroscopy could be particularly pertinent for this purpose. Among the techniques compatible with space exploration, it is one of the few capable of detecting living microorganisms at the surface of an extraterrestrial body without any sample preparation. The Raman spectrum of biological material consists of a superimposition of spectra of organic molecules, such as those described above in "organic molecules". Since biological materials are generally fluorescent, it is better

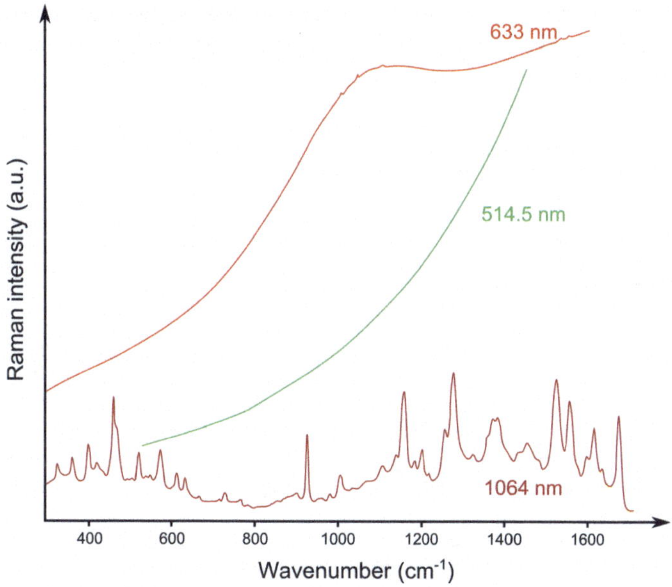

Fig. 13.8 Raman spectra of an Antarctic epilithic lichen, *Caloplaca saxicola*, from Beacon Sandstone, Mars Oasis, Antarctica, obtained using different laser wavelengths: 514.5, 633 and 785, and 1064 nm using FT Raman. Source: Adapted from Edwards (2004)

to use an IR excitation laser to detect living organisms, in particular in their natural environment as shown by Edwards (2004) (Fig. 13.8).

Observation of the Raman bands of organics in the range 400–1600 cm^{-1}, is generally difficult using a visible excitation laser, as displayed in Fig. 13.3. However, it is possible to detect the strong signal of CH and OH bands, in the 2800–3300 cm^{-1} and 3100–3650 cm^{-1} spectral regions, respectively, as well as the RR signal of pigments described above in "Pigments" (Edwards et al. 2013; Baqué et al. 2016; Jehlička et al. 2016).

13.4 Summary

Raman spectroscopy is particularly well-suited to the study of biosignatures, spanning a range from degraded biomolecules to living organisms. However, use of an inappropriate excitation laser wavelength may hamper the analyses. Most laboratory systems are equipped with several lasers, but this is not the case for the miniaturised system used for space exploration. In this case, the excitation wavelength must be chosen carefully in accordance with the aims of the analyses carried out and the corresponding targets (minerals, organics…). In any case, the high sensitivity of Raman spectrometers to carbonaceous matter will make it a key instrument for the detection of potential microfossils during the next *in situ* missions to Mars. In addition, with its capacity to identify minerals and to detect pigments, we can count on this technique to make important discoveries in astrobiology and planetology in the coming decades.

Acknowledgements I thank the *Centre National d'Etudes Spatiales* for funding. I thank Frances Westall and Keyron Hickman-Lewis for their useful comments.

References

Alleon J, Bernard S, Guillou CL et al (2016) Molecular preservation of 1.88 Ga Gunflint organic microfossils as a function of temperature and mineralogy. Nat Commun 7:11977

Altwegg K, Balsiger H, Bar-Nun A et al (2016) Prebiotic chemicals-amino acid and phosphorus-in the coma of comet 67P/Churyumov-Gerasimenko. Sci Adv 2(e1600285):1–5

Baqué M, Verseux C, Böttger U et al (2016) Preservation of biomarkers from cyanobacteria mixed with mars like regolith under simulated Martian atmosphere and UV flux. Orig Life Evol Biosph 46:289–310

Beegle LW, Bhartia R, DeFlores L et al (2014) SHERLOC: scanning habitable environments with raman & luminescence for organics & chemicals, an investigation for 2020. In: 45th Lunar and planetary science conference, Abstract 2835

Beyssac O, Goffé B, Chopin C et al (2002) Raman spectra of carbonaceous material in metasidiments: a new geothermometer. J Metamorph Geol 20:859–871

Beyssac O, Goffé B, Petitet J-P et al (2003) On the characterization of disordered and heterogeneous carbonaceous materials by Raman spectroscopy. Spectrochim Acta Part A 59:2267–2276

Brack A (2001) Water the spring of life. In: Baumstark-Khan C, Horneck G (eds) Astrobiology: the quest for the conditions of life. Springer, New York, pp 79–88

Brasier MD, Green OR, Jephcoat AP et al (2002) Questioning the evidence for earth's oldest fossils. Nature 416:76–81

Bustin RM, Ross JV, Rouzaud JN (1995) Mechanisms of graphite formation from kerogen: experimental evidence. Int J Coal Geol 28:1–36

Campbell KA, Lynne BY, Handley KM et al (2015) Tracing biosignature preservation of geothermally silicified microbial textures into the geological record. Astrobiology 15:858–882

Culka A, Jehlička J, Vandenabeele P et al (2011) The detection of biomarkers in evaporite matrices using a portable Raman instrument under alpine conditions. Spectrochim Acta Part A 80:8–13

Culka A, Jehlička J, Strnad L (2012) Testing a portable Raman instrument: the detection of biomarkers in gypsum powdered matrix under gypsum crystals. Spectrochim Acta Part A 86:347–350

Danger G, Orthous-Daunay F-R, de Marcellus P et al (2013) Characterization of laboratory analogs of interstellar/cometary organic residues using very high resolution mass spectrometry. Geochim Cosmochim Acta 118:184–201

De Gelder J, Gussem KD, Vandenabeele P et al (2007) Reference database of Raman spectra of biological molecules. J Raman Spectrosc 38:1133–1147

de Marcellus P, Meinert C, Myrgorodska I et al (2015) Aldehydes and sugars from evolved precometary ice analogs: importance of ices in astrochemical and prebiotic evolution. Proc Natl Acad Sci USA 112(4):965–970

Deing T, Hollricher O, Toporski J (2010) Confocal Raman spectroscopy, Springer series in optical sciences 158. Heidelberg, Berlin

Deldicque D, Rouzaud JN, Velde B (2016) A Raman - HRTEM study of the carbonization of wood: a new Raman-based paleothermometer dedicated to archaeometry. Carbon 102:319–329

Dickens J, Irvine W, Nummelin A et al (2001) Searches for new interstellar molecules, including a tentative detection of aziridine and a possible detection of propenal. Spectrochimica Acta Part A 57:643–660

Dubessy J, Caumon MC, Rull F (2012) Raman spectroscopy applied to earth sciences and cultural heritage, EMU notes in mineralogy 12. The Mineralogical Society of Great Britain and Ireland

Edwards HGM (2004) Raman spetroscopic protocol for the molecular recognition of key biomarkers in astrobiological exploration. Orig Life Evol Biosph 34:3–11

Edwards HGM, Hutchinson I, Ingley R et al (2013) Raman spectroscopic analysis of geological and biogeological specimens of relevance to the ExoMars mission. Astrobiology 13:543–549

Ferrari AC (2007) Raman spectroscopy of graphene and graphite: disorder, electron–phonon coupling, doping and nonadiabatic effects. Solid State Commun 143:47–57

Foucher F, Westall F (2013) Raman imaging of metastable opal in carbonaceous microfossils of the 700–800Ma old draken formation. Astrobiology 13:57–67

Foucher F, Ammar MR, Westall F (2015) Revealing the biotic origin of silicified Precambrian carbonaceous microstructures using Raman spectroscopic mapping, a potential method for the detection of microfossils on Mars. J Raman Spectrosc 46:873–879

Foucher F, Guimbretière G, Bost N et al (2017) Petrographical and mineralogical applications of Raman mapping. In: Maaz K (ed) Raman spectroscopy and applications. IntechOpen, London, pp 163–180

Jehlička J, Bény C (1999) First and second order Raman spectra of natural highly carbonified organic compounds from metamorphic rocks. J Mol Struct 480–481:541–545

Jehlička J, Urban O, Pokorny J (2003) Raman spectroscopy of carbon and solid bitumens in sedimentary and metamorphic rocks. Spectrochim Acta Part A 59:2341–2352

Jehlička J, Edwards HGM, Vitek P (2009) Assessment of Raman spectroscopy as a tool for the non-destructive identification of organic minerals and biomolecules for Mars studies. Planet Space Sci 57:606–613

Jehlička J, Culka A, Nedbalova L (2016) Colonization of snow by microorganisms as revealed using miniature Raman spectrometers—possibilities for detecting carotenoids of psychrophiles on Mars? Astrobiology 16:913–924

Lahfid A, Beyssac O, Deville E et al (2010) Evolution of the Raman spectrum of carbonaceous material in low-grade metasediments of the Glarus Alps (Switzerland). Terra Nova 22:354–360

Lazcano A (2011) Origin of life. In: Gargaud M, Amils R, Cernicharo Quintanilla J et al (eds) Encyclopedia of astrobiology, vol 2. Springer, Heidelberg, pp 1183–1190

Lopez-Reyes G, Rull F, Venegas G et al (2013) Analysis of the scientific capabilities of the ExoMars Raman laser spectrometer instrument. Eur J Mineral 25:721–733

Marshall CP, Emry JR, Marshall AO (2011) Haematite pseudomicrofossils present in the 3.5-billion-year-old apex chert. Nat Geosci 4:240–243

Marshall AO, Emry JR, Marshall CP (2012) Multiple generations of carbon in the apex chert and implications for preservation of microfossils. Astrobiology 12:160–166

Marshall CP, Marshall AO (2013) Raman hyperspectral imaging of microfossils: potential pitfalls. Astrobiology 13:920–931

Meinert C, Myrgorodska I, de Marcellus P et al (2016) Ribose and related sugars from ultraviolet irradiation of interstellar ice analogs. Science 352:208–212

Merlin JC (1985) Resonance Raman spectroscopy of carotenoids and carotenoid containing systems. Pure Appl Chem 57:785–792

Moreau JW, Sharp TG (2004) A transmission electron microscopy study of silica and kerogen biosignatures in ~1.9 Ga gunflint microfossils. Astrobiology 4:196–210

Mosier-Boss P, Putnam MD (2013) The evaluation of two commercially available, portable Raman systems. Anal Chem Insights 8:83–97

Pasteris JD, Wopenka B (2002) Images of the earth's earliest fossils? Nature 420:476–477

Pasteris JD, Wopenka B (2003) Necessary, but not sufficient: Raman identification of disordered carbon as a signature of ancient life. Astrobiology 3:727–738

Patel M, Bérces A, Kerékgyarto T et al (2004) Annual solar UV exposure and biological effective dose rates on the Martian surface. Adv Space Res 33:1247–1252

Pflug HD, Jaeschke-Boyer H (1979) Combined structural and chemical analysis of 3.800-Myr-Old microfossils. Nature 280:483–486

Poilblanc R, Crasnier F (2006) Spectroscopies Infrarouge et Raman, Collection Grenoble Science (ed) EDP Sciences, Grenoble
Qu Y, Engdahl A, Zhu S et al (2015) Ultrastructural heterogeneity of carbonaceous material in ancient cherts: investigating biosignature origin and preservation. Astrobiology 15(10):825–842
Quirico E, Montagnac G, Rouzaud JN et al (2009) Precursor and metamorphic condition effects on Raman spectra of poorly ordered carbonaceous matter in chondrites and coals. Earth Planet Sci Lett 287:185–193
Rividi N, van Zuilen M, Philippot P et al (2010) Calibration of carbonate composition using micro-Raman analysis: application to planetary surface exploration. Astrobiology 10:293–309
Rouzaud JN, Oberlin A (1989) Structure, microtexture, and optical properties of anthracene and saccharose-based carbons. Cent Eur J Phys 27:517–529
Ruiz-Mirazo K, Moreno A (2011) Life. In: Gargaud M, Amils R, Cernicharo Quintanilla J et al (eds) Encyclopedia of astrobiology, vol 2. Springer, Heidelberg, pp 919–921
Rull-Pérez F, Martinez-Frias J (2006) Raman spectroscopy goes to Mars. Spectroscopy Europe 18:18–21
Schmitt-Kopplin P, Gabelica Z, Gougeon RD et al (2010) High molecular diversity of extraterrestrial organic matter in Murchison meteorite revealed 40 years after its fall. Proc Natl Acad Sci USA 7 (7):2763–2768
Schopf JW, Kudryavtsev AB, Agresti DG et al (2002a) Laser-Raman imagery of earth's earliest fossils. Nature 416:73–76
Schopf JW, Kudryavtsev AB, Agresti DG et al (2002b) Images of the earth's earliest fossils? Schopf et al reply. Nature 420:477
Schopf JW, Kudryavtsev AB, Agresti DG et al (2005) Raman imagery: a new approach to assess the geochemical maturity and biogenecity of permineralized precambrian fossils. Astrobiology 5:333–371
Sforna MC, van Zuilen MA, Philippot P (2014) Structural characterization by Raman hyperspectral mapping of organic carbon in the 3.46 billion-year-old apex chert, Western Australia. Geochim Cosmochim Acta 124:18–33
Vandenabeele P, Jehlička J, Vitek P, Edwards HGM (2012) On the definition of Raman spectroscopic detection limits for the analysis of biomarkers in solid matrices. Planet Space Sci 62:48–54
Vitek P, Osterrothova K, Jehlička J (2009) Beta-carotene - a possible biomarker in the Martian evaporitic environment: Raman micro-spectroscopic study. Planet Space Sci 57:454–459
Vitek P, Jehlička J, Edwards HGM et al (2012) The miniaturized Raman system and detection of traces of life in halite from the atacama desert: some considerations for the search for life signatures on Mars. Astrobiology 12:1095–1099

Chapter 14
Searching for Signs of Life on Other Planets: Mars a Case Study

Jorge L. Vago, Frances Westall, Barbara Cavalazzi, and The ExoMars Science Working Team

Abstract Demonstrating the existence of simple life forms (past or present) on a cosmic body other than Earth is exceedingly challenging: (1) A naturally sceptic scientific community expects the evidence to be convincing—for example, several independent lines of analyses performed on a feature where the results can only be explained by a biological process. (2) Most bodies are difficult to explore in situ, just about the only way to achieve the above goal, and even then, typically, several missions are required to understand where to go and what to study. (3) Planets and moons that can only be observed remotely (e.g. exoplanets) or from orbit can at best provide some indirect hints of life potential. The actual verification of life would require studying samples containing biosignatures. With the exception of some active moons where jets and plumes may provide the means for satellites to analyse surface sourced material, most other cases require landing, exploring, collecting samples, and analysing them in situ—or bringing them back to Earth.

In this chapter we look at Mars as an example case and propose a scoring system for assigning a confidence value to a group of observations aiming to establish whether a location hosted (or still harbors) microbial life.

Life-seeking missions to other planets should target as many biosignatures as possible. Their discoveries cannot be conclusive unless they include powerful

---

The Exomars Science Working Team composed of ExoMars rover and surface platform instrument scientists as in Vago et al (2017).

J. L. Vago (✉)
European Space Agency, ESTEC, Noordwijk, The Netherlands
e-mail: jorge.vago@esa.int

F. Westall
Centre de Biophysique Moléculaire, CNRS, Orléans, France
e-mail: frances.westall@cnrs.fr

B. Cavalazzi
Dipartimento di Scienze Biologiche, Geologiche e Ambientali, Università di Bologna, Bologna, Italy

Department of Geology, University of Johannesburg, Johannesburg, South Africa
e-mail: barbara.cavalazzi@unibo.it

B. Cavalazzi, F. Westall (eds.), *Biosignatures for Astrobiology*, Advances in Astrobiology and Biogeophysics, https://doi.org/10.1007/978-3-319-96175-0_14

analytical chemistry instruments able to study biosignatures of biomolecules and their degradation products.

14.1 Introduction

Mars orbits close enough to Earth to have benefitted from a rich history of robotic exploration and discovery. It is thus the sole example we can justifiably invoke to illustrate the sustained effort required to search for signs of life elsewhere.

Based on what we knew about planetary evolution in the 1970s, many scientists regarded as plausible the presence of simple microorganisms on other planets, especially if their surface conditions could sustain at least episodically the presence of liquid water—which had been found not to be the case on Venus or the Moon. The 1976 Viking landers can be considered the first missions with a serious chance of discovering signs of life on Mars. That the landers did not provide conclusive evidence was not due to a lack of careful preparation. The Viking results were a consequence of the manner in which the question of life detection was posed, seeking to elicit signs of microbial activity from potential extant ecosystems within the samples analyzed (Klein et al. 1976). The twin Viking landers conducted the first in situ measurements on the martian surface. Their biology package contained three experiments, all looking for indications of metabolism in soil samples (Klein et al. 1976). One of them, the Labeled-Release Experiment, produced very provocative results (Levin and Straat 2016). If other information had not also been obtained, these data would have been interpreted as proof of biological activity. However, theoretical modelling of the martian atmosphere and regolith chemistry hinted at the existence of powerful oxidants that could, more or less, account for the results of the three biology package experiments (Klein 1999). The biggest blow was the failure of the gas chromatograph mass spectrometer (GCMS) to acquire evidence of organic molecules at the parts-per-billion level. With few exceptions, the majority of the scientific community concluded that the Viking findings did not demonstrate the presence of extant life (Klein 1998, 1999). As a consequence, our neighbour planet lost much of its allure and a multi-year gap in Mars exploration ensued.

During the 1990s and early 2000s, orbiters (Malin and Edgett 2000), landers, and the very successful Mars Exploration Rovers (Squyres et al. 2004a, b) focused on surface geology—searching for signs of life was not part of their objectives. However, this would change after findings by Mars Express 2003 and Mars Reconnaissance Orbiter 2005 revealed many instances of finely layered deposits containing phyllosilicate minerals that could only have formed in the presence of liquid water, thus reinforcing the hypothesis that early Mars had been wetter than today (Poulet et al. 2005; Bibring et al. 2006; Loizeau et al. 2010, 2012; Ehlmann et al. 2011; Bishop et al. 2013, 2018; Michalski et al. 2013). These discoveries rekindled the interest in young Mars as a potential abode for life.

More recently, two missions that have improved our understanding of chemical conditions on the martian surface are the 2007 Phoenix lander and the 2011

Curiosity rover. Phoenix included, for the first time, a wet chemistry analysis instrument that detected the presence of the perchlorate (ClO_4^-) anion in soil samples collected by the robotic arm (Hecht et al. 2009; Kounaves et al. 2010, 2014). Perchlorates are chemically inert at room temperature; however, if heated beyond a few hundred degrees, its four oxygen atoms are released, becoming very reactive oxidation vectors. It did not take long for investigators to recall that Viking had relied on thermal volatilization (TV; in other words heat) to release organics from soil samples (Navarro-González et al. 2010, 2011; Biemann and Bada 2011; Navarro-González and McKay 2011). If perchlorate had been also present in the soil at the two Viking lander locations, perhaps heating could explain the negative organic carbon results obtained? In fact, some simple chlorinated organic molecules (chloromethane and dichloromethane) had been detected by the Viking experiments (Biemann et al. 1977), but these compounds were interpreted to have resulted from a reaction between adsorbed residual methanol (a cleaning agent used to prepare the spacecraft) and hydrochloric acid (HCl). Today, the general consensus is that they were the outcome of heat-activated perchlorate dissociation and reaction with indigenous organic compounds (Steininger et al. 2012; Glavin et al. 2013; Quinn et al. 2013; Sephton et al. 2014; Goetz et al. 2016; Lasne et al. 2016).

This has been confirmed by measurements performed with the SAM (sample analysis at Mars) instrument on board Curiosity. The team detected oxygen (O_2) released by the thermal decomposition of oxychlorine species [i.e., perchlorates and/or chlorates (Archer et al. 2016)], as well as chlorine-bearing hydrocarbons attributable to the reaction of oxychlorine species with organics compounds, both when they analysed modern sand deposits as well as when they drilled into much older rocks (Glavin et al. 2013; Freissinet et al. 2015). The exogenous delivery of meteoritic organics (abiotic) to the martian surface has been estimated at ~10^5 kg C/year, mostly in the form of polycyclic aromatic hydrocarbons (PAHs) and kerogen that may undergo successive oxidation reactions. Therefore, a meteoritic source could have contributed the organic precursors needed for producing the observed chlorobenzene and dichloroalkanes (Freissinet et al. 2015).

Summarizing, as a result of painstaking research performed over many years and involving multiple missions, we can conclude that:

1. There was plentiful liquid water on early Mars (at least during its first billion years). Since life seems to have appeared on our planet as soon as the environment allowed it, sometime between 4.4–3.8 Ga ago, it is likely that conditions also existed for the emergence of life on Mars, even if it was colder than Earth (Solomon 2005; McKay 2010; Strasdeit 2010; Yung et al. 2010);
2. Today, it is still possible to find organic molecules close to the surface, though most probably they have a meteoritic (not biogenic) origin;
3. Since the martian atmosphere is more tenuous than Earth's, the ultraviolet (UV) radiation dose is higher than on our planet and will quickly damage exposed organisms or biomolecules; powerful oxidants exist that, when activated, can destroy the potential biosignatures we would like to study; and ionizing radiation penetrates into the uppermost meters of the planet's subsurface, causing a slow

Table 14.1 ExoMars Biosignature Score: A possible system to assign a confidence value (the score) to a group of robust observations aiming at establishing whether a location hosted life

	Score
Geological context	
Long-lived water or hydrothermal setting (morphology)	15–10
Long-lived water or hydrothermal setting (mineralogy)	15–10
Morphological biosignatures	
Multilayer organo-sedimentary structures (e.g. stromatolites)	20
Other candidate bio-mediated textures	10
Features suggestive of (fossil) microorganisms	20
Chemical biosignatures	
Detection of primary biomolecules or their degradation products	20
Enantiomeric excess (or other isomer selectivity)	30
Molecular weight clustering of organic compounds	20
Evidence of repeating constitutional subunits	20
Systematic isotopic ordering at molecular (group) level	20
Bulk isotopic fractionation	10

degradation process that, operating over many millions of years, can alter organic molecules beyond the detection sensitivity of analytical instruments.

These are the very specific boundary conditions that Mars missions must contend with. Orbiters and landers exploring other worlds, for example the moons of Jupiter or Saturn, would need to adapt their search-for-life strategy to their particular environments' geologic history—where are the potential biosignatures? how can we sample them? must we contend with radiation? But even if the details of the quest have to be tailored to the celestial body under investigation, it is possible to compile a somewhat universal list of observations having for objective to establish whether a location (on Mars or elsewhere) has hosted microbial life, past or present. In the work by Vago et al. (2017) we proposed such a list. We called this the *ExoMars Biosignature Score* because it is being developed while preparing for this mission. However, the list of biosignatures is quite exhaustive and encompasses more than those that the ExoMars rover will be able to assess (see Table 14.1).

Please note that Table 14.1 does not include morphological changes with time (e.g. colony growth), movement (e.g. creature displacement), or experiments designed to trigger and observe active metabolic responses (as in Viking). These "more dynamic" expressions of possible present life would not be easy to verify.

In this chapter we address each of the observations in Table 14.1, explain what their positive verification would entail, provide examples (or sketches) to illustrate what the data could look like, and justify the need to provide evidence that several of the principal biosignatures have been demonstrated before positing that life detection has been achieved.

14.2 Geological Context and Biosignatures

For planetary surface missions, typically the characterisation of geological context begins early, with landing site selection, as investigators canvas interesting locations searching for those that best fit the mission's scientific objectives.

Discovering candidate biosignatures embedded in a congruent geological landscape, that is, an environment that demonstrably possessed attributes conducive to the prosperity of microbial communities—for example, a long-lived, low-energy, aqueous or hydrothermal setting experiencing frequent fine sediment deposition—would help to increase substantially the confidence of any potential claim. For this reason we include "geological context" in the list of elements that bear investigation when searching for traces of life. However, some mission concepts do not afford this possibility. For example, a probe flying through an Enceladus plume seeking to trap and enrich organic molecule signatures would not have direct access to the geological setting where the organic molecules are being produced.

In general, microbial biosignatures can be grouped into three broad categories (Cady et al. 2003; Westall and Cavalazzi 2011) as follows: (1) cellular fossils that preserve organic remains of microbes and their extracellular matrices, as well their colonies and biofilms and mats, the study of which typically requires complex sample preparation and high-resolution instruments not currently available on landed space missions (Westall et al. 2011a); (2) bio-influenced fabrics and sedimentary structures (Noffke and Awramik 2013; Westall 2008, 2012; Davies et al. 2016), which provide a macroscale imprint of the presence of microbial biofilms that can be more readily identified, for example, laminated stromatolites; and (3) organic chemofossils preserved in the geological record (Parnell et al. 2007; Summons et al. 2008) that can be either primary biomolecules or diagenetically altered compounds known as biomarkers. Other, more tenuous or indirect information could complement the above, but we do not include them among the major biosignature categories. For example, metabolic effects preserved in the geologic record, such as mineral precipitation (e.g. magnetite by magnetotactic bacteria, Lefèvre and Bazilinski 2013, or carbonate, Dupraz et al. 2009), or leaching of elements in rocks (microbial corrosion, e.g. Foucher et al. 2010), or trace element concentration (e.g. Johannesson et al. 2014).

14.2.1 *Morphological Biosignatures*

The primordial types of microorganisms that could have existed on early Mars would have been prokaryotes similar to terrestrial chemotrophic microorganisms, i.e. of the order of a micron to a few microns in size. While the individual cells would be too small to distinguish, as on Earth, their permineralised or compressed microbial colonies and biofilms would be much larger.

In terrestrial marine (and other wet) environments, benthic microorganisms (e.g. those living in the seabed) form biofilms, highly organised microbial communities that are able to affect the accumulation of detrital sediments. Particle binding, bio-stabilisation, baffling, and trapping by biofilms can result in macroscopic edifices amenable to be recognised and studied with rover cameras and close-up imagers. In cases where sediment precipitation occurs in a repetitive manner, multi-layer constructions can ensue; for example, stromatolites constitute essential beacons of information, recording snapshots of microbial communities and environments throughout Earth's history (Allwood et al. 2006, 2009, 2013). In particular phototrophs produce large amounts of extracellular polymeric substances in the biofilm. If the biofilm covers a large enough area experiencing similar conditions, often multiple organo-sedimentary structures can arise in regularly spaced groups. But the presence of microbes does not always lead to the emergence of noticeable macroscale bio-sedimentary formations. An example of a less conspicuous expression is the layering found in some typical early Earth volcanic lithic environments, where organisms have colonised the surface of ashfall particles, creating visible, carbon-rich, biofilms in various sediment horizons (Westall et al. 2011b, 2015).

Hereafter we present a few examples illustrating evidence of microbial colonisation in ancient Earth rocks (dated at 3.5–3.3 Ga). We chose to depict them at the scales at which they would be observed during a mission: panoramic (tens of meters to a few centimetres), close-up range (a few centimetres to several tens of microns), and microscopic (submicron). We include the latter for reference, although it is not possible today for robotic missions to prepare and study samples at such high magnification.

14.2.1.1 Panoramic Scale

Although the observation of identifiable biosignatures at the panoramic scale is not expected on Mars (unless fortuitous evolution led to the emergence of phototrophic organisms forming three-dimensional stromatolite-like structures), documentation of fine-scale layering in sediments deposited under water that may or may not be associated with accumulations of carbon can be made at this scale. Figure 14.1a shows finely layered sediments from the ~3.6 Ga Gale Crater whose structural features, such as cross-bedding and channel bedding, suggest deposition of sediments in a shallow water environment (cf. Grotzinger et al. 2014). Sedimentary structures in Fig. 14.1b from the 3.33 Ga Barberton Greenstone Belt, including ripple bedding, are indicative of deposition in a shallow water tidal environment (Westall et al. 2015).

The rock from Josefsdal in Fig. 14.1c is a hydrothermal chert. It includes translucent layers of chemically precipitated silica alternating with matt black layers

Fig. 14.1 (**a**) Panoramic view of an outcrop of the ~3.6 Ga Shaler bed (Yellowstone Bay, Gale Crater, Mars), showing finely layered volcanic sediments deposited in a lake. Source: Adapted from NASA/JPL-Caltech/MSSS. (**b**) Finely laminated volcanic sediments from the Josefsdal Chert (Barberton Greenstone Belt, 3.33 Ga) deposited in shallow coastal waters. An important difference between the two rocks is that those from the early Earth were silicified by hydrothermal fluids after deposition and are therefore harder and more resistant to erosion than those of Gale Crater. Source: F Westall. (**c**) Hydrothermal facies from Josefsdal showing chemical siliceous sediment (translucent layers) alternating with highly silicified, black layers. Source: F Westall

that are faintly laminated (at the close-up scale). Observing sedimentary structures at this scale is important for interpreting the sedimentary context, but no further assertions can be made with respect to potential biosignatures.

14.2.1.2 Close-Up Scale

Close up observations are fundamental because they can provide more detailed information pertaining to the origin of the sediment particles as well as their mode of formation. For example, the size, sub-rounded habit and collision marks on the pebbles in Fig. 14.2a indicate strong aqueous transport (Edgar et al. 2017). Differential erosion of the fine, sandy component of the sediment has left the pebbles in relief. In comparison, Fig. 14.2b shows details of the Josefsdal Chert exposure shown in Fig. 14.1b exhibiting upwards grading of the sand, indicating passive settling of detrital particles deposited from airfall. Importantly, the close up images show that the black layers occur at the tops of the fining upwards cycles, indicating either that (1) they are composed of dark-coloured, hydrodynamically light detritus, and/or (2) the dark-coloured layers formed in situ on top of the sediment bedding planes, suggesting either diagenetic alteration of a precursor mineral into a dark phase (e.g. pyritisation) or the formation of microbial biofilms whose degraded organic remnants form dark-coloured kerogen.

In contrast, the coarsely laminated hydrothermal sediments in the Josefsdal outcrop in Fig. 14.1c exhibit faint internal lamination in close-up imagery (Fig. 14.2c, arrow). What is of additional interest is the mottled texture exhibited by the matt black layers (N.B. the image has been lightened) indicating the presence of layers of small-scale internal clotted features some 100 s μm in size.

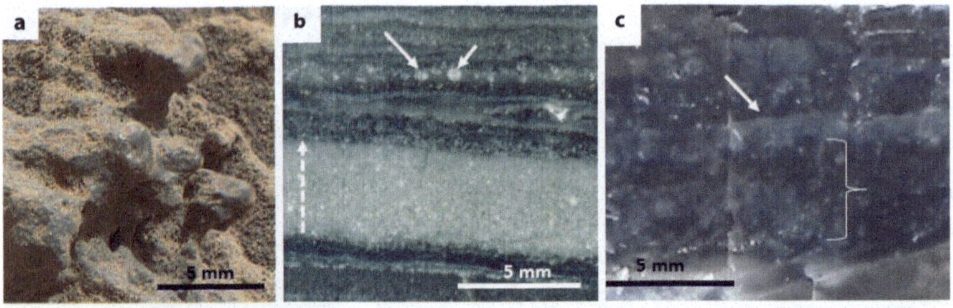

Fig. 14.2 (**a**) Close-up image of pebbles in a sandy matrix from the Shaler deposit at Yellowstone Bay, Gale Crater. Source: Adapted from Edgar et al. (2017). Surface textures on the pebbles and their sub-rounded aspect in a sandy matrix indicates subaqueous transport by relatively strong currents. (**b**) Close-up image of sedimentary textures, such as graded bedding (dotted arrow) in the finely laminated sediments in the Josefsdal Chert, Barberton, indicate airfall deposits into water. The black layers occur at the tops of the fining upwards sequences and could have an either detrital or in situ origin. Source: F Westall. (**c**) The hydrothermal sediments from the Josefsdal Chert are faintly laminated (arrow) but also show a mottled texture, e.g. bracketed area. Source: F Westall

14.2.1.3 Microscopic Scale

The MAHLI-microscope image of sand grains in a dune in Gale Crater (Fig. 14.3a) show that good textural detail can be potentially obtained at the microscopic scale. However, in this specific instance, such detail is either not available or the dust cover on the surface of the rocks obscures potential detail. Microscopic observation, however, is essential for further interpretation, especially associated with a compositional method, such as Raman or IR spectroscopy. The examples in Fig. 14.3b–d show that some sample preparation is necessary.

Thin sections (30 μm thick) of the rock in Fig. 14.2c were made and investigated under optical microscopy in transmitted light and Raman spectral mapping. The high spatial resolution study of details of the clotted texture (which were just faintly visible in close-up imagery) showed that the rock consists of volcanic particles largely coated with a black phase, which could be either kerogen or an oxide. The Raman mapping documented alteration of the underlying volcanic particle to anatase, a process that takes place in the presence of water. It also showed that the particle was thoroughly coated by kerogen. The irregular thickness of the coating (Westall et al. 2015) suggests in situ growth, such as colonisation by chemotrophic microbes, rather than coating by abiogenic carbon, which would produce a conformable layer.

In Fig. 14.3c, the rock surface was treated with hydrofluoric acid (HF) to highlight the differential mineralogy of the different components within the rock, thus preferentially corroding the silica matrix of the sediment and exposing the "dirty", carbon-rich components (i.e. carbonaceous microfossils). Visualisation of the <1 μm-sized microfossils is only possible using high powered electron microscopy. Even with this kind of compositional and textural data, interpretation of such features can be controversial.

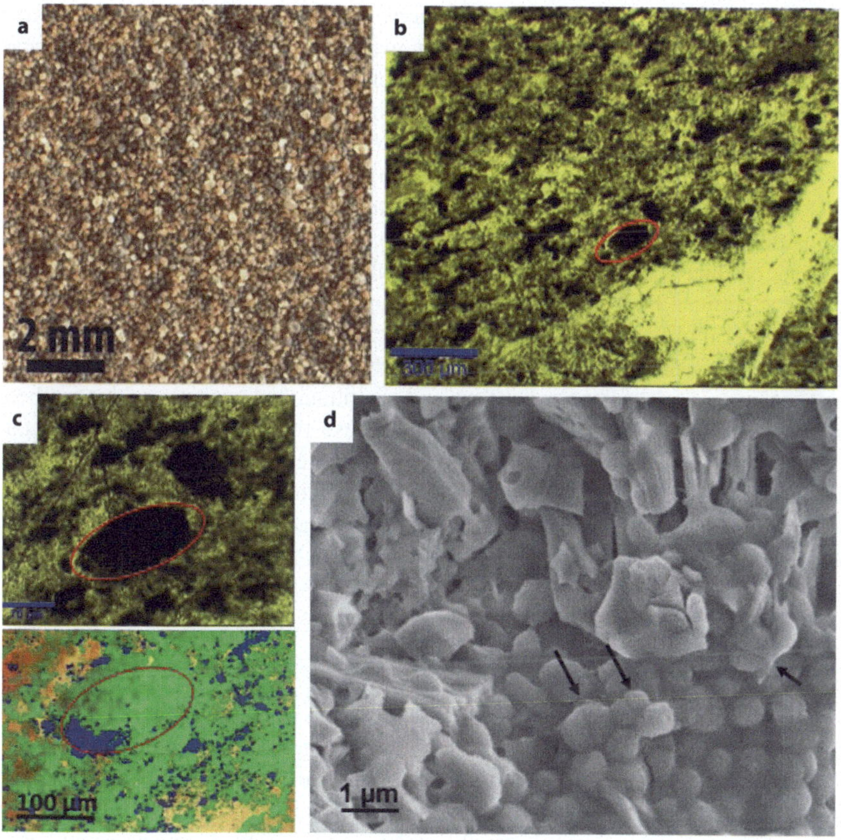

Fig. 14.3 (**a**) Sand grains on a dune in Gale Crater imaged by the instrument MAHLI. Individual grains are between 100–300 µm. Source: Adapted from Lapotre et al. (2017). (**b**) Optical microscope view in transmitted light of a sample from the Josefsdal Chert rock shown. Figure 14.2c showing that it contains particles 50–200 µm in size having a dark colour and producing a clotted texture. Source: courtesy F Foucher. (**c**) Optical detail of one particle (circled in **b**) with related Raman map (to the bottom) showing its mineralogy: anatase (blue), kerogen (green), silica (yellow–orange). Source: courtesy F Foucher. (**d**) SEM view of a colony of fossilised (silicified) microorganisms coating the surface of a similar volcanic grain from the 3.46 Ga Kitty's Gap Chert, Pilbara. Source: Adapted from Westall et al. (2006)

14.2.2 Chemical Biosignatures

When considering molecular biosignatures, the first obvious set of targets is the ensemble of primary biomolecules associated with active microorganisms, such as amino acids, proteins, nucleic acids, carbohydrates, some pigments, and intermediary metabolites. Detecting the presence of these compounds in high abundance would be diagnostic of extant life, but unfortunately on Earth they degrade quickly once microbes die.

Lipids and other structural biopolymers, however, are biologically essential components (e.g. of cell membranes) known to be stable for billions of years when buried (up to 1.64 Ga on Earth, Brocks et al. 2005). It is the recalcitrant hydrocarbon backbone that is responsible for the high-preservation potential of lipid-derived biomarkers relative to that of other biomolecules (Eigenbrode 2008).

Along the path from primary compound to molecular fossil, all biological materials undergo in situ chemical reactions dictated by the circumstances of transport, deposition, entombment, and post-depositional effects on the original organisms. The end product of diagenesis is macromolecular organic matter, which, through the loss of superficial hydrophilic functional groups, slowly degrades into the solvent-insoluble form of fossil carbonaceous matter called kerogen, as seen in the example from the Josefsdal Chert (Fig. 14.3c), The heterogeneous chemical structure of the kerogen matrix can preserve patterns and distribution diagnostic of biosynthetic pathways. Kerogen also possesses molecular sieve properties allowing it to retain diagenetically altered biomolecules (Tissot and Welte 1984).

Apart from the direct recognition of biomolecules and/or their degradation products (which is not illustrated here), other characteristics of bioorganic compounds include the following (Summons et al. 2008, 2011).

14.2.2.1 Enantiomeric Excess

In the case of chiral molecules (those that can exist in either of two non-identical mirror image structures known as enantiomers), life forms synthesise exclusively one enantiomer, for example, left-handed amino acids (L-amino acids) to build proteins and right-handed ribose (D-ribose) for sugars and the sugars within ribonucleic acid (RNA) and deoxyribonucleic acid (DNA). Opposite enantiomers (D-amino acids and L-ribose) are neither utilized in proteins nor in the genetic material RNA and DNA.

The use of pure chiral building blocks is considered a general molecular property of life. For example, gradual loss of enantiomeric excess in amino acids, i.e. racemisation, in fossil shells has been used for dating Quaternary fossils (called aminochronology, Wehmiller 1993).

14.2.2.2 Molecular Weight Clustering of Organic Compounds

Many important biochemicals exist in discrete molecular weight ranges (e.g. C_{14}–C_{20} lipid fatty acids). For this reason, the molecular weight distribution of biologically derived matter exhibits clustering; it is concentrated in discrete clumps corresponding to the various life-specialized families of molecules (Summons et al. 2008). This is in contrast to the molecular weight distribution for cosmic organics (Ehrenfreund and Charnley 2000; Ehrenfreund and Cami 2010): the relative abundance for abiotic volatiles is uniform and drops off as the carbon number increases. For example, ToF-SIMS analysis of well-preserved, silicified kerogen from the 3.33 Ga Josefsdal

Chert in South Africa, show just such clustering of the organic molecules (Westall et al. 2011b).

14.2.2.3 Repeating Constitutional Subunits

Many biological products (e.g. proteins and nucleic acids) are synthesized from a limited number of simpler units. This can leave an identifiable molecular weight signature even in fragments recovered from highly derived products, such as petroleum. For example, in the case of material containing fossil lipids we would expect to find a predominance of even-carbon numbered fatty acids (C_{14}, C_{16}, C_{18}, C_{20}). This is because the enzymes synthesizing fatty acids attach two carbon atoms at a time (in C_2H_4 subunits) to the growing chain. Other classes of biomolecules can also exhibit characteristic carbon chain-length patterns, for example, C_{15}, C_{20}, C_{25} for acyclic isoprenoids constructed using repeating C_5H_{10} blocks.

This phenomenon is illustrated by analysis of kerogen from the Bitter Spring Formation (850 My) in northern Australia where the aromatic molecules linked by short-chain *n*-alkyl moieties exhibit odd carbon number predominance, perhaps indicative of bacterial cell wall lipids (Imbus and McKirdy 1993).

14.2.2.4 Systematic Isotopic Ordering at Molecular (Group) Level

Biological molecule building blocks, in particular some functional groups, can show significant differences in their degree of ^{13}C incorporation relative to ^{12}C. The "repeating subunit" conformation of biomolecules can result in an observable isotopic ordering in the molecular fingerprint.

14.2.2.5 Bulk Isotopic Fractionation

The isotopic fractionation of stable elements such as C, H, O, N, S, and Fe can be used as a signature to recognize the action of biological pathways. Although the qualitative chemical behaviour of the light and the heavy isotope is similar, the difference in mass can result in dissimilar bond strength and reaction rates. Thus, the isotopic discrimination associated with organic biosynthesis (which alters the natural equilibrium between C isotopes in favour of the lighter variant) is principally responsible for determining the $^{13}C/^{12}C$ ratios in terrestrial organic and inorganic crustal reservoirs.

Although interesting, we do not consider bulk isotopic fractionation a robust biosignature when applied to locations or epochs for which we have scant knowledge of sources and sinks. In the specific case of carbon, $^{13}C/^{12}C$ ratios may serve as reliable biosignatures for past or present life only if the key components of the C-cycling system (applicable at the time of deposition and since then) are well constrained (Summons et al. 2011). This is certainly not the case for Mars (or other bodies), and one can also wonder to what extent we are sure about our own past carbon dynamics when analysing very ancient samples.

Despite the above reservations we are willing to include bulk isotopic fractionation in this list, but with the caveat that it should be used in association with other, less indirect, biosignatures.

On Earth, groundbreaking work on isotopic signatures and their preservation in ancient rocks was made by Manfred Schidlowski (1988) and ever since, the $\delta^{13}C$ signature is considered as a useful accompanying biosignature. Schidlowski (1988) documented isotopically light carbon in 3.8 Ga rocks containing carbon from the Isua Greenstone Belt in Greenland, although Westall and Folk (2003) noted the presence of recent (<8000 year-old) fossilised endolithic microorganisms in the rocks analysed. More recent studies of Isua rocks using the graphite combustion method do indeed confirm the presence of isotopically light carbon with $\delta^{13}C$ values between −10 and −20‰ (Ohtomo et al. 2014).

14.2.2.6 Preservation of Organic Matter

Effective chemical identification of biosignatures requires access to well-preserved organic molecules.

Because the martian atmosphere is more tenuous than Earth's, three important physical agents reach the surface of Mars (and other airless Solar System bodies of interest) with adverse effects for the long-term preservation of biomarkers: (1) The UV radiation dose is higher than on our planet and will quickly damage exposed organisms or biomolecules. (2) UV-induced photochemistry is responsible for the production of reactive oxidant species that, when activated, can also destroy chemical biosignatures. The diffusion of oxidants into the subsurface is not well characterized and constitutes an important measurement that future missions must perform. Finally, (3) ionizing radiation penetrates into the uppermost meters of the planet's subsurface. This causes a slow degradation process that, operating over many millions of years, can alter organic molecules beyond the detection sensitivity of analytical instruments. Radiation effects are depth dependent: the material closer to the surface is exposed to higher doses than that buried deeper. Therefore, the molecular record of ancient martian life, if it ever existed, is likely to have escaped radiation and chemical damage only if trapped in the subsurface for long periods. Studies suggest that a subsurface penetration in the range of 2 m is necessary to recover well-preserved organics from the very early history of Mars (Kminek and Bada 2006), assuming there has been some help from additional, recently eroded overburden (Dartnell et al. 2007, 2012; Parnell et al. 2007; Pavlov et al. 2012).

It is also important to notice that large diurnal temperature excursions are typical for most bodies lacking a dense atmosphere able to provide much thermal inertia.

For example, at midlatitudes on Mars, noon surface temperatures hover around 0 °C, but quickly plummet to −120 °C during the night. This can work to our advantage. Just 50 cm underground the temperature remains close to the average between day and night, some −60 °C. Thus, the martian subsurface effectively behaves like a freezer, slowing down the chemical reactivity of damaging oxidants, greatly contributing to the chemical preservation of organic molecules. The weaker gravity also means that diagenetic processes that would normally affect the preservation of molecules are comparably weaker on Mars. The same column of rock produces less pressure than on a higher gravity planet. It is therefore likely that fragile molecules that would quickly degrade on Earth could last much, much longer on Mars preserved sufficiently deep in the cold subsurface.

14.3 Conclusion

We have proposed a possible scoring system for assigning a confidence value to a group of observations aiming to establish whether a location hosted microbial life.

We find that there is value in defining a set of measurements and rules to guide our preparations and help with the interpretation of any findings once we have reached our destination, Mars or elsewhere.

A closer examination of Table 14.1 reveals that, if we could tick all possible biosignatures, assigning maximum points with a perfect chemical background, the score would be 200. However, in the work by Vago et al. (2017), we claimed we only need a value of 100 to establish that there was/is life. This indicates that it is not necessary to verify all possible biosignatures, but that it is mandatory to provide evidence that a few of the principal biogenicity criteria or indicators are indeed demonstrated. Chemical biosignatures are awarded a higher importance because they provide "more direct" evidence of biogenicity than the other categories for which bioinfluence is "inferred". Note, however, that in certain circumstances morphological biosignatures can be equally well-constrained.

Life-seeking missions to other planets should target as many biosignatures indicated in Table 14.1 as possible. We claim that their discoveries will not be conclusive unless such missions include powerful analytical chemistry capabilities that can allow for the unambiguous identification of key biosignatures of biomolecules and their degradation products.

Acknowledgements The Authors would like to thank the ExoMars project team and scientists (past and present), as well as our colleagues from industry. We would also like to recognize the help and support of ESA, Roscosmos, the European states and agencies participating in the ExoMars program, and NASA. Portions of this Chapter have appeared in Vago et al. (2017) and are included courtesy of Astrobiology.

References

Allwood AC, Walter MR, Kamber BS et al (2006) Stromatolite reef from the early Archaean era of Australia. Nature 441:714–718

Allwood AC, Grotzinger JP, Knoll AH et al (2009) Controls on development and diversity of early Archean stromatolites. Proc Natl Acad Sci USA 106:9548–9555

Allwood AC, Burch IW, Rouchy JM et al (2013) Morphological biosignatures in gypsum: diverse formation processes of Messinian (~6.0 Ma) gypsum stromatolites. Astrobiology 13:870–886

Archer PD, Ming DW, Sutter B et al (2016) Oxychlorine species on Mars: implications from Gale Crater samples. In: 47th Lunar and Planetary Science Conference, Abstract 2947

Bibring J-P, Langevin Y, Mustard JF (2006) Global mineralogical and aqueous Mars history derived from OMEGA/Mars express data. Science 312:400–404

Biemann K, Bada JL (2011) Comment on "Reanalysis of the Viking results suggests perchlorate and organics at midlatitudes on Mars" by Rafael Navarro-González et al. J Geophys Res 116 (E12):E12001

Biemann K, Oro J, Toulmin P et al (1977) The search for organic substances and inorganic volatile compounds in the surface of Mars. J Geophys Res 82:4641–4658

Bishop JL, Loizeau D, McKeown NK et al (2013) What the ancient phyllosilicates at Mawrth Vallis can tell us about possible habitability on early Mars. Planet Space Sci 86:130–149

Bishop JL, Fairén AG, Michalski JR et al (2018) Surface clay formation during short-term warmer and wetter conditions on a largely cold ancient Mars. Nat Astron 2:206–213

Brocks JJ, Love GD, Summons RE (2005) Biomarker evidence for green and purple sulphur bacteria in a stratified Palaeoproterozoic Sea. Nature 437:866–870

Cady SL, Farmer JD, Grotzinger JP et al (2003) Morphological biosignatures and the search for life on Mars. Astrobiology 3:351–368

Dartnell LR, Desorgher L, Ward JM, Coates AJ (2007) Modelling the surface and subsurface martian radiation environment: implications for astrobiology. Geophys Res Lett 34(2):L02207. https://doi.org/10.1029/2006GL027494

Dartnell LR, Page K, Jorge-Villar SE, Wright G, Munshi T, Scowen IJ, Ward JM, Edwards HGM (2012) Destruction of raman biosignatures by ionising radiation and the implications for life detection on mars. Anal Bioanal Chem 403(1):131–144. https://doi.org/10.1007/s00216-012-5829-6

Davies NS, Liu AG, Gibling MR et al (2016) Resolving MISS conceptions and misconceptions: a geological approach to sedimentary surface textures generated by microbial and abiotic processes. Earth Sci Rev 154:210–246

Dupraz C, Reid RP, Braissant O et al (2009) Processes of carbonate precipitation in modern microbial mats. Earth Sci Rev 96:141–152

Edgar LA, Gupta S, Rubin DM et al (2017) Shaler: *in situ* analysis of a fluvial sedimentary deposit on Mars. Sedimentology 65:96–122

Ehlmann BL, Mustard JF, Murchie SL et al (2011) Subsurface water and clay mineral formation during the early history of Mars. Nature 479:53–60

Ehrenfreund P, Cami J (2010) Cosmic carbon chemistry: from the interstellar medium to the early earth. Cold Spring Harb Perspect Biol 2(12):a002097–a002097. https://doi.org/10.1101/cshperspect.a002097

Ehrenfreund P, Charnley SB (2000) Organic molecules in the interstellar medium, comets, and meteorites: a voyage from dark clouds to the early earth. Annu Rev Astron Astrophys 38 (1):427–483. https://doi.org/10.1146/annurev.astro.38.1.427

Eigenbrode JL (2008) Fossil lipids for life-detection: a case study from the early earth record. Space Sci Rev 135:161–185

Foucher F, Westall F, Brandstatter F et al (2010) Testing the survival of microfossils in artificial martian sedimentary meteorites during entry into Earth's atmosphere: the STONE 6 experiment. Icarus 207:616–630

Freissinet C, Glavin DP, Mahaffy PR (2015) Organic molecules in the Sheepbed Mudstone, Gale Crater, Mars. J Geophys Res 120:495–514

Glavin DP, Freissinet C, Miller KE et al (2013) Evidence for perchlorates and the origin of chlorinated hydrocarbons detected by SAM at the Rocknest aeolian deposit in Gale Crater. J Geophys Res 118:1955–1973

Goetz W, Brinckerhoff WB, Arevalo R et al (2016) MOMA: the challenge to search for organics and biosignatures on Mars. Int J Astrobiol 15:239–250

Grotzinger JP, Sumner DY, Kah LC, Stack K, Gupta S, Edgar L, Rubin D, Lewis K, Schieber J, Mangold N, Milliken R, Conrad PG, DesMarais D, Farmer J, Siebach K, Calef F, Hurowitz J, McLennan SM, Ming D, Vaniman D, Crisp J, Vasavada A, Edgett KS, Malin M, Blake D, Gellert R, Mahaffy P, Wiens RC, Maurice S, Grant JA, Wilson S, Anderson RC, Beegle LW, Arvidson R, Hallet B, Sletten RS, Rice M, Bell J, Griffes J, Ehlmann B, Anderson RB, Bristow TF, Dietrich WE, Dromart G, Eigenbrode J, Fraeman A, Hardgrove C, Herkenhoff K, Jandura L, Kocurek G, Lee S, Leshin LA, Leveille R, Limonadi D, Maki J, McCloskey S, Meyer M, Minitti M, Newsom H, Oehler D, Okon A, Palucis M, Parker T, Rowland S, Schmidt M, Squyres S, Steele A, Stolper E, Summons R, Treiman A, Williams R, Yingst A, Team MS, Kemppinen O, Bridges N, Johnson JR, Cremers D, Godber A, Wadhwa M, Wellington D, McEwan I, Newman C, Richardson M, Charpentier A, Peret L, King P, Blank J, Weigle G, Li S, Robertson K, Sun V, Baker M, Edwards C, Farley K, Miller H, Newcombe M, Pilorget C, Brunet C, Hipkin V, Leveille R, Marchand G, Sanchez PS, Favot L, Cody G, Fluckiger L, Lees D, Nefian A, Martin M, Gailhanou M, Westall F, Israel G, Agard C, Baroukh J, Donny C, Gaboriaud A, Guillemot P, Lafaille V, Lorigny E, Paillet A, Perez R, Saccoccio M, Yana C, Armiens-Aparicio C, Rodriguez JC, Blazquez IC, Gomez FG, Gomez-Elvira J, Hettrich S, Malvitte AL, Jimenez MM, Martinez-Frias J, Martin-Soler J, Martin-Torres FJ, Jurado AM, Mora-Sotomayor L, Caro GM, Lopez SN, Peinado-Gonzalez V, Pla-Garcia J, Manfredi JAR, Romeral-Planello JJ, Fuentes SAS, Martinez ES, Redondo JT, Urqui-O'Callaghan R, Mier M-PZ, Chipera S, Lacour J-L, Mauchien P, Sirven J-B, Manning H, Fairen A, Hayes A, Joseph J, Sullivan R, Thomas P, Dupont A, Lundberg A, Melikechi N, Mezzacappa A, DeMarines J, Grinspoon D, Reitz G, Prats B, Atlaskin E, Genzer M, Harri A-M, Haukka H, Kahanpaa H, Kauhanen J, Paton M, Polkko J, Schmidt W, Siili T, Fabre C, Wray J, Wilhelm MB, Poitrasson F, Patel K, Gorevan S, Indyk S, Paulsen G, Bish D, Gondet B, Langevin Y, Geffroy C, Baratoux D, Berger G, Cros A, D'Uston C, Forni O, Gasnault O, Lasue J, Lee Q-M, Meslin P-Y, Pallier E, Parot Y, Pinet P, Schroder S, Toplis M, Lewin E, Brunner W, Heydari E, Achilles C, Sutter B, Cabane M, Coscia D, Szopa C, Robert F, Sautter V, Le Mouelic S, Nachon M, Buch A, Stalport F, Coll P, Francois P, Raulin F, Teinturier S, Cameron J, Clegg S, Cousin A, DeLapp D, Dingler R, Jackson RS, Johnstone S, Lanza N, Little C, Nelson T, Williams RB, Jones A, Kirkland L, Baker B, Cantor B, Caplinger M, Davis S, Duston B, Fay D, Harker D, Herrera P, Jensen E, Kennedy MR, Krezoski G, Krysak D, Lipkaman L, McCartney E, McNair S, Nixon B, Posiolova L, Ravine M, Salamon A, Saper L, Stoiber K, Supulver K, Van Beek J, Van Beek T, Zimdar R, French KL, Iagnemma K, Miller K, Goesmann F, Goetz W, Hviid S, Johnson M, Lefavor M, Lyness E, Breves E, Dyar MD, Fassett C, Edwards L, Haberle R, Hoehler T, Hollingsworth J, Kahre M, Keely L, McKay C, Bleacher L, Brinckerhoff W, Choi D, Dworkin JP, Floyd M, Freissinet C, Garvin J, Glavin D, Harpold D, Martin DK, McAdam A, Pavlov A, Raaen E, Smith MD, Stern J, Tan F, Trainer M, Posner A, Voytek M, Aubrey A, Behar A, Blaney D, Brinza D, Christensen L, DeFlores L, Feldman J, Feldman S, Flesch G, Jun I, Keymeulen D, Mischna M, Morookian JM, Pavri B, Schoppers M, Sengstacken A, Simmonds JJ, Spanovich N, de la Torre Juarez M, Webster CR, Yen A, Archer PD, Cucinotta F, Jones JH, Morris RV, Niles P, Rampe E, Nolan T, Fisk M, Radziemski L, Barraclough B, Bender S, Berman D, Dobrea EN, Tokar R, Cleghorn T, Huntress W, Manhes G, Hudgins J, Olson T, Stewart N, Sarrazin P, Vicenzi E, Bullock M, Ehresmann B, Hamilton V, Hassler D, Peterson J, Rafkin S, Zeitlin C, Fedosov F, Golovin D, Karpushkina N, Kozyrev A, Litvak M, Malakhov A, Mitrofanov I, Mokrousov M, Nikiforov S, Prokhorov V, Sanin A, Tretyakov V, Varenikov A, Vostrukhin A, Kuzmin R, Clark B,

Wolff M, Botta O, Drake D, Bean K, Lemmon M, Schwenzer SP, Lee EM, Sucharski R, de Pablo Hernandez MA, Avalos JJB, Ramos M, Kim M-H, Malespin C, Plante I, Muller J-P, Navarro-Gonzalez R, Ewing R, Boynton W, Downs R, Fitzgibbon M, Harshman K, Morrison S, Kortmann O, Williams A, Lugmair G, Wilson MA, Jakosky B, Balic-Zunic T, Frydenvang J, Jensen JK, Kinch K, Koefoed A, Madsen MB, Stipp SLS, Boyd N, Campbell JL, Perrett G, Pradler I, VanBommel S, Jacob S, Owen T, Savijarvi H, Boehm E, Bottcher S, Burmeister S, Guo J, Kohler J, Garcia CM, Mueller-Mellin R, Wimmer-Schweingruber R, Bridges JC, McConnochie T, Benna M, Franz H, Bower H, Brunner A, Blau H, Boucher T, Carmosino M, Atreya S, Elliott H, Halleaux D, Renno N, Wong M, Pepin R, Elliott B, Spray J, Thompson L, Gordon S, Ollila A, Williams J, Vasconcelos P, Bentz J, Nealson K, Popa R, Moersch J, Tate C, Day M, Francis R, McCullough E, Cloutis E, ten Kate IL, Scholes D, Slavney S, Stein T, Ward J, Berger J, Moores JE (2014) A habitable fluvio-lacustrine environment at yellowknife bay, gale crater, mars. Science 343 (6169):1242777–1242777. https://doi.org/10.1126/science.1242777

Hecht MH, Kounaves SP, Quinn RC et al (2009) Detection of perchlorate and the soluble chemistry of martian soil at the Phoenix lander site. Science 325:64–67

Imbus SW, McKirdy DM (1993) Organic geochemistry of Precambrian sedimentary rocks. In: Engel MH, Macko SA (eds) Organic Geochemistry. Plenum, New York, pp 657–684

Johannesson KH, Telfeyan K, Chevis DA et al (2014) Rare earth elements in stromatolites—1. evidence that modern terrestrial stromatolites fractionate rare earth elements during incorporation from ambient waters. In: Dilek Y, Furnes H (eds) Evolution of Archean Crust and Early Life, Modern Approaches in Solid Earth Sciences, vol 7. Springer Science+Business Media, Dordrecht, pp 385–410

Klein HP (1998) The search for life on Mars: what we learned from Viking. J Geophys Res 103 (E12):28463–28466

Klein HP (1999) Did Viking discover life on Mars? Orig Life Evol Biosph 29:625–631

Klein HP, Lederberg J, Rich A et al (1976) The Viking mission search for life on Mars. Nature 262:24–27

Kminek G, Bada J (2006) The effect of ionizing radiation on the preservation of amino acids on Mars. Earth Planet Sci Lett 245(1–2):1–5. https://doi.org/10.1016/j.epsl.2006.03.008

Kounaves SP, Hecht MH, Kapit J et al (2010) Wet chemistry experiments on the 2007 Phoenix Mars Scout Lander mission: data analysis and results. J Geophys Res 115(E1):E00E10

Kounaves SP, Chaniotakis NA, Chevrier VF et al (2014) Identification of the perchlorate parent salts at the phoenix Mars landing site and possible implications. Icarus 232:226–231

Lapotre MGA, Ehlmann BL, Minson SE et al (2017) Compositional variations in sands of the Bagnold Dunes, Gale Crater, Mars, from visible-shortwave infrared spectroscopy and comparison with ground truth from the curiosity rover. J Geophys Res Planets 122:2489–2509

Lasne J, Noblet A, Szopa C et al (2016) Oxidants at the surface of Mars: a review in light of recent exploration results. Astrobiology 16:977–996

Lefèvre CT, Bazylinski DA (2013) Ecology, diversity, and evolution of magnetotactic bacteria. Microbiol Mol Biol Rev 77:497–526

Levin GV, Straat PA (2016) The case for extant life on Mars and its possible detection by the Viking labeled release experiment. Astrobiology 16:798–810

Loizeau D, Mangold N, Poulet F et al (2010) Stratigraphy in the Mawrth Vallis region through OMEGA, HRSC color imagery and DTM. Icarus 205:396–418

Loizeau D, Werner SC, Mangold N et al (2012) Chronology of deposition and alteration in the Mawrth Vallis region, Mars. Planet Space Sci 72:31–43

Malin MC, Edgett KS (2000) Sedimentary rocks of early Mars. Science 290:1927–1937

McKay CP (2010) An origin of life on Mars. Cold Spring Harb Perspect Biol 2:a003509–a003509

Michalski JR, Niles PB, Cuadros J et al (2013) Multiple working hypotheses for the formation of compositional stratigraphy on Mars: insights from the Mawrth Vallis region. Icarus 226:816–840

Navarro-González R, McKay CP (2011) Reply to comment by Biemann and Bada on "Reanalysis of the Viking results suggests perchlorate and organics at midlatitudes on Mars". J Geophys Res 116(E12):E12002

Navarro-González R, Vargas E, de la Rosa J et al (2010) Reanalysis of the Viking results suggests perchlorate and organics at midlatitudes on Mars. J Geophys Res 115(E12):E12010

Navarro-González R, Vargas E, de la Rosa J et al (2011) Correction to "Reanalysis of the Viking results suggests perchlorate and organics at midlatitudes on Mars". J Geophys Res 116(E8): E08011

Noffke N, Awramik SM (2013) Stromatolites and MISS—differences between relatives. GSA Today 23:4–9

Ohtomo Y, Kakegawa T, Ishida A et al (2014) Evidence for biogenic graphite in early archaean isua metasedimentary rocks. Nat Geosci 7:25–28

Parnell J, Cullen D, Sims MR et al (2007) Searching for life on Mars: selection of molecular targets for ESA's aurora ExoMars mission. Astrobiology 7:578–604

Pavlov AA, Vasilyev G, Ostryakov VM, Pavlov AK, Mahaffy P (2012) Degradation of the organic molecules in the shallow subsurface of mars due to irradiation by cosmic rays. Geophys Res Lett 39(13). https://doi.org/10.1029/2012GL052166

Poulet F, Bibring J-P, Mustard JF et al (2005) Phyllosilicates on Mars and implications for early martian climate. Nature 438:623–627

Quinn RC, Martucci HFH, Miller SR et al (2013) Perchlorate radiolysis on Mars and the origin of Martian soil reactivity. Astrobiology 13:515–520

Schidlowski M (1988) A 3,800-million-year isotopic record of life from carbon in sedimentary rocks. Nature 333:313–318

Sephton MA, JMT L, Watson JS et al (2014) Perchlorate-induced combustion of organic matter with variable molecular weights: implications for Mars missions. Geophys Res Lett 41:7453–7460

Solomon SC (2005) New perspectives on ancient Mars. Science 307:1214–1220

Squyres SW, Arvidson RE, Bell JF (2004a) The Spirit Rover's Athena science investigation at Gusev Crater, Mars. Science 305:794–799

Squyres SW, Arvidson RE, Bell JF (2004b) The opportunity Rover's Athena science investigation at Meridiani Planum, Mars. Science 306:1698–1703

Steininger H, Goesmann F, Goetz W (2012) Influence of magnesium perchlorate on the pyrolysis of organic compounds in Mars analogue soils. Planet Space Sci 71:9–17

Strasdeit H (2010) Chemical evolution and early Earth's and Mars' environmental conditions. Palaeodiversity 3:107–116

Summons RE, Albrecht P, McDonald G et al (2008) Molecular biosignatures. Space Sci Rev 135:133–159

Summons RE, Amend JP, Bish D et al (2011) Preservation of Martian organic and environmental records: final report of the Mars biosignature working group. Astrobiology 11:157–181

Tissot BP, Welte DH (1984) Petroleum formation and occurrence. Springer, Heidelberg

Vago JL, Westall F, Pasteur Instrument Teams et al (2017) Habitability on early Mars and the search for biosignatures with the ExoMars Rover. Astrobiology 17:471–510

Wehmiller JF (1993) Applications of organic geochemistry for Quaternary research: aminostratigraphy and aminochronology. In: Engel MH, Macko SA (eds) Organic Geochemistry. Plenum, New York, pp 755–784

Westall F (2008) Morphological biosignatures in early terrestrial and extraterrestrial materials. Space Sci Rev 135:95–114

Westall F (2012) The early earth. In: Impey C, Lunine J, Funes J (eds) Frontiers of astrobiology. Cambridge University Press, Cambridge, p 331

Westall F, Cavalazzi B (2011) Biosignature in rocks. In: Thiel V, Reitner J (eds) Encyclopedia of geobiology. Springer, Berlin, pp 189–201

Westall F, de Ronde CE, Southam G, Grassineau N, Colas M, Cockell C, Lammer H (2006) Implications of a 3.472-3.333 Gyr-old subaerial microbial mat from the Barberton greenstone belt, South Africa for the UV environmental conditions on the early Earth. Philos Trans R Soc Lond B Biol Sci 361(1474):1857–1876. https://doi.org/10.1098/rstb.2006.1896

Westall F, Folk RL (2003) Exogenous carbonaceous microstructures in early Archaean cherts and BIFs from the Isua Greenstone Belt: implications for the search for life in ancient rocks. Precambrian Res 126:313–330

Westall F, Foucher F, Cavalazzi B et al (2011a) Volcaniclastic habitats for early life on earth and Mars: a case study from ~3.5Ga-old rocks from the Pilbara, Australia. Planet Space Sci 59:1093–1106

Westall F, Cavalazzi B, Lemelle L et al (2011b) Implications of in situ calcification for photosynthesis in a ~3.3Ga-old microbial biofilm from the Barberton greenstone belt, South Africa. Earth Planet Sci Lett 310:468–479

Westall F, Foucher F, Bost N et al (2015) Biosignatures on Mars: what, where and how? Implications for the search for Martian life. Astrobiology 15:998–1029

Yung YL, Russell MJ, Parkinson CD (2010) The search for life on Mars. J Cosmol 5:1121–1130

Part IV
Biosignatures in a Philosophical Perspective

Chapter 15
The History and Philosophy of Biosignatures

David Dunér

... all this universe is perfused with signs, if it is not composed exclusively of signs.
C. S. Peirce, "The Basis of Pragmaticism in the Normative Sciences" (1906, 1998, 394).

Abstract This chapter examines the human search, understanding, and interpretation of biosignatures. It deals with four epistemological issues in the search for signs of life in outer space: (1) conceptualization, how we form concepts of life in astrobiology, how we define and categorize things, and the relation between our concepts and our knowledge of the world; (2) analogy, how we see similarities between things, and with inductive, analogical reasoning go from what we know to what we do not know, from the only example of life here on Earth, to possible extraterrestrial life; (3) perception, how we interpret what our senses convey in our search for biosignatures, how the information we get from the surrounding world is processed in our minds; and (4) the semiotics of biosignatures, how we, as interpreters, establish connections between things, between the expression (the biosignature) and the content (the living organism) in various forms of semiosis, as icons, indices, and symbols of life. In all, it is about how we get access to the world, and how we interpret and understand it, for achieving a well-grounded knowledge about the living Universe.

D. Dunér (✉)
History of Science and Ideas, Lund University, Lund, Sweden

Cognitive Semiotics, Lund University, Lund, Sweden
e-mail: David.Duner@kultur.lu.se

B. Cavalazzi, F. Westall (eds.), *Biosignatures for Astrobiology*, Advances in Astrobiology and Biogeophysics, https://doi.org/10.1007/978-3-319-96175-0_15

15.1 Introduction

In the quest for life in Universe, the most likely scenario is that we one day might find signs of life, biosignatures, that indicate certain biochemical processes that could have their origin in extraterrestrial biological activity. Probably, this is what we can hope for, insomuch as we will not in the foreseeable future find ways of exploring, in situ, foreign worlds around other stars. With current technology no manned mission will take us there. The hope of finding life on other planets or moons in our own Solar System, though, has still not vanished. In some hidden corner of our Solar System some microbes or unicellular organisms might hide, but some more complex forms of life would be very unlikely to find. However, it becomes more and more clear, that among the hundreds of billions of stars in our galaxy there are perhaps many millions of Earth-like planets of which many might be habitable and have the right conditions for harbouring life. Even though we will never hold these actual life-forms in our hands, nor be able to construct an optical telescope that will let us get a glance of the planetary surface, it is fully conceivable that we in the near future will be able to refine our methods and observations in order to find signs of life.

On one hand we might come across observable and verifiable phenomena that we call "biosignatures," and on the other, we infer the existence of certain unknown instances of known biochemical processes that we call "life" that we suppose are the causes of the former. In other words, we make connections between the expression (the biosignature) and the content (the living organism). The ones who make this connection are we human beings, with our inventive minds that are a result of a particular bio-cultural coevolution of human cognition, of our species, here on Earth (Dunér and Sonesson 2016). This is what the history and philosophy of biosignatures is all about. We cannot, by no means, rule out ourselves in the inference. The data obtained, even if processed by a computer, must be comprehended and interpreted by the human mind to acquire significance. The interpreters of the "biosignatures" are and will be we. No one will help us with the interpretation. In that respect we are alone in the Universe and need to rely on our own interpretive capabilities. For us to be successful in our search for life beyond our Solar System depends on, first, obviously, that it really exists life to be discovered; secondly, that we have the technology to discover it. But, that is not enough. The discovery depends on—and that is the most critical thing—the capability of the human brain, the organization and efficiency of that systematic search for knowledge that we call science, which is a product of the socio-cultural history of our species (Dunér 2016a). Our endeavour depends on human cognition and our ability to understand and interpret what we observe in our surrounding world.

Here is not the place for discussing the actual scientific research about biosignatures per se; it is dealt with elsewhere in this volume. Anyhow, biosignatures concern various things, and refer to chemical substances (elements, molecules, etc.), but also physical features (structures, shapes, morphology, etc.), and physical phenomena (electromagnetic radiation, light, temperature, etc.). They can vary in scale from atomic to planetary magnitude, or perhaps even larger. They can be searched for both by in situ investigations and through remote indirect sensing, on our nearest planets and moons as well as in other Solar Systems. These signatures are meant to be evidence for either living life or dead life, present or past life, distinctive from an abiogenic background. What I aim at here is to examine the human endeavour to make sense of the world around us, how we think about biosignatures, rather than explaining what they are. In the following, I will delve into the epistemic questions that the search for biosignatures provokes. The epistemology of astrobiology is a less explored philosophical territory concerning the limits of astrobiological knowledge, i.e., what is known, what is knowable in practice or in principle, and what is knowably unknown (Dunér 2013a; cf. Persson 2013). The epistemological problems of astrobiology are somewhat similar to those of other branches of science, but with the exception that the limits of our astrobiological knowledge seem to be much more uncertain. In this chapter, I will discuss four epistemological issues in the search for biosignatures: (1) conceptualization, how we form concepts of life in science, how we define and categorize things, and the relation between our concepts and our knowledge of the world; (2) analogy, how we see similarities between things, and with inductive, analogical reasoning go from what we know to what we do not know; (3) perception, how we interpret what our senses convey, how the information we get from the surrounding world is processed in our minds; and (4) the semiotics of biosignatures, how we, as interpreters, establish connections between things, between the expression and the content in various forms of semiosis, as icons, indices, and symbols of life. In all, it is about how we get access to the world, and how we interpret and understand it, for achieving a well-grounded knowledge about the living Universe.

15.2 The Concept of Life

When searching for biosignatures, the first and foremost challenge is to determine if the sign is of a biotic or abiotic nature. Are there any ways of distinguish life from nonlife? This differentiation between biotic and abiotic leads us to the question "What is life?" If we are searching for something that we call "life," we should at least know what kind of phenomenon we have in mind. This seems perhaps obvious, but however, it is trickier than what we at first glance might expect. The definition of life is one of the most debated and discussed philosophical questions in astrobiology. No consensus has so far been established. In the course of the history of the philosophy of biology amounts of definitions have been put forward, one more inventive and clever than the other, but again, none seems to be exhaustive and indefectible (e.g., Luisi 1998; Cleland and Chyba 2002, 2007; Ruiz-Mirazo et al. 2004; Oliver and Perry

2006; Robus et al. 2009; Bedau and Cleland 2010; Gayon 2010; Pross 2012; Losch 2017). When examining Earth-like life, we find that it displays a number of characteristics inherited due to a common origin: it is carbon-based, uses a few specific organic molecules, and further more it is something that we perceive as alive, have some sort of energy consumption, metabolism, and that it to some extent grows, and transforms. One can make up a shortlist of ecological requirements for life (McKay 2007), e.g., energy, carbon, liquid water, and other elements such as nitrogen, phosphorus, and sulphur. Life is made of these and other chemical components, which are related to the surrounding environment, but in different proportions (Conrad 2007). Campbell and Reece (2002) and Domagal-Goldman et al. (2016) have listed a number of traits that are common to life on Earth: ordered structure, reproduction, growth and development, energy utilization, response to the environment, homeostasis (to maintain a steady internal environment regardless of the external environment), and evolutionary adaptation. But which are the most constitutive attributes of life, reproduction, evolution, metabolism, deoxyribonucleic acid, entropy resistance... or what? Definitions of life that have been put forward commonly combine especially metabolism, reproduction, and evolution as the most decisive attributes (Palyi et al. 2002). The most popular definition, NASA's "working definition," defines life as "a self-sustaining chemical system capable of Darwinian evolution" (Joyce 1994). As has been noted (Domagal-Goldman et al. 2016), its strength is that it stresses on life as an evolutionary process, rather than its chemical composition. It pinpoints life as an evolutionary process in contrast to the individual sample of life that does not undergo Darwinian evolution itself. However, one could question how such a definition would be useful when searching and analysing biosignatures. Could the life that the biosignatures refer to be tested and proven to be capable of Darwinian evolution? Perhaps more helpful for a hunter of biosignatures is the first part of the definition. A "self-sustaining chemical system" should in one way or another differ from its surroundings. In all, a definition should tell us if the phenomenon we encounter is life or not, but also be broad enough so we will not dismiss samples of "life" that are not similar or identical to terrestrial life.

The difficulty in arriving at an acceptable definition has to do with, among other things: (1) what we mean by "definition" and what such a definition should do for us in our search for life or what it should explain (cf. Persson 2013); (2) how we categorize things and how concepts are formed and used in our minds; and (3) that we know only one living planet in Universe, our own planet, and we still do not know how life emerged here on Earth. The aim of the following is not to fully scrutinize the philosophical question of the definition of life, even less to arrive at a definite definition. But before we go into the epistemic and semiotic questions involved in the search for biosignatures, which are the main target of this chapter, we need to delve into the very concept that is in the focus for our search, life, and how it is connected to our understanding of definitions, categorization, and our own ignorance.

15.2.1 The Definition of Definition

Science, as well as in the case of astrobiology, concerns concepts. We need names and abstract concepts in order to be able to talk and reason about objects, structures, processes, etc., that we gather from our senses, through observations and experiments. When we are using these terms, they need to be reasonably well defined, some sort of consent needs to be established, so we can agree on what we talk about. We need definitions. One might think that these concepts are already out there, just for us to discover, but the input we get from the surrounding world has to actually be processed by our brains and depends on the cognitive abilities we possess. The scientific concepts we use are not just dependent on the particular characteristics of human cognition that is a product a biological evolution of the human brain, it is also a product of a specific cultural evolution here on Earth and many generations of natural philosophers and scientists in the history of human, terrestrial science.

Constructing concepts in order to be able to think and talk about the new phenomena encountered is a major task for astrobiological research. Astrobiologists use a wide range of concepts, besides biosignature, for example life, habitability, habitable zones (e.g., Kane and Gelino 2012), Earth analogues, exoplanets, and other concepts and terms inherited from already well-established scientific disciplines, which together form that multidisciplinary field of research we call astrobiology. The most debated and discussed concept in astrobiology is, as said, the concept of life. However, my point here is not how we actually define life, but rather what we mean with "definition" and what it should do for us. There are a number of ingredients life needs to have in order to be life, as mentioned above, but if life is a recipe, what are the essential ingredients and which are optional ones? So far, the debate has intuitively employed an Aristotelian conception of definition (Aristotle 1966, 2.3.90b30–31), in which a "definition" is a limited list of characteristics that are necessary and/or sufficient for something to be of the type of object it is, and from which all the characteristics of the object originate. Many definitions of life tend to be a list of necessary and/or sufficient properties that something needs to have in order to be called "life," and further more, this list has the pretension to be complete. In our daily lives, however, we make relatively little use of Aristotelian definitions and depend much more on prototypes (Rosch 1975, 1978). Dogs, cats, and horses may seem to be more typical representatives for "life" than arsenic resistant microbes. In astrobiology, the prototype for life is terrestrial life, a self-sustaining chemical system capable of Darwinian evolution of that sort we find here on Earth. Another option is to comprehend concepts in Wittgensteinian terms, that there is a family resemblance among the essential features of the concept, a series of overlapping similarities, but no one common to all (Wittgenstein 1953; cf. Rosch 1987; Pennock 2012; Persson 2013), and in our case, no prototype, no typical representative of "life."

The search for life in the universe has to a great extent highlighted how strongly the evolution of life and environment are intertwined (Golding and Glikson 2011; Schulze-Makuch et al. 2015; Cabrol 2016), that the coevolution of life and environment determines the uniqueness of an extraterrestrial life form (Watson 1999; Irwin and Schulze-Makuch 2001; Kooijman 2004; Dietrich et al. 2006). A definition of life should not only encompass all the life we know, it should also be anticipatory and be prepared for putting future phenomena in one of the categories "life" and "nonlife." Future discoveries in astrobiology will most likely challenge our categorizations and definitions, that is to say, our preconception of what the world is and not is. So, we should be prepared to re-categorize and redefine our concepts. Future exobiological systematics and taxonomy will face problems for how we identify, describe, and categorize what we encounter. And in that respect, the taxonomy of future extraterrestrial life will be a product of the human mind.

15.2.2 The Categorization of Our Categories

A definition of life aims to delineate life and nonlife, and be able to deal with the intermediate evolutionary stages between nonlife and life. Our understanding of the origin of life as a bio-chemical evolutionary process supposes that there are intermediate stages from "dead" chemical elements to increasingly more complex forms of life. But where can we draw the line and say to each other that after this point the existing chemical congregations must be called life? Other entities appear as conceptual intermediaries, such as viruses, that possess many characteristics similar to organisms; they evolve, have an ordered structure, and go through maturation. But for most biologists, they could not be regarded as life in a proper sense, due to their lack of homeostasis, energy utilization, response to the environment, and perhaps most critically, they cannot reproduce independently of the metabolism of their infected hosts. Other biological entities, such as transposons, are also close to the blurred borderline between life and nonlife. The evolution of life seems to suggest that there is a continuum between life and nonlife, or grades of life, a series of increasing possession of "liveability." Could something be less or more alive? My tentative answer is yes.

It is our human way of understanding the world, by using categories, that makes problems for us when we are dealing with phenomena far beyond the well-known of our daily lives. As said, the problem of defining life has to do with how we categorize things (Lakoff 1990; Taylor 2003). All living creatures seem to categorize the environment in terms of edible *versus* inedible, benign *versus* harmful, and so forth. Categorization becomes more complex in human cognition. The human mind tends to categorize, to see hierarchies and similarities between things, such as species and genera in taxonomy (Berlin 1992; Dunér 2013b). When encountering new unfamiliar things, the old categories, systems, and beliefs are challenged and sometimes will fall short. Our thinking, science, and belief systems will have to be revised, which will lead to adjustments, adaptations, and compromises. But anyhow, we need these categories and invent new ones in order to handle the world around us; otherwise we would fall into a chaotic abyss of unsorted impressions.

15.2.3 Empirical Ignorance

Rather than defining "life," one might aim for understanding "life." In astrobiology there is an endeavour to achieve an objective concept of "life" that can guide scientific research. The question is, if such objectivity is reachable. The way towards an objective, scientific concept of life is not, by no means, straightforward. A major question concerns how to distinguish those general characteristics of "life" in Universe from those that are specific to our own life here on Earth (Gayon 2010). Without additional examples of life, as Cleland and Chyba (2002) argues, it would be impossible to know if our concept of life refers to a universal, objective natural phenomenon or is just a subjective category. Some characteristics of life might be universal in its proper sense, that there are certain necessary features that life needs to have in order to be alive. The lack of a second instance of life makes it hard to distinguish those characteristics that are universal of life from those that are inherited due to a common origin. This empirical ignorance prevents us from arriving at a definite definition. If future search for life will be successful, it will fundamentally change or concept of life.

Searching for biosignatures is to a large extent a hypothetico-deductive endeavour. We formulate a hypothesis, that certain signatures indicate biological activity, and then we search for these "biosignatures," make observations, and deduce that they indicate that life exists on that particular planet. However, only when all other hypotheses have been disproved, and only the habitation-hypothesis remains, then we can consider it to be established scientific knowledge. But in most cases biosignatures are not unambiguous. Both biological and abiotic processes can produce them. The question is how to distinguish true biogenic signatures from abiotic mimics. On one hand we need to avoid "false positives" that mimic life, and on the other hand we need to avoid "false negatives," that real signatures of life are overlooked (Tarter et al. 2007; Horneck et al. 2016). Researchers have faced this ambiguity of biosignatures in the search for the earliest traces of life on Earth (Pilcher 2003; Golding and Glikson 2011; Westall and Cavalazzi 2011) as well as in the analysis of Martian rocks (Westall et al. 2015). A famous example of ambiguous experimental results is the controversy of the Viking mission results (Klein et al. 1976; Levin and Straat 1976, 1979, 2016). Viking 1 landed in Chryse Planitia and Viking 2 in Utopia Planitia. Both had instruments, such as gas chromatograph-mass spectrometers, for performing experiments in situ to detect traces of life, to search for organic molecules that Earth-based organisms have, and gases that are consumed or produced in the metabolism of terrestrial organisms. Despite a well-equipped mission, the results became ambiguous. Partly, this ambiguity of biosignatures lies within the human mind, in our definitions, categories, and lack of empirical knowledge. Next, I will go on to another cognitive peculiarity of human reasoning that plays tricks on us in our endeavours: analogical arguments.

15.3 The Analogy of the Earth-Twins

Above, I have put forward some of the reasons why the scientific community has failed, and probably will fail, to come up with a consistent definition of life as a limited list of necessary or sufficient conditions. Such a definition might, though, be of heuristic value, i.e. as a practical, useful aid by analogy from the known. To begin with, we assume that there is only one physics, one chemistry in the Universe, and accordingly all phenomena will follow the same natural laws. We know just one inhabited planet, and everything that live on that planet is related, have one common origin. In astrobiology it is assumed that this particular planet, except that it is a living planet, has no exceptional characteristics. There are billions of stars of the same type as our Sun in the Universe, with presumably millions, billions of Earth-like planets that possess similar physical characteristics as our planet. Earth is a rather mediocre place in the Universe. This so-called mediocrity principle states that there is nothing remarkable with Earth. If this is true, it opens up for the quest for Earth-twins.

A common form of argumentation in astrobiology is analogical reasoning from what we know to what we do not know (Dunér 2013c). An analogical argument could be explained as a search for similarities, i.e., a way of selecting features in the source domain that are to be mapped onto the target domain, and of transferring relevant properties from the source to the target. If x has the properties $P_1, P_2, P_3, P_4 \ldots P_n$, and there is a y that has P_1, P_2, P_3, we may conclude that it also has P_4. If we know there is an x that has these qualities, and we discover a y that also has some of these qualities, then we conclude that all y also have the quality that we are seeking, P_4, or formulated in first-order logic: $\exists x(P_1x \wedge P_2x \wedge P_3x \wedge P_4x \ldots P_nx) \wedge \exists y(P_1y \wedge P_2y \wedge P_3y) \Rightarrow \forall y(P_4y)$. The challenge is then to select the correct and relevant salient features from an infinite number of possible ones in the source domain, features that then will be transferred to and mapped onto the target domain. In some sense, astrobiology as a whole is one single, great analogy. Starting from the one particular type of life we happen to know something about, namely life on Earth, we proceed to search for life on other planets. We predominantly look for life as we know it: something needing liquid water, being based mainly on carbon, inhabiting a planet of a certain magnitude, gravitation, atmosphere, chemistry, temperature, etc., that revolves within the habitable zone, at a certain distance, period, eccentricity, inclination, etc., of a solar-type star (a G2 main sequence star of 4.5 Gyr of age), which in turn has to be of a certain size and temperature, and so on. There are a number of problems with such an analogical argument: first of all, do we know all the necessary and sufficient conditions for life? Another problem is that we restrict ourselves to one particular sort of life we happen to know, life as we know it, and might overlook other forms of "weird life," life as we do not know (Davies et al. 2009). The third problem lies in the unpredictability of evolution, that there are some stochastic events involved in the evolutionary process. The question is if all ingredients of life are in place, will that inevitably, by necessity lead to the emergence of life? Is life a natural manifestation of matter? (cf. de Duve 1995).

Logically speaking, analogical arguments are invalid. The historian of science Michael Crowe pointed out that one of the most pervasive logical and/or methodological fallacies in the plurality debate is mistaking necessary conditions for sufficient conditions (Crowe 1986). If liquid water is a necessary condition for life as we know it, it is just one among numerous other necessary conditions, which all need to be present if a planet is habitable. Just an evidence of an atmosphere, even if it contains the right chemical components including water vapour, is not enough to prove that life exists on its surface. All too often, Crowe states, finding some few necessary conditions among a larger set of necessary conditions, has been taken as a proof of life. If all necessary conditions need to be in place, the question arises if we know the complete set of conditions, or are there unknown conditions still to be discovered? The conclusion is that, not until we know all necessary and sufficient conditions for life, we will only be able to infer that the planet is habitable, but not that it is actually inhabited.

The fallacious use of analogical arguments were also discussed by the philosopher and logician Charles Sanders Peirce, stating that "There is no greater nor more frequent mistake in practical logic than to suppose that things that resemble one another strongly in some respects are any more likely for that to be alike in others" (Peirce 1957, 134; Crowe 1986). However, Peirce admitted, analogical argumentation could be accepted as a method of discovery, but not a method of proof. In providing us with some point of departure, analogical arguments still might hold some heuristic advantage in the search for life. Astrobiology as a great analogy is an inductive argument that cannot logically prove anything, just propose that life in Universe is a theoretical probability, but not an inductive probability. Just one sample of life, our own here on Earth, is a rather poor, to say the least, start of an empirical-inductive argument. Nor are large numbers, for example of stars, exoplanets, Earth-twins, any conclusive ground for an argument, just a theoretical probability.

What we are actually looking for is something that reminds us of ourselves, something similar to us, to earthly life, whether microbes or more complex life. In fact, to stretch it a bit, we are searching for "ourselves." Though, life might be very different from what we imagine. The history of science is actually a history of surprises. The world we are living in turned out to be very different from what we first thought: richer, more complex, more peculiar and more astonishing than what we could dream of. This will also be true for astrobiology. Future discoveries in astrobiology will surprise us completely.

15.3.1 The Mountains of the Moon

Analogical reasoning is very common in the history of astrobiology and in the search for biosignatures. Many works have discussed the history of the extraterrestrial life debate and the emergence of the science of astrobiology (see e.g. Dick 1982, 1996; Guthke 1983; Crowe 1986, 2008; Sullivan and Carney 2007; Dunér 2012, 2016b; Briot 2013; Crowe and Dowd 2013; Dunér et al. 2016), but the concept of biosignature has not got particular attention. In the extraterrestrial life debate of the seventeenth and eighteenth centuries our closest celestial body, the Moon, was the prime candidate for life on other worlds. A number of scientists and scholars also speculated about life on Venus, Mars and on other planets, both within our Solar System and beyond its frontiers.

November 30, 1609 Galileo trained his telescope on the Moon and, like the Earth, found it to be rugged and uneven, perhaps even having similar mountains and oceans. The first images of the Moon seen through a telescope were published in his book *Sidereus nuncius* (1610)

In the autumn of 1609, Galileo Galilei made the first closer observations of extraterrestrial bodies through a telescope. In the *Sidereus nuncius* from 1610, he shows, based on his telescopic observations and analogical reasoning, that the Moon has mountains and therefore has the same solid, opaque and rugged nature as the Earth. The irregular border between its dark and illuminated parts is incompatible with the idea that it is a perfect spherical solid. Galileo wrote: "Anyone will then understand with the certainty of the senses that the Moon is by no means endowed with a smooth and polished surface, but is rough and uneven and, just as the face of the Earth itself, crowded everywhere with vast prominences, deep chasms and convolutions" (Galileo 1610; Spranzi 2004: 459). The Moon had a smooth appearance, though, in its contour, which he explained was because it might have an atmosphere. Galileo never stretched his analogical reasoning so far as he could by clearly claiming the existence of life on other planets. However, he did not consider it impossible that there were inhabitants on these spheres. And further more, we could not take it for granted that life elsewhere in the Universe must resemble our own. In 1612 Galileo wrote: "I agree with Apelles [the astronomer Christoph Scheiner] in regarding as false and damnable the view of those who put inhabitants on Jupiter, Venus, Saturn and the Moon, meaning by inhabitants, animals like ours, and men in particular" (Galileo 1957: 137; Dick 1982). Later in the *Dialogo … sopra i due massimi sistemi del mondo* (1632), he stated that there is no water, no humidity, no seas on the Moon, and therefore no life (Galileo 1632, 1953; Spranzi 2004).

In his *Cosmotheoros*, the Dutch scientist Christiaan Huygens (1698) expressed the view that it was highly probable that there was life on other planets. He noted that liquid water is necessary for life, and he saw darker and lighter spots on the surfaces of Mars and Jupiter that he interpreted as water and ice. Beyond our Solar System there are stars similar to our sun, and he asked why these stars could not have their own planets with their own moons. As for Venus, he empirically stated that a thick atmosphere surrounds it. He could not clearly detect any patches on the surface that might be signs of seas and mountains. Perhaps, he said, there are no seas on Venus, or, as he believed more probable, the air and clouds around Venus reflect nearly all the light from the Sun.

The first more certain telescopic observations of Venus, after Galileo's discovery of its phases, were made during the 1761 transit of Venus. Many observers reported certain phenomena during the transit that they believed to have been caused by an atmosphere surrounding Venus (Proctor 1874; Woolf 1959; Maor 2000; Sellers 2001; Sheehan and Westfall 2004; Aspaas 2012; Wulf 2012; Sterken and Aspaas 2013). Certain astrodynamic facts relating to Venus were well known to the astronomical community, for example, its orbit around the Sun, magnitude, and phases, etc. However, the question of the atmosphere and topography of Venus was still unresolved. New observations of Venus against the solar disc changed the situation. The analogy argument started from the general supposition that there were no actual differences between Earth and Venus. They were both planets that orbited the Sun, were of similar size, solid, and, as some astronomers claimed, Venus also possessed mountains and an atmosphere. If there is life on Earth, then one may ponder why it could not also exist on Venus. If we can estimate the axial rotation and detect an atmosphere and mountains on that planet, it might also be true that it harbours life. These questions were in fact those that were investigated during the seventeenth and eighteenth centuries, and they included the length of its period of rotation and whether it had mountains, an atmosphere and life. The Russian polymath Mikhail Lomonosov argued that his observations of the transit of Venus in 1761 supported the idea of a Venusian atmosphere: "Based on these observations I conclude that the planet Venus is surrounded by a distinguished air atmosphere similar (or even possibly larger) than that is poured over our Earth" (Marov 2005: 214f). Because an existing atmosphere had been proved, then it could be concluded that Venus is also inhabited.

Later Johann Elert Bode (1801) at the Berlin Observatory accepted astronomer Johann Hieronymus Schröter's (1793, 1796) claims about the existence of mountains and valleys on Venus (Crowe 1986). Bode applied an apparently analogical reasoning. He concluded that if Venus had land and sea, mountains and valleys, clearings and condensations occurred in its atmosphere, and it had a companion moon, then it was entirely similar to our Earth and consequently also habitable. The French populariser of astronomy Camille Flammarion (1862) considered in *La pluralité des mondes habités* that it was absurd that the Sun was employed solely to illuminate and heat our small world. This absurdity became even more striking when Venus was found to be a planet of the same dimensions as the Earth, with mountains and plains, seasons and years, and days and nights analogous to our own. That analogy was expanded to the conclusion, that, since they are alike in their physical characteristics, they must be alike also in their role in the Universe: "if Venus were without population, then the Earth must be similarly lacking, and reciprocally, if the Earth were populated, Venus must be populated also" (Flammarion 1862; Crowe 2008: 417f). Flammarion demonstrates here a characteristic analogical thinking, typical of the astrobiological search for an Earth analogue. In a later work on popular astronomy he says of the inhabitants of Venus: "this world differs little from ours in volume, in weight, in density, and in the duration of its days and nights. [. . .] It should, then, be inhabited by vegetable, human, and animal races but little different from those of our planet" (Flammarion 1880, 1907: 371; Sheehan and Westfall 2004).

Such analogical arguments could be summarized as a search for similarities in an inductive manner, in order to pinpoint as many as possible, especially those of a significant nature, i.e., those critical features that were believed to indicate a habitable environment. It was known that both Earth and Venus were planets of a similar size, both orbited the Sun, and were exposed to its light and heat, and that both globes were opaque and had a solid ground. These similarities could be extended, as some astronomers maintained, to both of them having nearly exactly the same period of axial rotation, as well as a companion moon, an atmosphere, mountains and seas. If Earth and Venus seemed to be perfect twins, then there must be life on Venus too.

The last decades, analogy reasoning lies behind a number of experiments on terrestrial life as preparation for in situ investigation on extraterrestrial bodies. Terrestrial biosignatures have been treated as analogical models for possible extraterrestrial biosignatures. If we turn to ourselves, are we detectable; is it possible to discover life on Earth from outer space? Based on observations made with the Galileo probe, Sagan et al. (1993) proposed what biosignatures might be possible to observe in the reflective light of Earth. According to the spectroscopic results, water, oxygen, ozone, carbon dioxide, carbon monoxide, nitrous oxide and methane were detected in the atmosphere of the Earth. The Very Large Telescope (VLT) of the European Southern Observatory (ESO) in the Atacama Desert, Chile, was used by Sterzik et al. (2012) for studying the Earthshine, the light from Earth that reflects on the surface of the Moon, seen as a greyish light on the lunar surface which is not lighten by the sun. The colour and polarization of the Earthshine showed that Earth's atmosphere contains clouds, that its surface is partially covered by a sea, and that it has vegetation. The idea was that by studying how Earth look like from space, one could get a reference for future analyses of exoplanet atmospheres. Analogy reasoning, helpful or not, rests on another cognitive process. In order to make an analogy between observations, one needs to interpret the perceptual information.

15.4 Epistemic Perception

The philosopher in Bernard de Fontenelle's *Entretiens sur la pluralité des mondes* (1686) states that all philosophy (understood as the natural sciences) is based on two things: curiosity and poor eyesight. We want to know more than what we can see. In contemporary astrobiology life seems always be beyond the fields we know. What the senses convey have to be interpreted through means of specific cognitive processes, and the interpretation of what has been observed is based on a preconceived understanding, concepts, and prior knowledge. Observations are not separate from theory. In contrary to the tendency to place excessive trust in "objective" observations, many philosophers of science, such as N.R. Hanson, Thomas Kuhn, and Michael Polanyi, have emphasized that observations are theory-laden, that there are no sharp dichotomy between observation and theory in scientific research (Crowe 1986). We need theories to understand what we see, and our preconceptions and expectations lead us in one or another direction, sometimes the wrong course. Not seldom we see what we expect to find.

In the optical observations of distant worlds, preconceived understanding often shapes the interpretation of what the observers see. Through their senses, the observers receive impressions from outer space, and they collect and collate information using their sight. What their senses convey have to be interpreted by means of specific cognitive processes before becoming reality. As observers, we do not just passively receive images and input from the world around us. Instead, the brain actively searches out patterns in what is conveyed to it through the senses, and interprets them through a process that is determined by both biological and cultural factors. Perception is not a neutral, objective, and realistic recording of reality. This conceptual or epistemic perception implies an identification of what is seen, and takes place by applying our concepts to visual perceptions, that is, concepts affect what we see, and, should we lack any concept of a specific phenomenon, then it will be difficult to distinguish it among all our impressions. The world distorts our concepts, and the concepts distort our world.

15.4.1 The Atmosphere of Venus

Astronomer William Herschel's lunar observations from 28 May 1776 highlight the difficulty to see and arrive at a conclusive interpretation of the seen. He saw growing substances: "My attention was chiefly directed to Mare humorum, and this I now believe to be a forest, this word being also taken in its proper extended signification as consisting of such large *growing substances*" (Crowe 1986, 63). This forest would require, he said, trees at least 4–6 times the height of ours. In 1778 he even suspected that lunar craters could be towns of the Lunarians. He saw circular buildings on the Moon: "I am almost convinced that those numberless small Circuses we see on the Moon are the works of the Lunarians and may be called their Towns" (Crowe 1986, 65). Those certain luminous spots that occasionally could be seen on the dark side of the Moon, Herschel explained as volcanoes in eruption. Church minister and science teacher Thomas Dick later wrote that a more pleasing idea would be that they were "some occasional splendid illuminations, produced by the lunar inhabitants, during their long nights" (Crowe 2008, 262f.).

Other striking examples of this epistemic perception are the maps of Venus that delineated the surface of the planet, which showed mountains and other geological features. Even a dim light, faint spots and lines, and a companion moon seemed to appear when Venus was viewed through a telescope. The astronomers interpreted their obscure observations in line with their prior knowledge and their ideas of the nature of the world, and they often found what they sought. If they believed in the existence of mountains and an atmosphere on Venus, then they duly found them. In 1645 the Neapolitan astronomer Francesco Fontana recorded "a dark patch in the centre of the disk" of Venus, which can be said to be the first attempt to note surface details there (Fontana 1646; Moore 1956; Cattermole and Moore 1997). In 1667, the astronomer Giovanni Domenico Cassini saw "various bright and dusky patches" from which he deduced the first estimated period of rotation of 23 h and 21 min (Cassini 1667). Another Italian astronomer, Francesco Bianchini, drew the first "map" in 1726 recording oceans and continents (Bianchini 1728; Sheehan and Westfall 2004). On mist-free days, at twilight, he saw rounded patches similar to lunar craters, and from their movements, he deduced that the period of rotation of Venus was 24 days and 8 hours. There is no doubt that these records of the surface features of Venus were purely optical. Beside the fact that the optical quality of the telescopes was not always reliable, and that weather conditions could considerably influence the quality of the observations, there is obviously also an epistemic perception that changes the interpretations of the seen. The uncertain observations by Fontana, Cassini, Bianchini and others were interpreted in a particular way. If they believed in the existence of oceans and continents on Venus, they searched for them and found them, because their prior knowledge and beliefs directed their attention towards certain interpretations. The illusion or fallacy in their perception did not lie primarily and only in the flaws in their optical equipment, but in their imaginative minds, the cognitive apparatus that processed their sensory impressions.

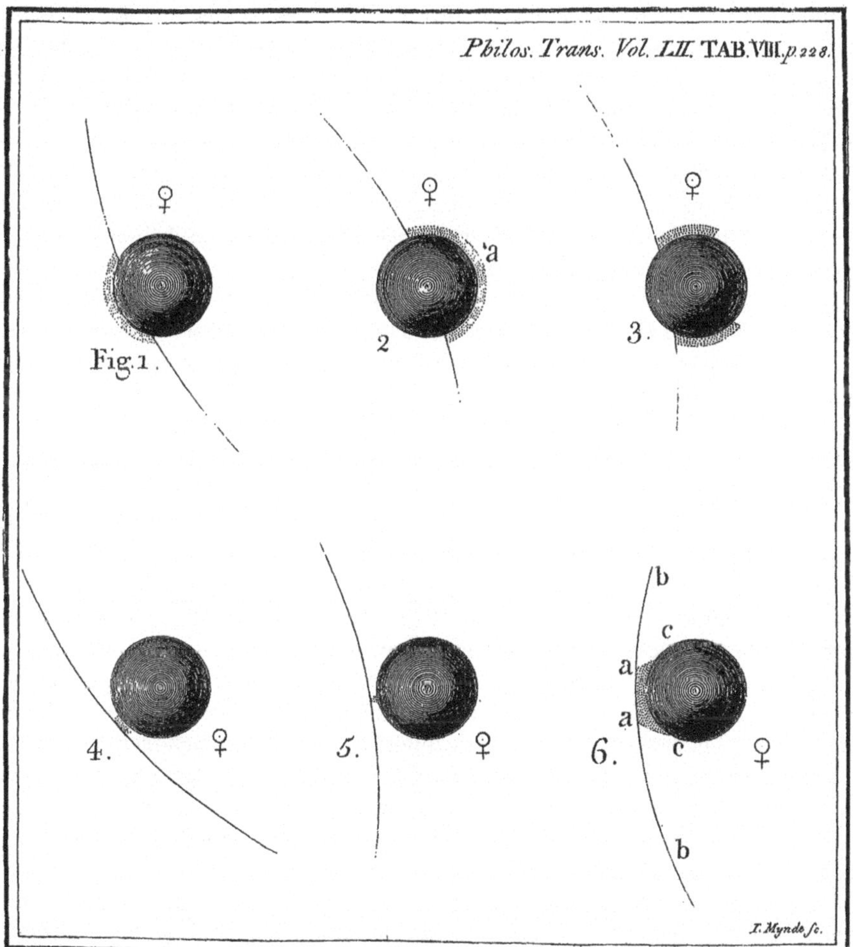

The bright ring and the black drop seen from Uppsala during the Venus Transit of 1761. From Torbern Bergman's letter published in *Philosophical Transactions* (1762)

During the transit of Venus of 6 June 1761, two surprising phenomena were observed: a bright ring around Venus and the "black drop" during the contacts. The ring was re-observed on 3 June 1769, and its causes were still being debated even then, but it was unanimously taken as proof of the existence of an atmosphere on Venus. Concerning the black drop, the astronomer Daniel Melanderhjelm explained it as caused by an atmosphere. The secretary of the Royal Swedish Academy of Sciences, Pehr Wilhelm Wargentin, on the other hand, saw it as a mere diffraction phenomenon (Dunér 2013c). Melanderhjelm's observations were, as he said, just "fallaciæ visus," optical fallacies. The physicist Johan Carl Wilcke performed a number of experiments during the summer of 1769 showing that the same phenomenon arises with a black body seen against a luminous body without any need to assume an atmosphere. Wargentin and Wilcke, however, did not disbelieve in the existence of an atmosphere, but the black drop could not provide proof of it.

Whether Venus has mountains and a surface similar to the Earth with valleys and seas had been debated ever since the first Venus maps appeared in the seventeenth century. Here, again, the conclusions were often a result of analogical reasoning and epistemic perception. The great observational astronomers Herschel and Schröter engaged in a heated argument as to the presence or absence of mountains on Venus. However, they both agreed that Venus has an atmosphere. It was well known in the era of Schröter and Herschel that Venus and the Earth, with regard to their size and mass, were almost perfect twins. It then became also reasonable to assume that their atmospheres were similar too, with regard to their extent and composition. In February 1788 Schröter perceived the ordinarily uniform brightness of the disk as being marbled by a filmy streak, and he concluded that what he was seeing was the outmost part of a dense, cloudy atmosphere (Cattermole and Moore 1997). Moreover, the horns of the crescent were seen to extend beyond the semi-circle, which could not be the case in the absence of an atmospheric mantle. On 28 December 1789 Schröter saw that the southern cusp of Venus was blunted, and that there was a small luminous speck beyond it (Baum 1973). He saw the same thing again in 1790 and 1791 and concluded from these observations that it must be a very lofty "enlightened mountain" that was catching rays of the Sun. Herschel re-observed Venus in 1793, and he questioned Schröter's findings. He agreed that Venus has an atmosphere, but he never found those high mountains that Schröter mentioned (Herschel 1912). In the *Philosophical Transactions* he states: "As to the mountains in Venus, I may venture to say that no eye, which is not considerably better than mine, or assisted by much better instruments, will ever get a sight of them" (Herschel 1793, 216; Moore 1956; Baum 1973).

The observers of Venus saw things that needed explanations and interpretations. They seemed to detect vague spots, streaks, lines, drops, a dim light, and a vague companion. Such optical misinterpretations, or rather what could be explained as an epistemic perception, were involved in the claims of a habitable Venus. The Ashen light, the dim visibility of the non-sunlit side of Venus, when it is in the crescent stage, was first reported in 1643 by the Italian Jesuit and astronomer Giovanni Battista Riccioli. This phenomenon was also the object of an epistemic perception leading to certain interpretations of the seen. The German astronomer Franz von Paula Gruithuisen in Munich declared that the light had been seen in 1759 and again in 1806, an interval of seventy-six Venusian years, and he wrote: "The observed appearance is evidently the result of general festival illumination in honour of the ascension of a new emperor to the throne of the planet" (Cattermole and Moore 1997, 17). However, Gruithuisen later modified his explanation and instead of a Venusian coronation, he suggested that the light might be caused by the burning of large areas of jungle to create new farmland. In a paper from 1824, "Discovery of Many Distinct Traces of Lunar Inhabitants, Especially of One of Their Colossal Buildings", Gruithuisen even reports that he had observed cities, forts, a temple, and animal trails on the Moon (Crowe 1986; Sheehan and Dobbins 2001). Later in the nineteenth century, the invent of spectroscopic analysis brought hope that lines for oxygen and water vapour in the atmosphere of Venus could be detected, and some preliminary results seemed to support that. It was only in 1966 and 1967 that the Russian space probes Venera 3 and Venera 4 dived into the cloud shield, and a very hostile Venusian environment was discovered (Harvey 2007).

15.4.2 The Canals of Mars

Perhaps the most famous example in the history of astrobiology of such epistemic perception is the debate concerning the canals of Mars. In 1877 Giovanni Schiaparelli at Brera Observatory in Milan recorded in detail the Martian network of canals for the first time. In the beginning other astronomers had problem to confirm these observations, but during the next decades many succeeded. Schiaparelli's findings were confirmed by the American astronomer Percival Lowell who detected hundreds of Martian canals that he interpreted as an artificial irrigation system (Hoyt 1976; Strauss 2001). In his book *Mars* (1895) he claimed that he now had conclusive evidence of an intelligent civilization on the dying planet of Mars. A change of Martian coloration was also observed, as evidence for seasons and vegetation. A darkening of the surface spread when the spring came on.

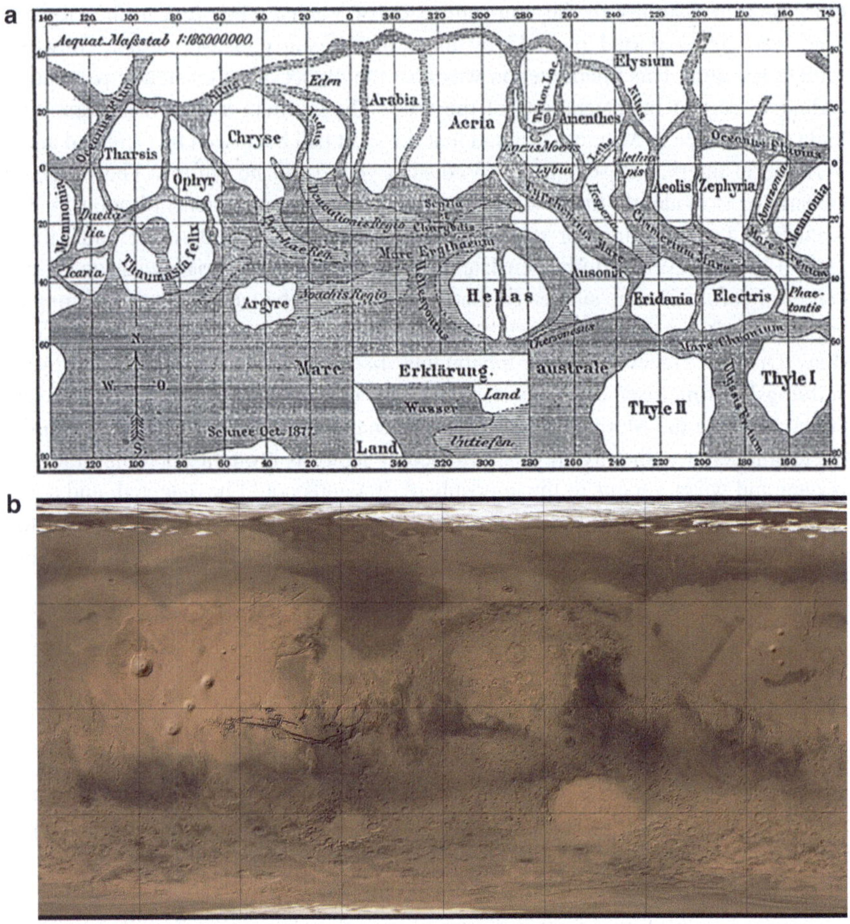

Giovanni Schiaparelli's map from 1888 of the canals of Mars compared to a modern satellite map

But there were sceptics. Edward Walter Maunder of Greenwich Observatory questioned the canal observations as mere optical illusions. The eye tends to integrate details, he explained, that are below the limits of vision. Eugène Antoniadi used the high quality telescope at Meudon Observatory, and made a drawing of a specific region of Mars, which he compared with Schiaparelli's drawing from the same area. Where Schiaparelli's drawing showed canals, Antoniadi's saw just diffuse details, no canals. Schiaparelli's canals had been, according to Antoniadi, created by connecting dots on the surface or on the border between darker and lighter regions. In the American magazine *Popular Science* 1912, the reader could by himself experience this optical illusion with a simple, practical test consisting of random dots (Hennessey 1999). By then most astronomers had abandoned this imaginative idea.

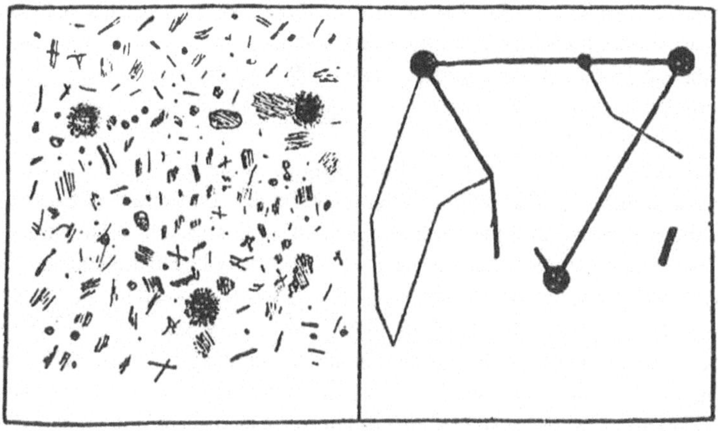

Is Professor Lowell deceived? Hold this 20 feet away, and the marks on the left look like the lines on the right. From *Popular Science* (1912)

Mapping the planets involves great patience, to monitor the shimmering images of the planets disturbed by the atmospheric conditions on Earth. The blurred images seen in the telescope were faint sensory impressions that needed interpretations by applying prior knowledge and conceptions to what the eyes perceived. The active mind searched for regularities, order, and comprehensibility in the observations. The story of human observations of the Moon, Venus, and Mars, tells us about the imprecision of the human eye-brain-hand system under difficult conditions (Sheehan 1988; Sullivan and Carney 2007). When searching for biosignatures, we are looking for regularities, order, and connections, based on previous knowledge, expectations, and theories. Next, I will turn to the question how we connect the expression (the sign of life) with its content (life).

15.5 Semiotics of Biosignatures

The semiotics of biosignatures is about the meaning-making processes of the human mind and its ability to make meaningful connections between things. The problem of biosignatures is very much a semiotic problem: how meaning can be discovered, invented, deciphered, and interpreted. One might say that the science of astrobiology "invents" connections between the signifier and the signified, expression and object, "signs of life" and "life." The first problem that arises in a situation of interpreting a biosignature is realising that it really is a sign at all. Some regularity and order, or finding a repetition in the pattern is not enough. The sign should be recognized as such by the interpreter, i.e., that it contains an expression that refers to a content, leading to an interpretive process by the interpreter. In other words, the interpreter needs to identify the physical phenomenon as containing semiotic meaning, something that can be a sign of life, a biosignature, that has a particular meaning by referring to its content "life." In our everyday lives as well as in well-established science, the expression of a phenomenon can easily be connected to its content. Seeing the footprints of an animal, we can infer—based on previous knowledge—what kind of animal it is, its weight and its direction. However, this semiotic confidence becomes much more uncertain when we turn to biosignatures of unknown forms of life. We are compelled to a number of assumptions that underlie the connection we try to establish between the expression and the content. Our assumptions might be mistaken, which leads to wrong interpretations, or even worse, we fail to make any assumptions at all, in other words, the phenomenon encountered, "the signature," would not be recognized as a meaningful sign whatsoever. Even though we have good reasons to believe that the connection we infer between the expression and the content, between the biosignature and the living organism, is scientifically correct, we need to rule out other explanations of the signrelation. The "biosignature" might not be a true biosignature at all, but instead is caused by an unknown or known abiotic process.

Before we go further into the semiotics of biosignatures, it would be in place to explain some fundamental semiotic concepts. As a study of signs, modern semiotics has a long prehistory that could be traced back to the philosophy of Plato and Aristotle, to John Locke and early modern medicine where semiotics was the interpretation of bodily signs, to draw conclusion from symptoms, to arrive at appropriate treatment of the patient. Particularly Peirce and Ferdinand de Saussure in the turn of the last century formed semiotics (or in the latter's case, semiology) of today. A certain sub-field of research, biosemiotics, has focused on the production and interpretation of signs in the biological realm (Wheeler 2006; Hoffmeyer and Favareau 2008; Romanini and Fernández 2014). Cognitive semiotics, on the other hand, studies the meaning-making structured by the use of different sign vehicles, and the properties of meaningful interactions with the surrounding environment, both with the physical and the social environment (Sonesson 2007, 2009; Zlatev 2015; Dunér and Sonesson 2016). Following attempt to uncover the meaning-making strategies in the search for biosignatures takes its departure from cognitive semiotics and related fields of research.

Semiotics is the study of meaning making. It concerns *signs*, which can be said to be something that we interpret as having meaning. According to Peirce, one of the main figures in semiotics, "A sign, or *representamen*, is something that stands to somebody for something in some respect or capacity" (Peirce 1932, 135; Saint-Gelais 2014). The sign, as *expression*, stands for something, its *object*. The sign does not include its meaning, rather the meaning is attributed through elaboration of an *interpreter*. So for something to be meaningful, an interpreter is needed, a human being (or other meaning-making creatures) who endows the sign a meaning. A physical phenomenon is meaningless so far as there is no one to recognize it as meaningful. Phenomena that we call biosignatures become meaningful phenomena, when we interpret them as containing a meaning by making a connection between the expression and the object, in other words, between the "biosignature" and "the living organism." The signs are the way we make sense of the world, to approach it, to get access to it and differentiate things from each other. Correspondingly, the biosignatures we detect are one way among many other ways of making sense of the data we receive from outer space. And these biosignatures exist so far as we find them meaningful. In that perspective, the biosignatures are not solely "out there," instead, they are to a great extent in our minds, in the interaction between our minds and the outer world. It is in our meaning-making practices the "biosignatures" become biosignatures. The *semiosis* is the sign process, how the signs operate in the production of meaning. In that sense, when the astrobiologist is interpreting biosignatures, he or she is involved in a meaning producing semiosis. This semiosis is triadic, it contains expression, object, and interpreter—which in our case respond to "biosignature," "life," and "astrobiologist." Depending on how the interpreter makes or interprets the connection between the expression and the object, we have basically three types of signrelations, *icon*, *index*, and *symbol*. But, let us return to these signrelations in more detail later on. For now, it is enough to conclude that the astrobiologist searching for biosignatures is a sort of semiotician, an astrobio-semiotician, trying to establish connections between expressions and objects in the Universe. Semiotics of biosignatures concerns qualities and categories, as well as the search for rules and regularities within such a nomothetic science as astrobiology concerned with generalities. More generally, semiotics of biosignatures concerns the meaning-making processes of astrobiology.

Finding the connection between expression and content are actually mental, in the mind of the interpreter. The search for biosignatures is based on the human endeavour of connecting things with each other, and of selecting the right elements for the connection among a wider range of possible elements. We ask ourselves, what are the meaningful properties of the information we gather through spectrometers, radio or optical telescopes, etc.? Which signatures (phenomena) have meaning and which are just meaningless noises? Hence, we are looking for the meaningful signs among a chaotic "noise" of data, the "biosignatures" that clearly "says" that they are signs of "life." The signifier is directly given, but the signified is only indirectly present, through the link with the signifier. "Life" (the signified) that we are searching for is just indirectly reachable for us, and we have to content ourselves with the only thing that is directly given, the biosignature (the signifier). As interpreters, we determine the relation between the signifier and the signified by picking out those elements we assume to be relevant. The challenge of the astrobiologist is to pick the right elements (properties) of the signifier. For example, when examining a Martian rock, we need to pick out those elements (shapes, molecules, etc.) that direct us to the signified, the living organism. And we need a reasonable explanation of the link between the signifier and the signified. Why, and how come, is this particular gas a result of a certain metabolism of a living organism? In what way is this shape a remnant of the morphology of a living organism? We need to know the physical processes that let us link the signifier with the signified.

In the following three sections, I will show that "biosignatures" is a very diverse category, not only in respect to its immense variety of expressions, but also in semiotic sense. It shows a great variation in signrelations that each has its particular epistemological problems. To begin with, in astrobiology the concept of "biosignature" is not completely unambiguous. The concept of "biomarker" is sometimes used interchangeably with "biosignature," but often restricted to refer to "an organic molecule whose origin can be directly related to an organic component of life" (Horneck et al. 2016, 227), or organic compounds characteristic of certain organisms, for example hopanoids in cell membranes of cyanobacteria. Biosignature, instead, refers to a broader range of signatures of life: morphological, geochemical, and organic—a diverse and ambiguous category of signatures related to either living organisms or fossilized remains. Biosignatures could be studied *in situ*, being of physico-chemical, geological, morphological, and mineralogical nature. Or they can be detected by remote methods, such as atmospheric spectroscopy, chemical disequilibrium, isotope ratios, etc. (Hegde et al. 2015). Physical microbial structures, stromatolites, mud mounds, atmospheric gases, etc., could be biosignatures, varying in scale from prebiotic molecular features to entire planets, referring to both life as we know and weird life, past and present life.

The following is a first attempt to bring some semiotic order in this chaotic variation of signs. Based on Peirce three signrelations—icon, index, and symbol—one could at least reveal some peculiarities of the semiosis of biosignatures. The meaning of the relation between expression and content, that the interpreter experiences, is based on either similarity (iconicity), proximity (indexicality), or habits, rules, or conventions (symbolicity). An icon is when the expression shares some *similarity* with the object. An index is when the expression has some *contiguity* with the object. And finally, a symbol is when the connection between expression and object is just a mere convention. Thus I put forward three kinds of biosignatures in semiotic sense: bioicons, bioindices, and biosymbols. However, in many cases a sign could be a combination of index, icon, and/or symbol, also as we will see in the case of biosignatures.

15.5.1 Icons of Life

Aristotle noticed similarities between certain seashell-shaped structures found in rocks and those that washed ashore on the beach. Are there a connection between the petrified seashells and the living ones? And how come? In the beginning of the eleventh century, the Persian polymath Ibn-Sīnā (Avicenna) put forward the theory of petrifying fluids as an explanation of the petrification. Through the ages, the question was debated, if these figure stones that resembled living organisms actually were fossilized seashells or just sports of nature, *lusus naturae*, if they grow in the bedrock or were traces of once living animals exterminated by the flood. In 1665, the Danish anatomist and geologist Nicolaus Steno found shark teeth in the Tuscan mountains, suggesting that where it is now high mountains, it had once been a sea (Cutler 2003). Other findings, however, seemed to have no counterparts in the living species. By the end of the eighteenth century, the French paleontologist Georges Cuvier began realizing that they actually were remnants of extinct species.

The idea of fossils was also combined with the idea of extraterrestrial life. If meteorites were coming from outer space, as it was realized in early nineteenth century, rather than being ejecta from volcanoes, these could be studied by chemical and geological methods. These meteorites from other worlds might contain evidence of extraterrestrial life, if not alive, in fossilized form. The analytical chemist Jöns Jacob Berzelius (1834) discovered that meteorites contained organic materials (hydrocarbons). He examined meteors from a meteor shower in November 1833, made a chemical analysis of a carbonaceous chondrite that had fallen in 1806, in Alais (Alès), France, but could not tell if it contained carboniferous of extraterrestrial origin (Crowe 1986). In the 1870s it was consensus that some meteorites contained organic materials, but no convincing evidence had been found that they contained extraterrestrial life forms. With Berzelius and others the idea of panspermia, that seed-bearing meteoric stones are moving around in outer space, became an increasingly plausible area of research, ending up in the physical chemist Svante Arrhenius's (1907) more elaborated panspermia hypothesis. And still, the hypothesis has not been completely ruled out. Current research has shown that microbial life could indeed travel between planets and survive in space (Horneck et al. 2010).

A theme in the history of palaeontology is the question of how to distinguish real remnants of living organisms from structures that just mimic living forms, to distinguish fossils from pseudofossils. These inorganic pseudofossils can be mistaken for fossils, for example branch-like structures like manganese dendrites in limestone, kidney ore, moss agates resembling moss leaves and other patterns in rock that arise through geological, not biological processes. This is still a challenge in the quest for microfossils for tracing the early history of life on Earth or in order to find fossilized life in Martian rocks. A famous example is the announcement in 1996, that fossilized life had been discovered in the Martian meteorite ALH84001 (McKay et al. 1996). Viewed under an electron microscope, certain tube-like structures in the meteorite resembled fossilized bacteria. It was a premature claim, abiological processes could in fact create these structures (Westall et al. 1998). This calls for new samples when claiming evidence for fossilized life in rocks on Mars or beyond. A second lot is needed to confirm or refute previous hypotheses based on the first batch of samples, or to continue the search.

Biosignatures in the form fossils are distinctly another thing than remote sensing of habitable atmospheres, not because how they are found, in situ, but in its signrelation. Biosignatures that share a similarity with living organisms, for example fossils, are in my terminology *bioicons*. In semiotics, an icon is a signrelation based on similarity, where the expression shares some of the object's properties. The similarity between properties is perceived on the background of other dissimilar properties. The most obvious examples of bioicons are body fossils, the imprints of the hard parts of animals and plants, where the imprints of skeletons or foliage let us, based on morphologic similarity, establish a link between the fossilized structure and the living thing. The very complexity of the expression (the fossil) directs us to the conclusion, based on the supposition that such a complexity cannot be the result of any known abiotic process. Microscopic fossils, microfossils, though, are more challenging. All life as we know it share the characteristics of having internal volumes isolated from the surrounding environment by a cell membrane. Based on this shared morphology, one could search for cellular structures. Well-preserved fossil cells can be identical in size, shape, and structure with living single-cell organisms. Their structures, that show less complexity, make it however more challenging to distinguish a biotic structure from an abiotic. On Earth, the Apex chert from Western Australia, dated at ~3.5 Ga, has been claimed either to be fossilized cells of filamentous bacteria or just a result of an abiotic process (Schopf 1993; Brasier et al. 2005). To be a true biosignature, it is not enough to notice a similarity between the expression and the object. We also need an explanation for how a living organism can become a fossil (a bioicon). If we find something that reminds us of a living thing, a microbe, a cell or that like, we need also a theory that links the living organism with the biosignature, establishing a physical correlation between the bioicon and the thing it signifies. Plenty of performed experimental fossilisation studies give us the right arguments to make this connection. Further more, fossils are not enough if we want to get a complete understanding of the life it refers to; they do not give us complete information of the biochemical nature of the living organism.

Bioicons are not just of visual nature, a similarity based on morphology or structure, they could exist in any sense modality. Based on chemical analyses, the researcher sees similarities between the expression and the content, not because of structural similarity, but because they share some chemical properties. In the case of chemical biosignatures, some are bioicons in that sense that the discovered biosignature has a chemical similarity or shared characteristics with the living organism, for example complex biological macromolecules, like carbohydrates, lipids, proteins, and nucleic acids (RNA and DNA). Most common biomolecules, however, usually modify and degrade, and the products (also called "molecular fossils") of this chemical breakdown (the diagenesis) have instead an indexical relation to the bioiconic biological macromolecules. For example the 2-methyl hopanes that are known to be the diagenetic products of 2-methylbacteriohopanepolyols are second order biosignatures, that is bioindices of bioicons that refer to its content, life in the form of cyanobacteria.

15.5.2 Indices of Life

"It happen'd one day about noon, going towards my boat, I was exceedingly surpris'd with the print of a man's naked foot on the shore, which was very plain to be seen in the sand: I stood like one thunder-struck, or as if I had seen an apparition" (Defoe 1719, 122). When Robinson Crusoe, shipwrecked and washed ashore on a seemingly uninhabited island, one day saw footprints in the sand, he knew that there was human life there—Friday. He came to the conclusion, not just because the footprint had a similarity (iconicity) with a human being, but because it had a causal link with the life-form who made it, as an index of life. An index is a sign caused by its object, it has an unintentional, causal link or contact with its content. "Smoke", for example, has this indexical relation to "fire," which it refers to and which we interpret as its cause. Indexicality is, in this respect, meaning by proximity or contiguity. This contiguity does not necessarily have to be of real physical causality, it could consist of the mere perceiving of two objects together in space. Indices could also be related to factorality, when seeing something as a part of something else (Sonesson 1994). The interpretation of indices requires empirical knowledge of the recurrent connection between the sign and what it refers to. The perceptual world consists of a profuse amount of potential indexicality, even though we do not yet recognize these indices as signs with meaning. But the human mind constantly searches for and infers causalities and meaning in things perceived.

Bioindices are thus biosignatures that have a connection to their objects (the living organisms) by contiguity. In other words, the connection between the expression and the content is not based on similarity, but on indexicality, and is in semiotic terms something distinctly different. Perhaps the clearest examples of bioindices are atmospheric, chemical biosignatures that refer to biological processes, such as the metabolism of living organisms. Homochirality and isotopic fractionation have been put forward as molecular evidence of metabolism. Biogenic minerals—deposits of calcium carbonate, calcium phosphate, iron oxides, manganese, and sulphur—could also be the products of microbial metabolic processes, and thus have this indexical relation, but are unfortunately very difficult to distinguish from minerals produced by mere abiotic processes. Fossils that record the behaviour or activity of an organism are another type of bioindices, in contrast to bioicons that has a similarity with the living thing. These artefacts of life, such as stromatolites formed by microbial mat communities, indicate a biotic origin. Other examples of bioindices that trace the behaviour of an organism are borings, burrows, footprints, etc. Again, the challenge here is to distinguish these bioindices from features that are a result of an abiotic process that mimic the biotic behaviour.

Remote sensing of planetary environments for habitability and biosignatures goes back to the nineteenth century. In his *Cours de philosophie positive* (1830–1842) the French positivist Auguste Comte said, concerning the celestial bodies, that "we will never by any means be able to study their chemical composition or their mineralogical structure" (Comte 1835, 2; Crowe 2008, 312). Some few decades later spectroscopy was developed. The turning point came with spectroscopic astronomy that gave a new powerful tool for searching extraterrestrial life. By analysing the spectra caused by the molecular absorption or emission at molecule-specific photon wavelengths, the spectroscopists could infer the chemical composition of the atmospheres of distant planets. The first spectroscopic observations aiming for detecting oxygen and water in the Martian atmosphere, were made by the astronomers William Huggins and Jules Janssen in the 1860s. By assuming water as a necessary condition for life, and by linking planetary environmental conditions (presence of water vapour in the atmosphere and liquid water on the surface) with the possibility for life to emerge and subsist, they got a clue. A detection of water vapour in the atmosphere of a planet would then be a crucial indication that it might be life on its surface. In 1867, Janssen claimed that he had discovered the presence of water vapour in the Martian atmosphere, but in fact it was probably terrestrial signatures, already refuted by the American astronomer William Wallace Campbell in 1894 (Raulin Cerceau 2013).

There are hopes that we in the future will be able to observe the absorption or emission properties of atmospheres of small, rocky exoplanets (Seager 2014; Seager and Bains 2015). A first step in the search for biosignatures of exoplanets would be to study the temperature, size, mass, density, gravitation, and light conditions of the exoplanet. Next, to search for indications of atmosphere, liquid water, clouds, surface, plate tectonics, daily rotation, seasons, and weather. The third step would be to look for bioindices. For sure, we will not be able observe any surfaces of exoplanets with current technology, but we might soon be able to detect certain gases that we connect with life by remote sensing, even though the interpretation of the spectra involves a number of difficulties. In the future, the European Extremely Large Telescope (E-ELT) will make it possible to perform spectroscopic analysis of the faint light of an exoplanet, and might result in the first exoplanet atmospheric biosignatures. Our hopes rest on the assumption that certain gases in the atmosphere are produced by life (as we know from studies of our own terrestrial atmosphere), such as oxygen, ozone, methane, and carbon dioxide. Oxygen enrichment in the atmosphere could indicate the presence of oxygenic photosynthesis. Ozone, which is produced photochemically from biologically produced oxygen, could be another indication of biological activity. And methane could likewise be connected to the metabolism of living organisms. However, these gases could also be produced by abiological processes and exist without any biological activity. Some gases that are products of life on Earth, such as CH_3Cl, CH_3SH, NO_2, NH_3, would not be detected with current technology, due to low amounts, others, such as water and carbon dioxide have significant abiotic sources, and are less suitable as conclusive signatures of life. To conclude, the argument starts from the premises: (P_1) that life produces certain gases as a by-product of metabolism; (P_2) some of theses gases will accumulate in the atmosphere; and (P_3) that these gases show a unique spectrum. From these premises—which we hold, to be true and to be sufficient for detection—we conclude that life could, in theory, be detected through spectroscopy. But if P_1 is false (there are metabolic processes that do not produce gases) or if P_2 is false (these gases do not leak into the atmosphere) or we do not recognise the unique spectrum (P_3), we will fail.

It might rather be the combination of gases and the quantity of them, that closer reveals if there are life on the planet. Life leads to disequilibria, for example in respect to atmospheric chemical composition, entropy, etc. Earth-like atmospheric biosignatures disappear relatively quickly on a planet where life has ceased to exist. If there is a certain amount of a biosignature gas, it needs to have a continuous source. That gases in disequilibrium could be diagnostic for life was first suggested by Joshua Lederberg and James Lovelock in 1965 (Lederberg 1965; Lovelock 1965; Catling and Kasting 2007). And this atmospheric disequilibrium is detectable by spectroscopy, as in the case of spectral analysis of Earthshine (Arnold et al. 2002; Arnold 2008). The simultaneous presence of oxygen and methane indicates an atmospheric disequilibrium that could be assumed as a spectral evidence of life. The sustainable source of these gases, in this ratio, is life. The discovery of significant amounts of methane in the atmosphere of Mars then implies that there must exist a recent or current source, otherwise the methane would rather quickly disappear. The source could be geological activity and water in the subsurface—or subsurface biology (Domagal-Goldman et al. 2016). As a recurrent theme in the search for life, the signs are ambiguous.

Another shipwrecked traveller in foreign territories, the ancient Greek philosopher Aristippus, was cast ashore on the Rhodian coast. But when he found geometrical figures in the sand, he became convinced that he had come to a land inhabited by civilized people (Vitruvius 1934). These Rhodian bioindices did not only indicate life, people, they indicated civilization. A certain class of bioindices could be categorized as *technoindices*, a second order index that indicate technology, which in its turn could indicate life. When analysing the spectra of exoplanets, one might find signs that do not have any known natural origin, such as industrial pollution, artificial molecules, for example pollutants like chlorofluorocarbons (CFCs), or other artificial traces of environmental disequilibrium that reverberates across the biosphere (Lin et al. 2014; Shostak 2015; Frank and Sullivan 2016), which we interpret as technoindices of advanced life forms that are able to artificially manipulate their environment. Monitoring the stars and planets in our galaxy we perchance come across signs of extraterrestrial civilizations revealed by their use of technology, for example radio emissions or other electromagnetic radiation leaking out from their planet voluntarily or involuntarily. Finding signs of technology does not necessarily lead to the conclusion that they originate from a civilization consisting of biological creatures, if one think of the highly hypothetic self-replicating "von Neumann machines" that replicate and disperse themselves without the dependence of biological creators.

The search for indices of life is a way of connecting phenomena around us, inferring that certain signs indicate a causal connection to their object and origin—life. This semiosis or meaning-making endeavour is however triadic, includes something more than expression and object, the biosignature and the living organism. As the astrophysicist Arthur Eddington (1920) touched upon: "We have found a strange footprint on the shores of the unknown. We have devised profound theories, one after another, to account for its origins. At last, we have succeeded in reconstructing the creature that made the footprint. And lo! It is our own" (Sullivan and Baross 2007, 6). It is ourselves, the interpreters that make this connection between expression and content. Searching for indices of life may reveal some knowledge about the living Universe wherein we live, but also an understanding of how we search for meaning in the seemingly chaotic world around us.

15.5.3 *Symbols of Life*

August 15, 1977, the Big Ear radio telescope in Ohio received a very strong narrowband radio signal that lasted for 72 seconds. While reviewing the record date, the astronomer Jerry R. Ehman was stunned and wrote the comment "Wow!" on the computer printout. This anomaly has not been confirmed nor repeated (Kraus 1979). The first problem one faces in such a situation is to determine if it is a natural or an artificial signal, if one could rule out all known natural causes of the signal and conclude that it is an artificial signal caused by an intelligent civilization with advanced technology. This *technosignature* might indicate the existence of technology, as such a technoindex. The next problem that arises is to determine if it is something more than just an index of technology, but actually contains a message, a content that is meant to be communicated, deciphered, and understood by the receiver. Probably, we would not be satisfied with a mere conclusion that it is a technoindex, but that it contains symbolical information, that it is a *technosymbol.*

Searching for extraterrestrial intelligence by means of radio astronomy has been an exciting challenge ever since the start of Project Ozma in 1960 (Sagan 1973; Weston 1988; Tarter 2001; Drake 2011; Schuch 2011; Dunér 2015, 2017; Traphagan 2015; Vakoch and Dowd 2015; Cabrol 2016). The starting point of the argument is plausible. Electromagnetic leakage from Earth is detectable from outer space, and likewise, if an extraterrestrial intelligence is engaged in radio communication, we would be able, at least in theory, to detect its voluntary or involuntary broadcasts. The problem of interstellar communication lies not so much in the physical or technological constraints, even though they very much challenge our scientific and technological skills, but in the cognitive and semiotic problems that interstellar message decoding provoke (Vakoch 1998; Dunér 2011b, 2014; Sonesson 2013).

Intelligence could be seen as an evolved mental gymnastics required to survive and reproduce within its specific environment. This includes the capability of representing activities and being able to make inner models of reality. By using symbols an intelligent creature could engage in abstract thinking detached from the environment, by which they can reason about things not existent; things that are not right in front of them, in a specific moment in time. Very effective tools for symbolizing thought are our communicational devices. According to the cognitive linguist John Taylor (2002), language can be understood as a set of resources that are available to the language user for the symbolization of thought, and for the communication of these symbolizations.

The problem with symbols is that they are conventional, or arbitrary, as the founder of semiology Ferdinand de Saussure (1916) called them. Icons and indices are signs that have some *non-arbitrary* similarity or contiguity with the signified, in contrast to the symbols' completely *arbitrary* relation. For example the word "life" has no causal link to what it stands for, nor does it resemble what it signifies. There are no intrinsic relationship between the expression and content whatsoever. It is the interpreters (the ones that construct the message and the ones that decode them, respectively) that joins them together and establish the connection between the expression and the content. And the matching between the transmitters' and the receivers' interpretation of the symbols is by no means self-evident. We may figure out the reference of the signal, but will probably have severe problems understanding extraterrestrial symbols. It is not impossible to imagine that the aliens would have certain knowledge about their environment that in its content is similar to our own knowledge of mathematics, physics, and chemistry. But their expression of it would most likely be very different from ours. It is the message's expression rather than its content that becomes the difficulty for the interpreter. In symbols, there is a gap between the sign and meaning. Nothing in the physical appearance of the sign gives any clue to its object; they are instead linked by an arbitrary correlation. In fact, most attempts at interstellar message constructions violate this basic semiotic understanding of signs that distinguishes between expression and content. Symbols are detached representations and, as such, dependent on cultural and social interactions that create some specific regularities that have their origin in more or less stochastic habits, conventions, etc., of the species (Sonesson and Dunér 2016). Our communication and symbolization have evolved through an evolutionary and cultural-historical process here on Earth, and are thereby constrained by our human bodies, terrestrial environment, and the socio-cultural characteristics of our species. And likewise, a potential information transfer containing a symbolic message from an alien civilization would be constrained by the bio-cultural coevolution of the extra-terrestrial intelligence that coded it.

15.6 Conclusion

So far, we have no conclusive evidence of the existence of extraterrestrial life. But could we ever be hundred percent sure that we are alone? One might object that the plurality of life hypothesis labours under a major problem: unfalsifiability. According to the philosopher of science Karl Popper (1963), theories that are unfalsifiable and rich in explanatory power are often known to be wrong. In our case, there is no method or way of proving that life does not exist and cannot exist elsewhere in Universe. Astrobiology has an attractive flexibility and opens up for a richness of various explanations for the failure to prove the existence or for future success in discovering life. If we do not find life on the surface of Mars, we continue searching under its surface. If we do not find life in our Solar System, we continue searching in other Solar Systems. If one exoplanet does not show any signs of life, we go on to the next, and so on. When will we give up and conclude that life most likely does not exist elsewhere? After hundred thoroughly studied exoplanets, after thousand, millions? We will never be able to search for life in every corner of the Universe, doing in situ investigations on every exoplanet in all galaxies in the entire Universe. We will never know empirically that life does not exist on other planets. We can just move on, refining our methods, observations, theories, etc., but never reach a final conclusion beyond uncertainty. It is just a question of probability. But to verify the hypothesis, it is just enough with one single sample. Finding life is not empirically falsifiable, but, in principle, verifiable.

Anyhow, one day we might encounter signs of another living planet. We will then get some potential knowledge, among other things, about its chemical composition and environmental circumstances. But above all, the descriptions, interpretations, and conclusions concerning this new world will also tell something about ourselves and our place in the Universe, and how we interpret and understand that reality we experience. There are things we know. Even though life might not exist out there, it is we human beings with our brains, bodies, and cultures who are searching for it. The history and philosophy of biosignatures is centred on humans, or more specifically, on the scientific endeavour's dependence on the human mind and human culture (Dunér 2011a, 2013a; Dunér et al. 2013). Astrobiologists have brains, for sure; they are using cognitive tools that are a result of the bio-cultural coevolution of human cognitive abilities. Certain cognitive processes are at work when astrobiologists encounter unknown things, when interpreting potential signs of life, when they gather and classify information, and make conclusions from the observational data. This does not go on in subjective isolation. Astrobiologists live in a culture, in a certain time in history, in a specific research environment, and collaborate with other thinking beings. In this chapter, I have touched upon some epistemological issues in the search for biosignatures, how we conceptualize things, make analogies, how we perceive our surrounding world and endow it meaning. The quest for signs of life rests on the cognitive and socio-cultural capabilities of that human species that makes this lonely planet alive and thinking.

References

Aristotle (1966) Posterior analytics. In: Tredennick H, Forster ES (eds) Topica. Heinemann, London

Arnold L (2008) Earthshine observation of vegetation and implication for life detection on other planets: a review of 2001–2006 works. Space Sci Rev 135:323–333

Arnold L, Gillet S, Lardière O et al (2002) A test for the search for life on extrasolar planets: looking for the terrestrial vegetation signature on the earthshine spectrum. Astron Astrophys 392:231–237

Arrhenius S (1907) Panspermy: the transmission of life from star to star. Sci Am 196:196

Aspaas PP (2012) Maximilianus Hell (1720–1792) and the eighteenth-century transits of Venus: a study of Jesuit science and Central European contexts. Tromsø University, Tromsø

Baum R (1973) The planets: some myths & realities. David & Charles, New York

Bedau MA, Cleland CE (2010) The nature of life: classical and contemporary perspectives from philosophy and science. Cambridge University Press, Cambridge

Berlin B (1992) Ethnobiological classification: principles of categorization of plants and animals in traditional societies. Princeton University Press, Princeton, NJ

Berzelius JJ (1834) Ueber meteorsteine. Ann Phys Chem 33:113–148

Bianchini F (1728) Hesperi et Phosphori nova phænomena, sive observationes circa planetam Veneris. Rome

Bode JE (1801) Allgemeine Betrachtungen über das Weltgebäude. Berlin

Brasier MD, Green OR, Lindsay JF et al (2005) Critical testing of Earth's oldest putative fossil assemblage from the ~3.5 Ga Apex chert, Chinaman Creek, Western Australia. Precambrian Res 140:55–102

Briot D (2013) Elements for the history of a long quest: search for life in the Universe. Int J Astrobiol 12:254–258

Cabrol NA (2016) Alien mindscapes: a perspective on the search for extraterrestrial intelligence. Astrobiology 16:1–16

Campbell NA, Reece JB (2002) Biology, 6th edn. Benjamin Cummings, New York

Cassini G (1667) Touchant la découuerte qu'il a faite du mouuement de la Planete de Venus à l'entour de son axe. Journal des Scavans

Catling D, Kasting JF (2007) Planetary atmospheres and life. In: Sullivan WT, Baross JA (eds) Planets and life: the emerging science of astrobiology. Cambridge University Press, Cambridge, pp 91–116
Cattermole P, Moore P (1997) Atlas of Venus. Cambridge University Press, Cambridge
Cleland CE, Chyba CF (2002) Defining "life". Orig Life Evol Biosph 32:387–393
Cleland CE, Chyba CF (2007) Does 'life' have a definition? In: Sullivan WT, Baross JA (eds) Planets and life: the emerging science of astrobiology. Cambridge University Press, Cambridge, pp 119–131
Comte A (1835) Cours de Philosophie Positive: Tome 2, Contenant la Philosophie Astronomique et la Philosophie de la Physique. Bachelier, Paris
Conrad PG (2007) Instruments and strategies for detecting extraterrestrial life. In: Sullivan WT, Baross JA (eds) Planets and life: the emerging science of astrobiology. Cambridge University Press, Cambridge, pp 473–482
Crowe MJ (1986) The extraterrestrial life debate 1750–1900: the idea of a plurality of worlds from Kant to Lowell. Cambridge University Press, Cambridge
Crowe MJ (2008) The extraterrestrial life debate, antiquity to 1915: a source book. University of Notre Dame, Notre Dame, IN
Crowe MJ, Dowd MF (2013) The extraterrestrial life debate from antiquity to 1900. In: Vakoch DA (ed) Astrobiology, history, and society: life beyond earth and the impact of discovery. Springer, Berlin, pp 3–56
Cutler A (2003) The seashell on the mountaintop: a story of science, sainthood and the humble genius who discovered a new history of the earth. Dutton, New York
Davies PCW, Benner SA, Cleland CE et al (2009) Signatures of a shadow biosphere. Astrobiology 9:241–249
de Duve C (1995) Vital dust: life as a cosmic imperative. Basic Books, New York
de Fontenelle B (1686/1767) Entretiens sur la pluralité des mondes. Amsterdam: Mortier; trans, Conversations on the plurality of worlds. Thomas Caslon, London
de Saussure F (1916) Cours de linguistique générale. Payot, Paris
Defoe D (1719/2001) Robinson Crusoe. ed. Richetti J. Penguin, New York
Dick SJ (1982) Plurality of worlds: the origins of the extraterrestrial life debate from Democritus to Kant. Cambridge University Press, Cambridge
Dick SJ (1996) The biological universe: the twentieth-century extraterrestrial life debate and the limits of science. Cambridge University Press, Cambridge
Dietrich LEP, Michael M, Newman DK (2006) The co-evolution of life and earth. Curr Biol 16: pR395–pR400
Domagal-Goldman SD, Wright KE, Adamala K et al (2016) The astrobiology primer v2.0. Astrobiology 16:561–653
Drake F (2011) The Search for extra-terrestrial intelligence. Philos Trans R Soc A Math Phys Eng Sci 369:633–643
Dunér D (2011a) Astrocognition: prolegomena to a future cognitive history of exploration. In: Landfester U, Remuss N-L, Schrogl K-U, Worms J-C (eds) Humans in outer space – interdisciplinary perspectives. Springer, Wien, pp 117–140
Dunér D (2011b) Cognitive foundations of interstellar communication. In: Vakoch DA (ed) Communication with extraterrestrial intelligence. State University of New York Press, Albany, NY, pp 449–467
Dunér D (2012) The history and philosophy of astrobiology. Astrobiology 12:901–905
Dunér D (2013a) Extraterrestrial life and the human mind. In: Dunér D, Parthemore J, Persson E, Holmberg G (eds) The history and philosophy of astrobiology: perspectives on extraterrestrial life and the human mind. Cambridge Scholars Publishing, Newcastle-upon-Tyne, pp 7–31
Dunér D (2013b) The language of cosmos: the cosmopolitan endeavour of universal languages. In: Rydén G (ed) Sweden in the eighteenth-century world: provincial cosmopolitans. Ashgate, Farnham, pp 41–65

Dunér D (2013c) Venusians: the planet Venus in the 18th-century extraterrestrial life debate. In Sterken C, Aspaas PP (eds) Meeting Venus: a collection of papers presented at the Venus transit conference in Tromsø 2012. J Astronom Data 19:145–167
Dunér D (2014) Interstellar intersubjectivity: the significance of shared cognition for communication, empathy, and altruism in space. In: Vakoch DA (ed) Extraterrestrial altruism: evolution and ethics in the cosmos. Springer, Dordrecht, pp 139–165
Dunér D (2015) Length of time such civilizations release detectable signals into space, L, pre-1961. In: Vakoch DA, Dowd MF (eds) The Drake equation: estimating the prevalence of extraterrestrial life through the ages. Cambridge University Press, Cambridge, pp 241–269
Dunér D (2016a) Science: the structure of scientific evolutions. In: Dunér D, Sonesson G (eds) Human lifeworlds: the cognitive semiotics of cultural evolution. Peter Lang, Bern, pp 229–266
Dunér D (2016b) Swedenborg and the plurality of worlds: astrotheology in the eighteenth century. Zygon J Relig Sci 51:450–479
Dunér D (2017) On the plausibility of intelligent life on other worlds: a cognitive-semiotic assessment of $f_i \cdot f_c \cdot L$. Environ Humanit 9(2):433–453
Dunér D, Sonesson G (2016) Human lifeworlds: the cognitive semiotics of cultural evolution. Peter Lang, Bern
Dunér D, Parthemore J, Persson E, Holmberg G (2013) The history and philosophy of astrobiology: perspectives on extraterrestrial life and thehuman mind. Cambridge Scholars Publishing, Newcastle-upon-Tyne
Dunér D, Malaterre C, Geppert W (2016) The history and philosophy of the origin of life. Int J Astrobiol 15:1–2
Eddington AS (1920) Space, time, and gravitation: an outline of the general relativity theory. Cambridge University Press, Cambridge
Flammarion C (1862) La Pluralité des Mondes Habités. Mallet-Bachelier, Paris
Flammarion C (1880/1907) Astronomie populaire: description générale du ciel. Paris. In: Marpon C, Flammarion E, trans. Gore JE, Popular astronomy. Appleton, New York
Fontana F (1646) Novæ observationes terrestriumq[ue] rerum observationes, et fortasse hactenus non uulgatae. Neapoli
Frank A, Sullivan WT III (2016) A new empirical constraint on the prevalence of technological species in the universe. Astrobiology 16:359–362
Galileo G (1610/2009) Siderevs nvncivs magna, longeqve admirabilia spectacula pandens, ... Venice; trans. Shea WR, Galileo's Sidereus nuncius, or, a sidereal message. Science History Publications, Sagamore Beach, MA
Galileo G (1632/1953) Dialogo ... sopra i due massimi sistemi del mondo ... Firenze; trans. de Santillana G, Dialogue on the great world systems. Chicago University Press, Chicago, IL
Galileo G (1957) Discoveries and opinions of Galileo, ed. & trans. Drake S. Doubleday, Garden City, NY
Gayon J (2010) Defining life: synthesis and conclusions. Orig Life Evol Biosph 40:231–244
Golding SD, Glikson M (2011) Earliest life on Earth: habitats, environments and methods of detection. Springer, Dordrecht
Guthke KS (1983) Der Mythos der Neuzeit: das Thema der Mehrheit der Welten in der Literatur- und Geistesgeschichte von der kopernikanischen Wende bis zur Science Fiction. Francke, Bern
Harvey B (2007) Russian planetary exploration: history, development, legacy, prospects. Praxis, Chichester
Hegde S, Paulino-Lima IG, Kent R, Kaltenegger L, Rothschild L (2015) Surface biosignatures of exo-earths: remote detection of extraterrestrial life. Proc Natl Acad Sci USA 112:3886–3891
Hennessey R (1999) Worlds without end: the historic search for extraterrestrial life. Tempus, Stroud
Herschel W (1793) Observations on the planet Venus. Philos Trans 83
Herschel W (1793/1912) Observations on the planet Venus. In: Dreyer JLE, Herschel W (ed) The Scientific Papers of William Herschel, London
Hoffmeyer J, Favareau D (2008) Biosemiotics: an examination into the signs of life and the life of signs. University of Scranton Press, Scranton

Horneck G, Klaus DM, Mancinelli RL (2010) Space microbiology. Microbiol Mol Biol Rev 74:121–156

Horneck G et al (2016) AstRoMap European astrobiology roadmap. Astrobiology 16:201–243

Hoyt WG (1976) Lowell and Mars. University of Arizona Press, Tuscon, AZ

Huygens C (1698) Kosmotheōros, sive de terris cœlestibus, earumque ornatu, conjecturæ. Den Haag

Irwin NI, Schulze-Makuch D (2001) Assessing the plausibility of life on other worlds. Astrobiology 1:143–160

Joyce GF (1994) Foreword. In: Deamer DW, Fleischaker GR (eds) Origins of life: the central concepts. Jones & Bartlett, Boston, MA, pp xi–xii

Kane SR, Gelino DM (2012) The habitable zone and extreme planetary orbits. Astrobiology 12:940–945

Klein HP, Horowitz NH, Levin GV, Oyama VI, Lederberg J, Rich A, Hubbard JS, Hobby GL, Straat PA, Berdahl BJ, Carle GC, Brown FS, Johnson RD (1976) The Viking biological investigation: preliminary results. Science 194:99–105

Kooijman SALM (2004) On the co-evolution of life and its environment. In: Schneider SH, Miller JR, Crist E et al (eds) Scientists debate Gaia: the next century. MIT Press, Cambridge, MA, pp 343–351

Kraus J (1979) We wait and wonder. Cosmic Search 1:31–34

Lakoff G (1990) Women, fire, and dangerous things: what categories reveal about the mind. University of Chicago Press, Chicago, IL

Lederberg J (1965) Signs of life: criterion system of exobiology. Nature 207:9–13

Levin GV, Straat PA (1976) Viking labeled release biology experiment: interim results. Science 194:1322–1329

Levin GV, Straat PA (1979) Completion of the Viking labeled release experiment on Mars. J Mol Evol 14:167–183

Levin GV, Straat PA (2016) The case for extant life on Mars and its possible detection by the Viking labeled release experiment. Astrobiology 16:798–810

Lin HW, Abad GG, Loeb A (2014) Detecting industrial pollutants in the atmospheres of Earth-like exoplanets. Astrophys J 792

Losch A (2017) What is life?: on Earth and beyond. Cambridge University Press, Cambridge

Lovelock JE (1965) A physical basis for life detection experiments. Nature 207:568–570

Lowell P (1895) Mars. Houghton, Mifflin, Boston, MA

Luisi P (1998) About various definitions of life. Orig Life Evol Biosph 28:613–622

Maor E (2000) June 8, 2004: Venus in transit. Princeton University Press, Princeton, NJ

Marov MY (2005) Mikhail Lomonosov and the discovery of the atmosphere of Venus during the 1761 transit. In: Kurtz DW (ed) Transits of Venus: new views of the solar system and galaxy. Cambridge University Press, Cambridge

McKay CP (2007) How to search for life on other worlds. In: Sullivan WT, Baross JA (eds) Planets and life: the emerging science of astrobiology. University Press Cambridge, Cambridge, pp 461–472

McKay DS, Gibson EK Jr, Thomas-Keprta KL et al (1996) Search for past life on Mars: possible relic biogenic activity in Martian meteorite ALH84001. Science 273:924–930

Moore P (1956) The planet Venus. Faber and Faber, London

Oliver J, Perry RS (2006) Definitely life but not definitively. Orig Life Evol Biosph 36:515–521

Palyi G, Zucchi C, Caglioti L (2002) Fundamentals of life. Elsevier, New York

Peirce CS (1932) Collected papers 2: elements of logic. Belknap Press of Harvard University Press, Cambridge, MA

Peirce CS (1957) Essays in the philosophy of science. In: Tomas V (ed), Liberal Arts Press, New York

Peirce CS (1906/1998) The basis of pragmaticism in the normative sciences. In: The essential Peirce: selected philosophical writings, vol 2. Indiana University Press, Bloomington, IN, pp 1893–1913

Pennock RT (2012) Negotiating boundaries in the definition of life: Wittgensteinian and Darwinian insights on resolving conceptual border conflicts. Synthese 185:5–20
Persson E (2013) Philosophical aspects of astrobiology. In: Dunér D, Parthemore J, Persson E et al (eds) The history and philosophy of astrobiology: perspectives on extraterrestrial life and the human mind. Cambridge Scholars Publishing, Newcastle-upon-Tyne, pp 29–48
Pilcher CB (2003) Biosignatures of early earths. Astrobiology 3:471–486
Popper K (1963) Conjectures and refutations: the growth of scientific knowledge. Routledge, London
Proctor R (1874) Transits of Venus: a popular account of past and coming transits from the first observed by Horrocks A.D. 1639 to the transit of A.D. 2012. Longmans, Green, London
Pross A (2012) What is life?: how chemistry becomes biology. Oxford University Press, Oxford
Raulin Cerceau F (2013) Pioneering concepts of planetary habitability. In: Vakoch DA (ed) Astrobiology, history, and society: life beyond earth and the impact of discovery. Springer, Berlin, pp 115–129
Robus O, Haydon N, McGlynn S et al (2009) Life as a functional concept: functionalism as a robust framework for theories and definitions of multi-realized living systems. Orig Life Evol Biosph 39:390
Romanini V, Fernández E (2014) Peirce and biosemiotics: a guess at the riddle of life. Springer, New York
Rosch E (1975) Cognitive representations of semantic categories. J Exp Psychol Gen 104:192–233
Rosch E (1978) Principles of categorization. In: Rosch E, Lloyd BB (eds) Cognition and categorization. Erlbaum, Hillsdale, NJ, pp 27–48
Rosch E (1987) Wittgenstein and categorization research in cognitive psychology. In: Chapman M, Dixon R (eds) Meaning and the growth of understanding: Wittgenstein's significance for developmental psychology. Erlbaum, Hillsdale, NJ, pp 151–166
Ruiz-Mirazo K, Peretó J, Moreno A (2004) A universal definition of life: autonomy and open-ended evolution. Orig Life Evol Biosph 34:323–346
Sagan C (1973) Communication with extraterrestrial intelligence. MIT Press, Cambridge, MA
Sagan C, Thompson WR, Carlson R, Gurnett D, Hord C (1993) A search for life on Earth from the Galileo spacecraft. Nature 365:715–721
Saint-Gelais R (2014) Beyond linear B: the metasemiotic challenge of communication with extraterrestrial intelligence. In: Vakoch DA (ed) Archaeology, anthropology, and interstellar communication. NASA, Washington, DC, pp 78–93
Schopf JW (1993) Microfossils of the early Archean Apex chert: new evidence of the antiquity of life. Science 260:640–646
Schröter JH (1793) Neuere Beobachtungen der Venuskugel. Astronomisches Jahrbuch für 1793
Schröter JH (1796) Aphroditographische Fragmente, zur genauern Kenntniss des Planeten Venus. Helmstedt
Schuch HP (2011) Project Ozma: the birth of observational SETI. In: Schuch HP (ed) Searching for extraterrestrial intelligence: SETI past, present, and future. Springer, Berlin, pp 13–18
Schulze-Makuch D, Irwin LN, Fairén AG (2015) Extraterrestrial life: what are we looking for? In: Kolb VM (ed) Astrobiology: an evolutionary approach. CRC Press, Boca Raton, FL, pp 399–412
Seager S (2014) The future of spectroscopic life detection on exoplanets. Proc Natl Acad Sci USA 111:12634–12640
Seager S, Bains W (2015) The search for signs of life on exoplanets at the interface of chemistry and planetary science. Sci Adv 1:1–11
Sellers D (2001) The transit of Venus: the quest to find the true distance of the Sun. MagaVelda, Leeds
Sheehan W (1988) Planets and perception: telescopic views and interpretations, 1609–1909. University of Arizona Press, Tucson
Sheehan W, Dobbins T (2001) Epic Moon: a history of lunar exploration in the age of the telescope. Willmann-Bell, Richmond, VA

Sheehan W, Westfall J (2004) The transits of Venus. Prometheus Books, Amherst, NY
Shostak S (2015) Fraction of civilizations that develop a technology that releases detectable signs of their existence into space, f_c, 1961 to the present. In: Vakoch DA, Dowd MF (eds) The drake equation: estimating the prevalence of extraterrestrial life through the ages. Cambridge University Press, Cambridge, pp 27–240
Sonesson G (1994) Prolegomena to the semiotic analysis of prehistoric visual displays. Semiotica 100:267–332
Sonesson G (2007) From the meaning of embodiment to the embodiment of meaning. In: Zimke T, Zlatev J, Frank R (eds) Body, language and mind. Vol. 1: embodiment. Mouton, Berlin
Sonesson G (2009) The view from Husserl's lectern: considerations on the role of phenomenology in cognitive semiotics. Cybern Human Knowing 16:107–148
Sonesson G (2013) Preparations for discussing constructivism with a Martian (the second coming). In: Dunér D, Parthemore J, Persson E et al (eds) The history and philosophy of astrobiology: perspectives on the human mind and extraterrestrial life. Cambridge Scholars Publishing, Newcastle-upon-Tyne, pp 189–204
Sonesson G, Dunér D (2016) The cognitive semiotics of cultural evolution. In: Dunér D, Sonesson GP (eds) Human lifeworlds: the cognitive semiotics of cultural evolution. Peter Lang, Bern
Spranzi M (2004) Galileo and the mountains of the moon: analogical reasoning, models and metaphors in scientific discovery. J Cogn Cult 4:451–483
Sterken C, Aspaas PP (2013) Meeting Venus: a collection of papers presented at the Venus transit conference in Tromsø 2012. J Astronom Data 19:145–167
Sterzik MF, Bagnulo S, Palle E (2012) Biosignatures as revealed by spectropolarimetry of earthshine. Nature 483:64–66
Strauss P (2001) Percival Lowell: the culture and science of a Boston Brahmin. Harvard University Press, Cambridge, MA
Sullivan WT, Baross JA (2007) Planets and life: the emerging science of astrobiology. Cambridge University Press, Cambridge
Sullivan WT, Carney D (2007) History of astrobiological ideas. In: Sullivan WT, Baross JA (eds) Planets and life: the emerging science of astrobiology. Cambridge University Press, Cambridge, pp 9–45
Tarter JC (2001) The search for extraterrestrial intelligence (SETI). Annu Rev Astron Astrophys 39:511–548
Tarter JC, Backus PR, Mancinelli RL et al (2007) A reappraisal of the habitability of planets around M dwarf stars. Astrobiology 7:30–65
Taylor JR (2002) Cognitive grammar. Oxford University Press, Oxford
Taylor JR (2003) Linguistic categorization. Oxford University Press, Oxford
Traphagan J (2015) Extraterrestrial intelligence and human imagination: SETI at the intersection of science, religion, and culture. Springer, New York
Vakoch DA (1998) Constructing messages to extraterrestrials: an exosemiotic perspective. Acta Astronaut 42:697–704
Vakoch DA, Dowd MF (2015) The Drake equation: estimating the prevalence of extraterrestrial life through the ages. Cambridge University Press, Cambridge
Vitruvius (1934) On architecture: Books VI–X, ed. Granger F. Harvard University Press, Cambridge, MA
Watson AJ (1999) Coevolution of the earth's environment and life: Goldilocks, Gaia and the anthropic principle. Geol Soc Spec Pub 150:75–88
Westall F, Cavalazzi B (2011) Biosignatures in rocks. In: Thiel V, Reitner J (eds) Encyclopedia of geobiology. Springer, Berlin, pp 189–201
Westall F, Gobbi P, Gerneke D, Mazzotti G (1998) Ultrastructure in the carbonate globules of Martian meteorite ALH84001. In: Chela-Flores J, Raulin F (eds) Exobiology: matter, energy, and information in the origin and evolution of life in the universe. Kluwer, Amsterdam, pp 245–250

Westall F, Foucher F, Bost N et al (2015) Biosignatures on Mars: what, where, and how? Implications for the search for Martian life. Astrobiology 15:998–1029
Weston A (1988) Radio astronomy as epistemology: some philosophical reflections on the contemporary search for extraterrestrial intelligence. Monist 71:88–100
Wheeler W (2006) The whole creature: complexity, biosemiotics and the evolution of culture. Lawrence and Wishart, London
Wittgenstein L (1953) Philosophical investigations. Blackwell, Oxford University Press, Oxford
Woolf H (1959) The transits of Venus: a study of eighteenth-century science. Princeton University Press, Princeton, NJ
Wulf A (2012) Chasing Venus: the race to measure the heavens. William Heinemann, London
Zlatev J (2015) Cognitive semiotics. In: Trifonas P (ed) International handbook of semiotics. Springer, Berlin, pp 1043–1068

Index

A
Abiogenesis, 183
Abiogenic carbon, 290
Abiotic, 276
Abiotic CH_4, 257
Abiotic minerals, 139
Abiotic origin, 196
Abiotic processes, 124, 168
Absorption band, 254
Adaptations, 104
Aerobic methane oxidizers, 75
Aerosols, 12
Akilia, 75
Algae, 69
ALH 84001, 168, 176
Allende, 31
Amino acids, 6, 7, 11, 15, 30, 101, 184, 186, 218
Amorphous carbons, 36
Amorphous water ice, 191
Anaerobic archaeal consortia, 75
Anaerobic chemotrophic colonies, 160
Anaerobic process, 59
Anoxygenic photosynthetic bacteria, 61
Antarctic, 168, 176, 178
Antarctica, 16, 176
Antoniadi, E., 320
Apatite, 133
Apex chert, 326
Aqueous alteration (AA), 27
Aragonite, 129
Archaea, 58, 103
Archaean, 72, 74
Archaean atmosphere, 184
Aristotle, 322
Artefacts, 93
Ashfall particles, 288
Asteroids, 11, 23
Astrobiologist, 323
Astrobiology, 23, 30, 112, 183, 242, 267, 274, 307
Atacama Desert, 25
Atmosphere, 5, 39, 72, 229, 230, 232, 239, 258, 263
Atmospheric evolution, 237
Atmospheric oxygen, 231
Atomic hydrogen, 236
Atomic structure, 123
Australia, 60, 62, 74, 326
Authigenic mineral, 133
Autotrophic bacteria, 74
Autotrophy, 66

B
Bacteria, 13, 57, 58, 177
Bacterial colonies, 178
Bacteriomorph, 153, 154
Baja California, 57
Baja California Sur, 57
Barberton Greenstone Belt, 288
Bioapatites, 133
Biochemical markers, 168
Biochemical processes, 304
Biochemicals, 66
Biochemical signatures, 184
Biochemistry, 4, 174, 185, 188, 199
Biofilms, 155, 288
Biogenic, 150, 178
Biogenic magnetite, 172

B. Cavalazzi, F. Westall (eds.), *Biosignatures for Astrobiology*, Advances in Astrobiology and Biogeophysics, https://doi.org/10.1007/978-3-319-96175-0

Biogenic minerals, 168, 178
Biogenic origin, 157
Biogenic sedimentary fabrics, 55
Biogenicity, 123, 134, 146, 160, 275
Biogeochemical, 98, 132
Biogeochemical cycles, 99
Bioicons, 325
Bioindices, 327
Biological activity, 90, 124
Biological carbon fixation, 67
Biological environments, 229
Biological isotopic fractionation, 229
Biological material, 278
Biomarkers, 58, 63, 64, 70, 175, 195, 294, 323
Biomass, 12, 57, 94, 102
BIOMEX project, 115
Biomineralization, 129
Biominerals, 55, 123, 125, 132, 133, 137, 277
Biomolecules, 7, 54, 56, 69, 112, 119, 184, 185, 191, 199, 291, 294
Biomorphs, 137
BIOPAN, 212
Biopolymers, 4, 13
Bio-sedimentary formations, 288
Biosignatures, 51, 52, 61, 66, 67, 113, 114, 116, 117, 146, 157, 160, 240, 242, 267, 277, 286, 304, 307
BIOSIGN project, 115
Biospheres, 67, 98, 104, 113, 146
Biosynthesis, 56, 58, 68, 187
Biosynthetic, 58
Biota, 70
Biotic, 276
Biotopes, 4
Bode, J.E., 314
Bona fide, 149
Bones, 133
Brines, 4
Buck Reef Chert, 155
Building blocks, 15
Building blocks of life, 184
Bulk isotopic fractionation, 293, 294

C
δ13C, 70
Calcite, 134, 170
Calvin cycle, 67, 74
Canada-France Hawaii Telescope, 253
Canals of Mars, 319, 321
Canyon Diablo, 65
Carbohydrates, 66, 101
Carbon, 193
Carbonaceous chondrites, 8, 9, 16
Carbonaceous films, 153
Carbonaceous laminae, 153
Carbonaceous material, 152
Carbonaceous matter, 274, 276, 277
Carbonaceous meteorites, 258
Carbonaceous structures, 275
Carbonate dissolution, 127
Carbonate precipitation, 126
Carbonates, 5, 126, 128, 168, 171, 174, 175
Carbon atoms, 30
Carbon cycle, 72, 95
Carbon dioxide, 5, 193, 238
Carbon fixation, 67
Carbon isotope signature, 158
Carbon isotopic composition, 171
Carbonization, 275
Carbon-oxygen-nitrogen cycles, 229
Carbon precursor, 150
β-Carotene, 274
Carotenoids, 56
Cassini, 113
Catagenesis, 62
Catalyse, 37
Catalysts, 191
Cell activity, 99
Cell division, 99
Cell-like, 160
Cells, 68, 98, 101
Cellular membranes, 55
Cellular systems, 54
Cell wall, 176
CH_4, 255, 263
Chemical biosignatures, 3, 291, 292
Chemical compositions, 125
Chemical compounds, 199
Chemical product, 196
Chemical tracers, 92
Chemisorption, 189
Chemistry, 191
Chemocline, 64
Chemoorganotrophs, 156
Chemotrophic cells, 160
Chemotrophic colonies, 156, 160
Chemotrophic microorganisms, 287
Chert matrix, 152
Chlorinated organic molecules, 285
Chlorine-bearing hydrocarbons, 285
Chlorophyll, 228
C, H, N, O, P and S, 54
Chondritic organic matter, 38
Chrondrite, 33
Classical HZ, 226

Clathrates, 263
Clay minerals, 11
Clays, 191
Climate, 237
Clots, 156
Clotted layers, 156
Clotted texture, 291
Coccoids, 154
Cometary grains, 27
Comet 67P/Churyumov–Gerasimenko, 9
Comets, 8, 11, 23, 184
Concept of life, 52, 305–307
CONCORDIA, 27
Condensation-sublimation cycle, 260
Conditions, 136
Cosmic dust, 8
Cosmic rays, 6, 231
Cosmic spherules, 27
Cosmomaterials, 24
Coupled climate-photochemistry, 238
Cryosphere, 159
Crystal growth, 137
Crystalline, 263
Crystalline minerals, 38
Crystallinity, 133, 135
Crystallites, 150, 171
Crystallization, 125
Crystals, 13, 128
CSHELL spectrometer, 253
Cube satellite, 115
Curiosity, 193, 255, 285
Curiosity rover, 78, 284
Cyanobacteria, 58, 60, 61, 63, 129, 153, 158, 239
Cyanobacterial ma, 64
Cytoplasmic, 56

D
Darwinian evolution, 3, 52, 53, 306
Dead Earths, 224, 240
Decomposition, 174
Deep biosphere (DB), 88
Deep ocean, 88, 117
Definition, 307, 308
Definitions of life, 306
Degassing, 126
Degree of disorder, 32
Desiccation, 60
Deuterium, 38
Diagenesis, 57, 62, 186
Diagenetic, 64
Disk, 37
DNA, 93, 116
DNA sequences, 89
DNA sequencing, 97
Doppler shift, 255
Dormant life, 158
Dresser Formation, 76
Dwarf planets, 119
Dynamical processes, 232

E
Early Earth, 159, 183, 224, 230, 233, 234
Early life, 60, 62, 160
Early Mars, 285
Earth system, 229
Earth, 5, 24, 26, 52, 54, 114, 119, 139, 146, 160, 184, 185, 218, 224, 239, 284, 310
Earth-based life, 52
Earth-like, 225
Earth-like exoplanets, 230, 235–239
Earth-like planets, 304
Earthshine, 229
Earth-twins, 310, 311
Ecological niches, 88
Ecological systems, 55
Ecosystems, 64, 74
Eeathering, 231
Electric fields, 262
Empirical ignorance, 309, 310
Enantiomeric excess, 292
Enantiomers, 59
Enceladus, 113–115
Environmental conditions, 41
Enzymes, 76, 101, 105
Eoarchaean, 146, 160
Epifluorescence, 93
Epistemic perception, 315, 316
Epistemology of astrobiology, 305
ESA Rosetta, 9
Eta-Earth, 225
Eukaryotes, 68, 69
Eukaryotic cytoplasm, 158
Europa, 113–115
Evaporating brine, 172
Evaporation, 126
Excitation laser wavelength, 268, 270
Exobiological systematics, 308
Exogenous methane, 258
ExoMars 2020, 193, 268
ExoMars biosignature, 286
ExoMars rover, 286
Exoplanetary, 223–242
Exoplanetary biosignature, 224

Exoplanetary missions, 242
Exoplanetary science, 230
Exoplanets, 113, 119, 224
Exploration missions, 185
Exportadora de Sal, 57, 60
EXPOSE, 115
EXPOSE facility, 217
Extracellular polymeric substances (EPS), 128, 288
Extraterrestrial, 17
Extraterrestrial analogues, 199
Extraterrestrial biological activity, 304
Extraterrestrial bodies, 199
Extraterrestrial environments, 111
Extraterrestrial life, 160, 168, 186, 267, 324
Extraterrestrial origin, 325
Extraterrestrial samples, 168
Extreme, 146
Extreme environmental, 146
Extremophiles, 89, 147

F
False negatives, 309
False positives, 309
Fatty acids, 62, 127
Field investigation, 268
Filaments, 154
Film-like substance, 154
Fischer-Tropsch reactions, 31
Fischer Tropsch-type (FTT), 206
Flammarion, C., 314
Fluid flows, 89
Fluorescence, 28, 270
Formation, 137
Fossil bones, 134
Fossilized organisms, 186
Fossil meteorites, 25
Fossils, 267
Fossil traces, 132
FOTON/BIOPAN, 117
Fourier Transform Spectrometer (FTS), 253
Framboids, 136
Franz von Paula Gruithuisen, 319
FT-Raman systems, 271
Functional groups, 128, 191
Fungi, 89, 98, 99, 105
Fusion crust, 169

G
Gale Crater, 78, 193, 194, 259, 288
Galilean and Saturnian moons, 235
Galilei, G., 312
Galileo spacecraft mission, 239
G- and D-bands, 32
Gas Chromatograph Mass Spectrometer (GCMS), 12, 192
Gas giants, 236
Gas hydrates, 101
Gas-phase, 197
Gas-phase biosignature, 224
Gas-phase species, 230
Genes, 105
Genomes, 63, 97
Genomic fragments, 102
Geobiology, 139
Geochemical environment, 267
Geochemistry, 77
Geohopanoids, 64
Geological context, 287
Geological environment, 192
Geological settings, 146
Geological time, 231
Geothermometry, 171
Glass, 172
Global climate model (GCM), 259
Globules, 170
Glycerol, 56
Glycine, 13
Graphene, 191
Graphite, 147, 150, 175, 191, 274
Graphitization, 275
Great Oxidation Event (GOE), 124, 233
Greenland, 75
Guerrero Negro, 57
Gypsum, 60

H
H2O, 262
Habitability, 111, 158, 159, 224, 227, 307
Habitable, 53
Habitable zones (HZ), 113, 225, 226, 236, 307, 310
Habitats, 90, 96, 149
Hadean, 146, 160, 233
Halophiles, 4
Hazes, 239
Heavy metals, 103, 104
Herschel, W., 316
Hesperian, 194
Heterogeneity, 36
HNOPS, 158
Homochirality, 13
Homochiral sequences, 11
Hopanes, 186
Hopanoid molecules, 59
Hopanoids, 58
Huygens, C., 313

Huygens probe, 12
Hydrated minerals, 194
Hydrocarbons, 4, 56, 184, 186
Hydrogen peroxide (H2O2), 262
Hydrosphere, 72
Hydrothermal activity, 147, 152, 258
Hydrothermal chert vein, 146
Hydrothermalism, 27
Hydrothermal sediments, 290
Hydrothermal systems, 7
Hydrothermal vents, 76, 104, 156
Hydroxyapatite, 133, 134, 191
Hypersaline mats, 58
Hypersaline sabkhas, 60
Hyper-thermophiles, 90

I
Icons of life, 324–326
Icy moons, 113, 116, 119
Icy ocean worlds, 119
Igneous crust, 89
Impact craters, 24, 172, 178
Impactors, 8
Indices of life, 326–329
Infrared Astronomical Satellite, 11
Inner-sphere, 191
Insoluble organic matter (IOM), 28
Intelligence, 330
Interactions, 188
International Space Station (ISS), 115, 217
Interplanetary Dust Particles (IDPs), 24
Interstellar Medium (ISM), 30
Interstellar origin, 37
Ionic exchange, 190
Ionizing radiation, 294
Iron isotopes, 77
Iron oxide, 199
Irradiation, 42
Isomers, 31
Isoprenoids, 56, 69, 186
Isotopes, 193
Isotopic compositions, 31, 37, 65, 134
Isotopic fractionations, 37
Isotopic ratios, 229, 230
Isotopic signals, 228
Isotopic signatures, 89, 149

J
Josefsdal Chert, 155, 289
Jupiter, 112

K
Kaapvaal, 74
Kerogens, 28, 150, 274, 285, 290
Ketone, 36
Kitty's Gap Chert, 155

L
Labeled-Release experiment, 284
Laminae, 60
Late heavy bombardment, 184, 185
Life, 3, 17, 52, 101, 119, 124, 139, 146, 160, 231, 237, 284, 305, 306, 310, 323
Life detection, 114, 199
Life forms, 149, 160
Limits of life, 168
Lipids, 56, 57, 66, 101
Liquid phase, 189
Liquid water, 4, 12, 90, 192, 285, 310
Lithopanspermia, 186
Living organisms, 327
Living system, 4
Lomonosov, M., 314
Low Earth Orbit (LEO), 118
Lowell, P., 319
Lunar landers, 118
Lunar regolith, 40
Lunar soils, 40

M
Macromolecules, 104, 133, 184
Magnetic field, 16, 113
Magnetite crystals, 131
Magnetites, 131, 173, 178
Magnetotactic bacteria (MTB), 130, 172
Marine sedimentary, 150
Marine water, 129
Mars, 26, 39, 40, 52, 78, 113, 116, 123, 139, 159, 160, 185, 186, 192, 196, 235, 253–264, 284, 312
Mars 2020, 268
Mars Exploration Rovers, 284
Mars Express 2003, 284
Mars reconnaissance orbiter, 194
Mars Reconnaissance Orbiter 2005, 284
Martian atmosphere, 196, 263, 294
Martian landers, 41
Martian meteorite, 146
Martian meteorite Allan Hills (ALH) 84001, 168
Martian meteorite ALH84001, 136, 325

Martian origin, 168
Martian polar regions, 114
Martian soil, 196, 263
Martian surface, 263
Mass spectrometry (MS), 28, 93
Mats, 160
M-dwarf stars, 236
Membrane composition, 104
Membranes, 4, 56, 116, 176
Mercury, 5, 185
Mesoarchaean, 146, 156, 157
Mesoproterozoic, 75
Mesosphere, 232
Metabolic activity, 129, 186
Metabolic processes, 55
Metabolic signatures, 99
Metabolism, 53, 126, 127, 176, 184, 229, 284
Metabolites, 105, 114, 116
Metagenomic analyses, 95
Metal oxide, 191
Metals, 191
Metamorphism, 186, 275
Meteorites, 9, 25, 38, 169, 176, 184, 275
Meteoroids, 39
Methane, 91, 94, 113, 193, 234, 263, 264
Methane oxidation, 258
Methanogenic, 58
Methanogenic bacteria, 239
Microbes, 105, 149
Microbial biominerals, 125
Microbial biosignatures, 287
Microbial carbonates, 126
Microbial colonies, 154
Microbial communities, 104, 287
Microbial degradation, 186
Microbial diversity, 93, 97
Microbial ecology, 98, 229
Microbial growth, 90
Microbialites, 126
Microbial life, 160, 192
Microbial life, past or present. In the work by Vago et al (2017) we proposed such a list, 286
Microbially induced sedimentary structures (MISS), 153, 156
Microbial mats, 59, 153, 327
Microbial metabolisms, 91, 149
Microbial mineralization, 124
Microbial moulds, 157
Microbial populations, 65
Microbiologists, 92
Microbiota, 98, 100
Microcosm, 58
Microcrystals, 136
Microfossils, 55, 93, 152, 154, 160, 168, 176–178, 274, 276, 290
Micrometeorites, 8, 10, 39, 240
Microniches, 90
Microorganisms, 4, 55, 62, 93, 94, 112, 146, 156, 174, 178
Micropalaeontology, 267
Microstructures, 138
Middle Marker, 155
Miller, 32
Miller-Urey, 32
Miller-Urey experiment, 206
Miller-Urey reactions, 32
Mineral formation, 124
Mineral grains, 42
Mineralization, 61
Mineral matrix, 274
Mineral morphologies, 125
Mineral nucleation, 129
Mineralogical traces, 124
Mineralogical traces of life, 139
Minerals, 123, 127, 137, 188, 195, 270
Missions, 112
Molecular adsorption, 187
Molecular complexity, 30, 53
Molecular composition, 267
Molecular isotopic abundances, 66
Molecular synthesis, 206
Molecular weight clustering, 292, 293
Molecule, 30
Molecule-mineral interactions, 187, 191
Molecules, 128, 175, 176, 191, 197, 274
Moodies Group, 155–157
Moon, 5, 40, 112, 118, 236, 240, 284, 312
Morphological biosignatures, 287, 288
Morphological fossils, 156
Mountains of the Moon, 312–315
Murchison, 16, 28, 30, 31, 212
Murchison meteorite, 9
Murchison-type carbonaceous micrometeorites, 258
Murray, 16

N

Nannobacteria, 177
Nanoparticles, 125
Nanophase, 174
Nanophase magnetites, 173
Nanostructures, 150
Neoarchaean, 60
Neoproterozoic, 74

Nitrous oxide (N2O), 232, 234
NMR spectra, 31
Noachian, 168
Nonlife, 308
Nucleation, 128
Nucleic acids, 4, 66, 93

O

Ocean planets, 159
Oceanic crust, 89
Oligotrophy, 105
OMEGA, 195
Opal, 194
Open ocean, 99
Optical microscopy, 290
ORGANIC and AMINO experiments, 217
Organic biosignatures, 116
Organic compounds, 6, 176, 270
Organic laminae, 152
Organic materials, 115, 174
Organic matter (OM), 41, 91, 133, 178
Organic molecules, 4, 6, 41, 101, 125, 128, 184, 186, 220, 278, 285
Organic polymers, 128
Organic precursors, 285
Organics, 41, 61, 197
Organic solids, 38
ORGANICS experiment, 214
Organisms, 55, 176, 267
Organo-sedimentary structures, 59, 288
Orgueil, 31
Orgueil chondrite, 37
Osmotic pressure, 90
Oxidants, 294
Oxidation processes, 42
Oxidation reactions, 285
Oxidative stress, 105
Oxide minerals, 191
Oxides, 240, 290
Oxygen, 193, 239
Oxygenic photosynthesis, 60
Oxygen isotopic composition, 168
Oxygen isotopic fractionation, 230
Ozone (O3), 262
Ozone, 232, 234, 238

P

PAHs, 285
Palaeoarchaean, 156, 160
Palaeoredox, 136
Paleoenvironmental, 132
Parent body, 27, 30, 37
Payloads, 194
PCR amplification, 93
Peirce and Ferdinand de Saussure, 322
Perchlorates, 263, 285
Perseus, 210
Perseus mission, 212
pH, 77, 190
Phanerozoic, 70
Phenomena, 323
Phoenix lander, 284
Phosphates, 69, 191
Phosynthetic biofilms, 160
Photic zone, 61
Photocatalysis, 192
Photochemical and spectral responses, 224
Photochemical responses, 226
Photochemistry, 16, 206–220, 262, 294
Photochemistry models, 235
Photodissociation, 258
Photolysis, 192, 236
Photon capsules, 219
Photons, 268
Photoreactions, 220
Photosynthesis, 98, 192, 229, 274
Photosynthetic mats, 153
Phototrophic biofilms, 156
Phototrophic layers, 277
Phototrophs, 60
Photsynthetic microbial biofilms, 160
Phyllosilicate minerals, 284
Phyllosilicates, 26, 37, 153, 194
Phylogeny, 95
Physicochemical extremes, 89
Pigments, 274, 275
Pilbara, 60, 74
Pilbara Craton, 60
Planetary atmospheres, 191, 226
Planetary conditions, 117
Planetary exploration, 268
Planetary Fourier Spectrometer (PFS), 253
Planetary habitability, 226
Planetary sciences, 23
Planets, 139, 227
Plasma, 31
Plate tectonics, 5
Plumes, 113
Polyaromatic organic solid, 28
Polyaromatic structure, 32
Polycyclic aromatic hydrocarbons (PAHs), 174
Polymerization, 15
Polysaccharides, 128
Pore-water, 101

Prebiotic, 6, 184
Prebiotic and space-like conditions, 199
Prebiotic chemistry, 185
Prebiotic processes, 188
Precambrian, 60, 72, 176
Preservation of organic matter, 294
Primitive Earth, 7
PROCESS experiment, 217
Prokaryotes, 4, 99, 287
Prokaryotic, 99
Proteins, 4, 66, 116
Proterozoic, 70
Protocols, 62
Protosolar, 37
Protosolar disk, 31
Psychro/meso/thermophiles, 89
Pyrite, 136, 191
Pyrolysis, 31, 194, 258
Pyruvate, 70

Q

Quartz, 134, 209, 263

R

Radiation chemistry, 206
Radiolysis, 31
Raman, 32, 158
Raman effect, 268, 270–272
Raman Laser Spectrometer (RLS), 268
Raman map, 272
Raman mapping, 278
Raman scattering, 270
Raman spectroscopy, 156, 267–280
Raman spectrum, 274
Rare earth elements (REE), 134
Rayleigh effect, 270
Record of life, 160
Recrystallization, 133
Red Planet, 123
Redox disequilibrium, 239
Redox reactions, 55
Regolith, 42, 263
Renazzo, 32
Renazzo-type (CR) chondrites, 16
Repeating constitutional subunits, 293
Replication, 53
Rock record, 52
Rocks, 132
Rocky planets, 236
Rock/water interface, 147
Rods, 154
Rosetta Orbiter Spectrometer for Ion and Neutral Analysis (ROSINA), 9
Rosettes, 170–172
Rover, 255

S

Sample analysis at Mars (SAM), 285
Satellites, 113
Saturated solutions, 128
Saturn, 112
Scanning electron microscopy, 177
Scanning Habitable Environments with Raman & Luminescence for Organics & Chemicals (SHERLOC), 268
Schiaparelli, G., 319
Scientific concepts, 307
Seawater, 4, 7, 230
Second Oxidation Event (SOE), 233
Secondary Ion Mass Spectroscopy (SIMS), 32
Sedimentary rocks, 66, 90
Sedimentary textures, 290
Self-organization, 191
Self-replication, 53
Self-sustaining chemical system, 306
Semiosis, 323
Semiotics of biosignatures, 305
Serpentinisation, 91, 101, 257
Shock metamorphism, 27
Signs, 322
Signs of Life, 284–295
Silica, 157, 172, 191
Silicates, 36
Skeletal tissues, 134
Skeletons, 58
Snowball Earth, 233
Solar System, 7, 112, 117, 158, 224, 234, 235, 304, 305
Solar System atmospheres, 224
Solar-type star, 310
Solar wind, 27
Soluble organic matter (SOM), 28
Source rocks, 64
South Africa, 74
Space, 185
Space environment, 115
Space exploration, 111
Space-expose experiments, 209
Spectroscopic exoplanetary biosignatures, 224, 228–240
Spectroscopy, 32, 197, 229
Spheroidal microstructures, 157
Stable isotopes, 65

Stalks, 137
Stardust mission, 9, 184
Stratosphere, 237
Strelley Pool Chert, 60, 150
Strelley Pool Formation, 152
Stromatolites, 55, 59–61, 150, 152, 153, 267, 277
Stromatolitic layers, 152
Subseafloor, 89, 90, 94, 99, 101
Sulfate reducing bacteria (SRB), 76
Sulfate reduction, 75
Sulfurization, 64
Sulphates, 102, 194
Sulphides, 7, 174
Sulphur, 240
Sulphur-oxidizing bacteria, 137
Sun, 5
Sunlight, 42
SuperCam, 42, 272
Supernova, 16
Surface charge, 190
Symbol, 323
Symbols of life, 329–331
Syngenicity, 147, 150, 155, 160
Synthesis, 206
Synthetic membranes, 59
System, 234, 235
Systematic isotopic ordering, 293

T
Tagish Lake, 33
Tanpopo, 115
Taphonomy, 62
Taxa, 60
Taxonomy, 308
Technosignature, 330
Technosymbol, 330
Telescopes, 227
Terrestrial analogue, 117
Terrestrial biosignatures, 314
Terrestrial life, 13, 306, 314
Terrestrial planets, 119, 227
The Raman bands, 279
Thermal emission, 11
Thermal Emission Spectrometer (TES), 254
Thermal metamorphism (TM), 27
Thermophiles, 90
Thiophenes, 64
Tidal flats, 60
Titan, 12
To sustain life, 53
Trace elements, 134, 169
Trace gases, 196
Trace Gas Orbiter (TGO), 264
Traces of fossil life, 160
Traces of life, 146
Transit of, 318
Transit spectrophotometry, 228
Troilite meteorite, 65
Trophic level, 58
Troposphere, 232, 237
Tunable Laser Spectrometer (TLS), 255

U
Ultramafic rocks, 101
Universe, 191, 304
Ultraviolet (UV), 32, 42, 139
UV photons, 42
UV radiation, 112, 218, 285
UV Raman spectroscopy, 270

V
Vacuum UV conditions, 219
Vega 1 and 2, 8
Vegetation red edge, 228
Venera 3, 319
Venera 4, 319
Venting plumes, 115
Venus, 159, 185, 235, 284, 312–314, 318
Vibrational energy, 268
Viking, 10, 192, 284, 285, 309
Viking landers, 284
Viking mission, 309
Volcanic particles, 290

W
Water, 232
Water-ice crust, 113
Water vapour, 193
Weathering, 25
Western Australia, 76

X
XANES, 35
X-Ray, 40

The manufacturer's authorised representative in the EU is Springer Nature Customer Service Centre GmbH, Europaplatz 3, 69115 Heidelberg, Germany. If you have any concerns regarding our products, please contact ProductSafety@springernature.com

Printed and bound by CPI Group (UK) Ltd, Croydon, CR0 4YY
23/09/2026
02221733-0007